Igor G. Kuleyev, Ivan I. Kuleyev, Sergey M. Bakharev, Vladimir V. Ustinov
Phonon Focusing and Phonon Transport

Texts and Monographs in Theoretical Physics

Edited by
Michael Efroimsky, Bethesda, Maryland, USA
Leonard Gamberg, Reading, Pennsylvania, USA

Igor G. Kuleyev, Ivan I. Kuleyev,
Sergey M. Bakharev, Vladimir V. Ustinov

Phonon Focusing and Phonon Transport

In Single-Crystal Nanostructures

DE GRUYTER

Physics and Astronomy Classification 2010
65.80.-g; 62.23.Hj; 66.70.Df; 62.30.+d

Authors

Dr. Igor G. Kuleyev
Institute of Metal Physics
Ekaterinburg
Russian Federation
kuleev@imp.uran.ru

Dr. Sergey M. Bakharev
Institute of Metal Physics
Ekaterinburg
Russian Federation
bakharevsm@imp.uran.ru

Dr. Ivan I. Kuleyev
Institute of Metal Physics
Ekaterinburg
Russian Federation
kuleyevII@imp.uran.ru

Dr. Vladimir V. Ustinov
Institute of Metal Physics
Ekaterinburg
Russian Federation
ustinov@imp.uran.ru

ISBN 978-3-11-067039-4
e-ISBN (PDF) 978-3-11-067050-9
e-ISBN (EPUB) 978-3-11-067057-8
ISSN 2627-3934

Library of Congress Control Number: 2020931493

Bibliographic information published by the Deutsche Nationalbibliothek
The Deutsche Nationalbibliothek lists this publication in the Deutsche Nationalbibliografie; detailed bibliographic data are available on the Internet at http://dnb.dnb.de.

Cover image: Filograph/iStock/Getty Images Plus
Typesetting: Integra Software Services Pvt. Ltd.
Printing and binding: CPI books GmbH, Leck

www.degruyter.com

Contents

Abstract

The monograph is devoted to the investigation of physical processes that govern phonon transport in bulk and nanoscale samples of cubic symmetry. A series of works the authors have carried out over the past six years underlies the monograph. The main attention has been paid to the study of phonon, focusing in cubic crystals and its influence on boundary scattering and lattice thermal conductivity of bulk materials and nanostructures. An analytical solution to the problem of the Knudsen phonon gas flow in finite-length samples with circular, square-shaped, and rectangular cross-sections is given. Phonon transport is described for nanostructures with different types of elastic energy anisotropy at low temperatures when diffuse phonon scattering at the boundaries is the dominant relaxation mechanism. The influence of the phonon focusing on the dependencies of phonon mean free paths on the geometric parameters of nanostructures for different directions of phonon heat flux and film orientations relative to the axes of the crystals is accounted for. The film orientations and directions of the heat flux, which provide maximum or minimum heat removal from elements of semiconductor chips, are determined. The temperature dependencies of the thermal conductivity of silicon nanowires and thin films are calculated within the three-mode Callaway model.

The monograph is addressed to researchers in the field of condensed matter physics, as well as teachers, graduate students, and students of physical and technical universities.

Translation by
Yu G. Gorelykh, PhD, Physical and Mathematical Sciences

https://doi.org/10.1515/9783110670509-203

Introduction

The monograph is devoted to the study of physical processes that underlie phonon transport in bulk and nanoscale monocrystalline samples of cubic symmetry. An important aspect of this problem is to investigate the influence of focusing on the propagation of phonon modes and the thermal conductivity of monocrystalline films and nanowires. Moreover, close attention is required for research on dispersion and relaxation mechanisms of the phonon momentum, which govern the dependencies of kinetic characteristics on temperature and geometric parameters for different directions of heat flux in nanowires and the orientation of the film planes.

In light of the development of technology on manufacturing nanofilms and nanowires and of their widespread use in microelectronics, interest in the study of their heat-conducting properties has recently grown dramatically [1–7]. Peculiarities of phonon transport in such structures are due to the fact that the mean free paths of phonons in a wide temperature range turn out to be greater than or comparable with the characteristic sizes of samples in the nanometre scale. Therefore, phonon scattering at the boundaries plays an important role in the thermal resistance of nanoscale materials in the temperature range from helium to room temperature. In this case, the mean free path of phonons is determined by the nature of the phonon-surface interaction. Such a situation, when diffuse phonon scattering at the boundaries is a dominant relaxation mechanism, is commonly called the phonon-boundary scattering regime. According to [8, 9], the diffuse nature of phonon scattering occurs if the geometric parameters of the roughness at the sample boundary are greater than or comparable with the phonon wavelength. In bulk materials, the phonon-boundary scattering regime takes place at rather low temperatures, when the phonon–phonon relaxation mechanisms are frozen out. In this case, the propagation of heat flux along a dielectric rod can be treated as the flow of a rarefied gas through a tube. In the event of passing a rarefied molecular gas through a tube, collisions of the gas molecules with each other can be neglected and the flow magnitude is assumed to be controlled by the nature of the molecule–surface interaction (see, e.g., [10, 11]). In 1909, M.H.C. Knudsen solved the problem of the flow of a rarefied gas through a circular cross-section infinite-length tube and showed that, with diffuse scattering of the gas molecules at the boundaries, the mean free path of the molecules is equal to the diameter of the tube [12].

In 1938, H. Casimir was the first to pose the problem of thermal conductivity of a thin dielectric infinite-length rod in the model of an isotropic medium [13]. He analysed the case of diffuse phonon scattering at the sample boundaries. Casimir suggested that, when impacted with the surface, all phonons are absorbed and then re-emitted isotropically into half-space towards inside the sample, with the intensity depending on the surface temperature and following the theory of black-body radiation. He revealed that the mean free path of phonons in a cylindrical infinite-length

https://doi.org/10.1515/9783110670509-001

rod is equal to its diameter. In this regard, the mean free paths of phonons in infinite-length samples are referred to as the Casimir lengths. The findings obtained coincide with the result of Knudsen [12] for the flow of a rarefied molecular gas through a circular cross-section infinite tube. Therefore, the phonon-boundary scattering regime is called the Knudsen phonon-gas flow regime. Later, Berman and colleagues examined the influence of partially mirror reflection of phonons from the surface of the sample, as well as the influence of finite length on conductivity in the boundary scattering regime [14, 15]. It was shown that the thermal conductivity of the finite-length sample diminishes by a magnitude of the order of the ratio of the sample width to its length, compared to an infinite sample. However, no analytical expressions for corrections to the thermal conductivity due to the finite length of the sample were derived.

Anisotropy of the elastic properties of crystals poses a number of new effects in phonon transport. Among them is phonon focusing. In [16–18], it has been shown that due to the non-collinearity of the phase and group velocities, phonon flux emitted by a point heat source focuses along certain directions in the crystal lattice. Experimental studies carried out by McCurdy et al. [19] have proved that phonon focusing causes two effects concerning the thermal conductivity of cubic crystals in the boundary scattering regime. The former exhibits the $k(T)$ dependence of the thermal conductivity on the direction of the temperature gradient T relative to the crystallographic axes: for square-shaped cross-section Si samples, the thermal conductivity in the [001] direction at low temperatures turned out to be 40% and 50% higher than in the [011] and [111] directions. For CaF_2 crystals, the opposite is true: the thermal conductivity in the [001] direction is 40% less than that in the [111] direction. The latter makes itself felt in the dependence of the thermal conductivity of rectangular cross-section samples on the orientation of the side faces at low temperatures. For two samples having the same geometric parameters and the [110] temperature gradient direction, the thermal conductivity of the sample with {001} wide and narrow {110} faces appears to be 33% higher than that of a sample with {110} wide and {001} narrow faces. At temperatures above the maximum of $k(T)$, the phonon mean free path becomes less than the transverse dimensions of the sample, the thermal conductivity of cubic crystals becomes isotropic, and it does not depend on the direction of the heat flux in the crystal.

In [19], the Casimir theory has been generalized to the case of elastically anisotropic crystals. It was assumed that the heat flux and temperature distribution are uniform along the sample length, with a plane of mirror symmetry being perpendicular to the direction of the heat flux. Under these assumptions, the mean free paths of phonons in Si and CaF_2 samples were calculated for symmetric directions at a temperature of 3K. The resulting values were consistent with the experimental data of [19] at T = 3K, with an error not exceeding 8%. As to finite-length samples, the authors of [19] failed to deduce analytical expressions for the relaxation times of phonons involving in diffuse boundary scattering within the Casimir theory [13] and to establish the temperature dependencies of thermal conductivity. Therefore, over the 40 years since the publication of [19], none of the works devoted to study the phonon focusing effect on the temperature dependencies of the thermal conductivity of dielectric crystals has been done. Such a calculation was performed in [20, 21]. This made it possible to find the relaxation times of phonons of various polarizations during diffuse phonon-boundary scattering for finite-length samples with circular, square-shaped, and rectangular cross-sections. For this purpose, piecewise smooth functions for various angle intervals were utilized through the relations between the group velocity components and the geometric parameters of the samples.

To correctly explain the phenomena discovered in [19], two orientation parameters that take into account the dependencies of the kinetic characteristics on the direction of heat flux [*I*] and the orientation of the film plane {*J*} (or a wide face of the sample) should be introduced for calculating the thermal conductivity of monocrystalline samples. Bearing in mind phonon focusing, it was shown that the orientation parameters {*I*} and {*J*} can be determined through the group velocity components parallel and perpendicular to the heat flux direction [22]. This method and analytical solutions in [20, 21] allowed one to adequately describe the experimental data on the thermal conductivity of silicon crystals [19] over the entire temperature range for all directions of the temperature gradient and all orientations of the side faces for rectangular cross-section samples [22].

Until now, to analyse the temperature dependencies of thermal conductivity for phonon relaxation rates at the boundaries and in anharmonic scattering processes, fitting parameters have always been introduced to compensate for the flaws of theoretical models underlying their calculations. As to phonon-boundary scattering, these parameters were earlier necessary to take into account the influence of geometric parameters and orientations of the side faces of samples, as well as heat flux directions on the thermal conductivity of semiconductor and dielectric crystals. However, with a view to getting rid of them, the problem of the Knudsen phonon gas flow was analytically solved. It has been shown that, in the temperature range from 3 to 15 K, when phonon scattering dominates at the boundaries and by isotopic disorder, our theory holds true. It quantitatively describes the temperature dependencies of Si thermal conductivity for all heat flux directions in square-shaped cross-section samples and the dependence on the orientation of the side faces for rectangular cross-section samples, without using the fitting parameters. In calculating the relaxation characteristics of phonons and phonon transport in elastically anisotropic crystals, the method of accounting for phonon focusing [20–22] was used in [23, 24] to build the temperature dependencies of thermal conductivity of silicon nanowires with diameters of 115 and 56 nm and films with thicknesses of 1.6, 0.83, 0.42, 0.1, and 0.02 μm. The three-mode Callaway model agrees well with the experimental data of [4–7] for the thermal conductivity of films and nanowires in the temperature range from 20K to 350K.

The method proposed is relevant, since a significant number of publications devoted to the study of phonon transport in films and nanowires ignore the effects caused by phonon focusing. For example, in [25, 26], the Casimir theory [13] is used for the boundary scattering in silicon- and diamond-based nanostructures. However, it is valid only for the model of an isotropic medium. In turn, the isotropic medium model is also applied for interpreting the experimental results in the reviews [1–3]: the directions of the heat flux and the orientation of the film planes relative to the crystallographic axes are omitted. In the works [23, 27], it is shown that, depending on the temperature gradient direction, the thermal conductivity values for nanowires can vary by 1.5–2 times and a change in the orientation of the film plane can lead to an increase in the thermal conductivity by 2–2.5 times. Therefore, neglecting of orientation data makes the results uninformative. Obviously, the isotropic medium model utilized in interpreting experimental data on the thermal conductivity of nanostructures contributes to errors of up to 250%.

The papers [5–7, 28, 29] explore the phonon-boundary scattering in sufficiently thin films (the film thickness is much smaller than its width) within the isotropic medium model in the same way as was done in the works of Fuchs [30] and Sondheimer [31] in analysing the conductivity of thin metal films. In doing so, the mean free paths and phonon relaxation times in the films were assumed to depend only on their thicknesses [5–7, 28–31]. This result is incorrect. According to [25, 32], the thermal conductivity of the films and phonon mean free paths within the isotropic medium model substantially depends on their geometric dimensions. Even more, the Casimir lengths diverge logarithmically as the film width tends to infinity. As noted in [24, 31], it is phonons that propagate almost parallel to the film plane bring to this divergence. Previously, the issue of the influence of the finite width of films on the divergence of the Casimir lengths has left aside. Also, many important problems such as the influence of elastic anisotropy of cubic crystals on the dependencies of thermal conductivity on the geometric parameters of nanostructures, as well as on the heat flux directions and the film plane orientations relative to the crystal axes have escaped the attention of researchers. A solution to these problems would help determine the optimal orientations of the film planes and the directions of heat flux to provide the maximum or minimum heat removal from the elements of microcircuits. These challenging objectives are as relevant as ever not only for silicon films widely used in microelectronics but also for other semiconductor microstructures to manufacture [1–7, 28, 29]. Therefore, the discussion of these difficult tasks should be dwelled on. The work [27] examines the influence of elastic anisotropy of cubic crystals on the dependence of the thermal conductivity of films on their geometric parameters. It is shown that the divergence above specified can be eliminated through making allowance for the finite length of the films.

Research of the dependence of the thermal conductivity of nanostructures on geometric parameters is of great interest for technical applications of the theory elaborated in this monograph. One of the important goals of the present work is to analyse the influence of focusing on the propagation of phonon modes and the anisotropy of the thermal conductivity of monocrystalline films and nanowires within the framework of the Casimir-McCurdy phenomenological method [13, 19]. This method offers to exploit a three-dimensional phonon spectrum. Therefore, there arise natural restrictions on the diameters of nanowires and film thicknesses, under which the influence of spatial confinement on the spectrum acoustic modes can be neglected. An analysis performed in [4–7, 23, 24] showed that for nanowires with diameters greater than 50 nm and films with thicknesses greater than 20 nm in the temperature range from 20 to 50 K, the thermal conductivity repeats the dependence $k(T)\sim T^3$, as the heat capacity of bulk samples does it in the Debye theory. Therefore, the restrictions afore-indicated give the right to ignore the influence of spatial confinement on the spectrum of acoustic modes. Calculations of the temperature dependencies of the thermal conductivity of silicon films and nanowires in [23, 24] confirm this conclusion.

It should be noted that the numerical method such as the molecular dynamics method ensures no sufficiently infallible results in calculating the temperature dependencies of the thermal conductivity when accounting for phonon focusing. The calculation of thermal conductivity in [33, 34] led to mutually opposite findings for the thermal conductivity anisotropy in symmetrical directions. It was found in [34] that the thermal conductivity of diamond nanowires in the [011] directions is much larger than that in the [001] and [111] directions. This result contradicts the experimental data of McCurdy et al. [19] and our analysis. According to [19–22], the thermal conductivity maxima for Ge, Si crystals, and diamond in the low-temperature region should be observed in the [001] directions. They are due to a slow transverse mode, which focuses precisely in these directions. The results of [33] for the thermal conductivity anisotropy are in qualitative agreement with the calculations of [20, 21] and experimental data in [19].

However, the calculations in [33] claim that significant anisotropy of thermal conductivity occurs at temperatures well above the temperature of maximum thermal conductivity. This contradicts the data of [19]. The latter holds that at temperatures above the thermal conductivity maximum, a transition to bulk relaxation mechanisms occurs, and the thermal conductivity anisotropy quickly disappears.

The first chapter analyses the Christoffel equation and calculates the dynamic characteristics of phonons in cubic crystals within the anisotropic continuum model. It has been shown that all cubic crystals can be divided into two types depending on the value of the anisotropy parameter: crystals with positive and negative anisotropy of second-order elastic moduli. A method is proposed for approximating the phonon spectrum of cubic crystals, which is obtained from data on inelastic neutron scattering for symmetric directions over the entire Brillouin zone. A study of phonon focusing in cubic crystals with various types of anisotropy of elastic energy has shown that focusing directions in the first type crystals become defocusing directions in the second type crystals and vice versa. The influence of focusing on the angular distribution of the density of phonon states (DPS) is analysed. It has been shown that, in elastically anisotropic crystals, the maximum and minimum DPS values are achieved in phonon focusing and defocusing regions, respectively. Therefore, the DPS maxima directions in type-I crystals turn into the minima directions in type-II crystals. Peculiarities of the propagation of phonon pulses in cubic crystals with various types of anisotropy of elastic energy are investigated. An exact analytical expression for the enhancement factor of the phonon flux is derived. Also, the behaviour features of the coefficient are analysed depending on the type of curvature of the isoenergetic surface.

The second chapter presents an analytical solution to the problem of the Knudsen phonon gas flow in circular, square-shaped, and rectangular cross-section finite-length samples. The relaxation times for diffuse phonon scattering at the boundaries are represented in the form of piecewise-smooth functions for various intervals of angles. The latter are determined by the relations between the group velocity components and the geometric parameters of the samples. It has been shown that square-shaped and circular cross-section samples produce the maximum phonon mean free paths for each vibrational mode in the focusing directions, exceeding those of the other vibrational modes in these directions. An analytical solution to the problem of diffuse phonon scattering at the boundaries and determination of the phonon relaxation times make it possible to get rid of fitting parameters in calculating the thermal conductivity of bulk single crystals and nanostructured samples.

The third chapter develops a method of accounting for phonon focusing to compute the thermal conductivity of monocrystalline samples. The method proposed and the phonon relaxation times calculated allow one to adequately describe the temperature dependencies of the thermal conductivity of square-shaped and rectangular cross-section silicon crystals to explain both effects discovered by McCurdy for thermal conductivity. The theory [20–22] quantitatively sets forth the temperature dependencies of the thermal conductivity of silicon crystals within the experimental error for all samples and both effects discovered by McCurdy [19] without applying fitting parameters. This is true for the temperature range from 3 to 15 K when scattering at the boundaries and by isotopic disorder dominates.

The fourth chapter investigates phonon transport in nanostructures with various types of anisotropy of elastic energy at low temperatures when diffuse phonon scattering at the boundaries is a dominant relaxation mechanism. A correlation between the angular dependences of the densities of phonon states (DPS) and mean free paths of quasi-transverse modes is established. It has been shown that, thanks to the focusing effect, in elastically anisotropic crystals there arise regions of directions of the maximum mean free path and maximum DPS values. The influence of phonon focusing on the dependencies of phonon mean free paths on the geometric parameters of nanostructures for different directions of heat flux and orientations of film planes relative to the crystal axes is explored. The orientations of the film planes and the heat flux directions that provide the maximum or minimum heat removal from the elements of semiconductor microcircuits are identified.

The fifth chapter discusses the influence of anisotropy of elastic energy on phonon transport in monocrystalline films in the regime of the Knudsen phonon-gas flow. The influence of focusing on the propagation of phonons and thermal conductivity of films with various types of anisotropy of elastic energy is explored. The physical causes leading to the anisotropy of thermal conductivity for heat flux in the planes of films with various types of anisotropy of elastic energy and various ratios of geometric parameters are found. It has been shown that the anisotropy of the thermal conductivity of the films substantially depends on their orientation.

The sixth chapter deals with calculating the temperature dependencies of the thermal conductivity of silicon nanowires and thin films in the framework of the three-mode Callaway model. The method proposed and the phonon relaxation times calculated make it possible to reconcile the results of calculating the temperature dependencies of the thermal conductivity of larger than 50-nm-diameter silicon nanowires and larger than thickness 20-nm films with experimental data. It has been shown that, both at low and at room temperatures, the {100} and {111} orientations of Si films generate maximum and minimum thermal conductivity, respectively. The role of boundary scattering at room temperature is investigated. The contribution of boundary scattering to the heat resistance reaches 60% for nanowires with a diameter of 56 nm and films with a thickness of 20 nm, respectively.

The seventh chapter discusses unresolved problems of the influence of phonon focusing on phonon transport and thermoelectric effects due to electron-phonon drag in elastically anisotropic crystals and nanostructures based on them.

Chapter 1
Propagation of Elastic Waves and Phonon Focusing in Cubic Crystals

A significant number of publications devoted to analysing thermal conductivity, both in bulk materials [9, 13, 35, 36] and nanostructures [5, 6, 7, 25, 37], have used an isotropic medium model for the spectrum and phonon polarization vectors. Through isotropic media, only purely longitudinal and purely transverse vibrational modes propagate, with the latter being degenerate. This model is not adequate for conducting the analysis of phonon transport and absorption ultrasound in elastically anisotropic crystals (see, for example [19, 27, 38–40]). In the long-wave approximation, when the wave vector of a phonon is much smaller than the Debye wave vector, the anisotropic continuum model is a suitable approximation. This model allows one to calculate the elastic energy of a cubic crystal using three second-order elastic moduli [41–43]. For most crystals, they are experimentally determined.

Anisotropy of elastic energy leads to a number of new effects in the dynamic characteristics of elastic waves and phonon transport of cubic crystals. Firstly, quasi-longitudinal or quasi-transverse vibrations can propagate in cubic crystals, whereas pure modes [27, 40–43] propagate only in symmetrical directions, such as [100], [110], and [111]. Secondly, the phonon spectrum becomes anisotropic, and the degeneracy of the transverse vibrational modes is removed. A detailed analysis of the dynamic characteristics of elastic waves in cubic crystals was carried out in [44] (see also Section 1.1). It shows that the contribution of the transverse component to quasi-longitudinal vibrations in cubic crystals is small and can be left aside. On the contrary, the contribution of the longitudinal components to quasi-transverse modes is not small and, therefore, it needs to be accounted for calculating the relaxation characteristics of phonon systems. Due to their elastic properties, cubic crystals can be divided into two types: crystals with positive (type-I) and negative (type-II) anisotropy of second-order elastic moduli [44]. The shape of the spectrum of vibrational branches for such crystals differs qualitatively compared to the phonon spectra within one type of crystal where they differ quantitatively. Thirdly, the anisotropy of the elastic properties of cubic crystals leads to non-collinearity of the group and phase velocities of phonons and, accordingly, to focusing or defocusing of vibrational modes [16–18, 45]. Phonons mainly propagate in the focusing directions: their density of states can be significantly larger than in the isotropic medium model. Whereas in the directions close to the defocusing directions, the density of states of vibrational modes can be much lower than in the isotropic medium model. We have shown that apart from qualitative differences in not only the spectrum anisotropy and the behavior of the polarization vectors for the first- and second-type crystals, the effects of phonon focusing can also emerge in different ways. For crystals of one

https://doi.org/10.1515/9783110670509-002

type, the directions of focusing and defocusing of the vibrational modes coincide. As for crystals of various types, these directions are opposite: the focusing directions in the first-type crystals turn into the directions of defocusing in the second-type crystals. At low temperatures when the phonon mean free path exceeds the transverse size of the sample, a qualitative difference in the anisotropy phonon transport in cubic crystals of various types can appear.

Section 1.1 deals with the calculation of the dynamic characteristics of phonons in cubic crystals within the anisotropic continuum model. Section 1.2 offers a method that allows approximating the phonon spectrum found from inelastic neutron scattering data for symmetric directions, over the entire Brillouin zone for cubic crystals. Section 1.3 evaluates the heat capacity of Si and Ge crystals using the approximation phonon spectrum. Section 1.4 analyses the influence of focusing on group velocities and features of phonon propagation in cubic crystals with various types of anisotropy of elastic energy. Section 1.5 covers the investigation of the influence of the elastic energy anisotropy on the density of phonon states in cubic crystals with various types of anisotropy of elastic energy. Section 1.6 discusses the enhancement factor of phonon flux.

1.1 Dynamic Characteristics of Phonons in Cubic Crystals within the Anisotropic Continuum Model

The phenomenological theory of elasticity regards a crystal in an undeformed equilibrium state as a continuum with a constant density ρ. When exposed to an external force, the crystal becomes deformed, and the relative positions of the particles change. In this case, the state of the crystal is described by the vector displacement field $\mathbf{u}(\mathbf{r},t)=(u_1,u_2,u_3)$, which defines the displacement at time t of some point having the coordinate $\mathbf{r}=(x_1,x_2,x_3)$ in equilibrium. Such a description is applicable when the displacements of neighbouring atoms in a crystal are much smaller than the distance between them. Consequently, they can be represented as a continuously and slowly changing displacement field [41–43, 46, 47]. To characterize the continuum deformation, a symmetric strain tensor $\eta_{ik}(i,j=1,2,3)$ should be entered [41–43]:

$$\eta_{ik}=\frac{1}{2}\left(\frac{\partial u_i}{\partial x_k}+\frac{\partial u_k}{\partial x_i}+\sum_j\frac{\partial u_j}{\partial x_i}\frac{\partial u_j}{\partial x_k}\right)\equiv\frac{1}{2}\left(\xi_{ik}+\xi_{ki}+\sum_j\xi_{ji}\xi_{jk}\right), \tag{1.1}$$

where $\xi_{ij}=\partial u_i/\partial x_j$ are the components of the distortion tensor. In the case of small strains ($\xi_{ij}<<1$), we neglect the terms of the second order of smallness. Then, the strain tensor takes the form:

$$\eta_{ik} = \frac{1}{2}(\xi_{ik} + \xi_{ki}). \tag{1.2}$$

The theory of elasticity (stress) represents forces as surface ones that must be described by a symmetric stress tensor σ_{ij} [41–43]. The symmetry of σ_{ij} arises from the requirement of vanishing the force moments at equilibrium. In the case of small stresses, the relationship between stresses and strains, according to Hooke's law, are linear and expressed through a fourth-rank tensor or an elastic modulus tensor c_{ikmn} [41–43, 46, 47]:

$$\sigma_{ik} = \sum_{m,n} c_{ikmn}\varepsilon_{mn}. \tag{1.3}$$

The density of elastic energy U per unit volume of a non-deformable crystal has the form [41–43, 46, 47]:

$$U = \frac{1}{2}\sum_{i,k,m,n} \eta_{ik}c_{ikmn}\eta_{mn}. \tag{1.4}$$

The tensor of elastic moduli is symmetric relative to both the permutation of the indices inside the first or second pair and the permutations of the first pair and the second one:

$$c_{ikmn} = c_{kimn} = c_{iknm} = c_{mnik}. \tag{1.5}$$

The first property follows from the symmetry of the strain and stress tensors, the second property comes from the existence of an energy density in the form (1.4). In most cases, instead of the fourth-rank tensor c_{ikmn}, a simpler second-rank tensor $c_{\alpha\beta}(\alpha, \beta = 1, 2, ..., 6)$ is used. The latter was introduced by Voigt [48] according to the scheme presented in Table 1.1. According to (1.5), the matrix $c_{\alpha\beta}$ is symmetric $(c_{\alpha\beta} = c_{\beta\alpha})$ and contains 21 independent quantities.

Table 1.1: Determination of elastic constants by Voigt [48] (relationship between c_{ikmn} and $c_{\alpha\beta}$).

$c_{ik,mn}$	11	22	33	23	13	12
11	c_{11}	c_{12}	c_{13}	c_{14}	c_{15}	c_{16}
22	c_{21}	c_{22}	c_{23}	c_{24}	c_{25}	c_{26}
33	c_{31}	c_{32}	c_{33}	c_{34}	c_{35}	c_{36}
23	c_{41}	c_{42}	c_{43}	c_{44}	c_{45}	c_{46}
13	c_{51}	c_{52}	c_{53}	c_{54}	c_{55}	c_{56}
12	c_{61}	c_{62}	c_{63}	c_{64}	c_{65}	c_{66}

This table can be represented in a more compact form [41]:

ik* or *mn	**11**	**22**	**33**	**23**	**13**	**12**
α or *β*	1	2	3	4	5	6

Let us go over to Voigt's notation to demonstrate the simplest examples of an isotropic medium and a cubic crystal. An isotropic medium possesses the highest degree of symmetry. For it, the tensor c_{iklm} is invariant with respect to all rotations and reflections in three-dimensional space. This condition is satisfied only by the unit tensor (up to a scalar factor). Therefore, the components of the tensor c_{iklm} must be expressed through the combinations of the components of the unit tensor δ_{ij} with some coefficients. It is easy to make sure that taking into account the symmetry properties, we can write down the elastic modulus tensor c_{iklm} for an isotropic medium as follows [42]:

$$c_{iklm} = c\delta_{ij}\delta_{lm} + a(\delta_{il}\delta_{km} + \delta_{im}\delta_{kl}). \tag{1.6}$$

This expression has only two independent constants a and c. They can be written in terms of the elastic moduli c_{11} and c_{12}: $c = c_{12}$, $a = (c_{11} - c_{12})/2$. We use Table 1.1 for finding the Voigt matrix $c_{\alpha\beta}$ for an isotropic medium, corresponding to the tensor (1.6):

$$c_{\alpha\beta} = \begin{pmatrix} c_{11} & c_{12} & c_{12} & 0 & 0 & 0 \\ c_{12} & c_{11} & c_{12} & 0 & 0 & 0 \\ c_{12} & c_{12} & c_{11} & 0 & 0 & 0 \\ 0 & 0 & 0 & (c_{11}-c_{12})/2 & 0 & 0 \\ 0 & 0 & 0 & 0 & (c_{11}-c_{12})/2 & 0 \\ 0 & 0 & 0 & 0 & 0 & (c_{11}-c_{12})/2 \end{pmatrix}. \tag{1.7}$$

A cubic crystal is characterized by three mutually perpendicular axes of symmetry of the second and fourth orders and the axis of symmetry of the third order, directed along the spatial diagonal of the cube [42]. When the coordinate system rotates, the quantities c_{ikmn} are modified according to the rules of transformation of the tensor components. If D_{ik} is an orthogonal matrix that describes the rotation of the coordinate system ($\boldsymbol{r}' = D\boldsymbol{r}; x'_i = \sum_l D_{il}x_l$), the altered system appears as:

$$c'_{ikmn} = \sum_{i',k',m',n'} D_{ii'}, D_{kk'}, D_{mm'}, D_{nn'}\, c_{i'k'm'n'}. \tag{1.8}$$

Let us look at the transformations of the components of the tensor c_{ikmn} during rotation relative to the second-order axes. It follows from relations (1.8) that the invariance of the elastic modulus tensor relative to such rotations results in non-vanishing only components of the elastic modulus tensor c_{ikmn}, in which each index occurs an even number of times [42]. Now, we employ the conditions (1.5) and Table 1.1 to obtain the elastic modulus tensor $c_{\alpha\beta}$ in terms of Voigt's notation, with it being invariant with respect to 180° rotations around the X, Y, and Z axes:

$$c_{\alpha\beta} = \begin{pmatrix} c_{11} & c_{12} & c_{13} & 0 & 0 & 0 \\ c_{12} & c_{22} & c_{23} & 0 & 0 & 0 \\ c_{13} & c_{23} & c_{33} & 0 & 0 & 0 \\ 0 & 0 & 0 & c_{44} & 0 & 0 \\ 0 & 0 & 0 & 0 & c_{55} & 0 \\ 0 & 0 & 0 & 0 & 0 & c_{66} \end{pmatrix}. \tag{1.9}$$

The invariance of the elastic modulus tensor with respect to rotations around the third-order axis gives rise to the equivalence of all three coordinates x, y, z, which pass one into another when turning around the spatial diagonal of the cube. Therefore, the elements of the matrix c_{ikmn} should not change for any permutation of indices 1, 2, and 3.

Hence, we arrive at the following relations [42]

$$c_{1111} = c_{2222} = c_{3333}, \quad c_{1122} = c_{2233} = c_{3311}, \quad c_{2323} = c_{3131} = c_{1212}, \tag{1.10}$$

or in Voigt's notation:

$$c_{11} = c_{22} = c_{33}, \quad c_{12} = c_{13} = c_{23}, \quad c_{44} = c_{55} = c_{66}. \tag{1.11}$$

Finally, we can come up with the elastic modulus tensor $c_{\alpha\beta}$ in the form [41–43, 46, 47]:

$$c_{\alpha\beta} = \begin{pmatrix} c_{11} & c_{12} & c_{12} & 0 & 0 & 0 \\ c_{12} & c_{11} & c_{12} & 0 & 0 & 0 \\ c_{12} & c_{12} & c_{11} & 0 & 0 & 0 \\ 0 & 0 & 0 & c_{44} & 0 & 0 \\ 0 & 0 & 0 & 0 & c_{44} & 0 \\ 0 & 0 & 0 & 0 & 0 & c_{44} \end{pmatrix}. \tag{1.12}$$

Exactly the same form of the tensor $c_{\alpha\beta}$ can be derived using the invariance of the elastic modulus tensor c_{ikmn} with respect to rotations around the fourth-order X, Y, and Z axes. So, the elastic properties of cubic crystals are governed by three elastic moduli: c_{11}, c_{12}, and c_{44}.

Applying expressions (1.4), (1.12) and reading Table 1.1 allow one to represent the elastic energy density of cubic crystals through the strain tensor components in the form:

$$U = \frac{1}{2} c_{11} \left(\eta_{11}^2 + \eta_{22}^2 + \eta_{33}^2\right) + c_{12} \left(\eta_{11}\eta_{22} + \eta_{22}\eta_{33} + \eta_{33}\eta_{11}\right) + 2c_{44} \left(\eta_{12}^2 + \eta_{23}^2 + \eta_{13}^2\right). \tag{1.13}$$

As expected, the anisotropic continuum model offers the elastic energy of a cubic crystal through three second-order moduli c_{11}, c_{12}, and c_{44}. Expressing the strain tensor components as components of the distortion tensor ξ_{ij}, we can restrict ourselves to the linear approximation (1.2) in the harmonic approximation. Then, the elastic energy density (1.13) can be modified:

$$U = \frac{1}{2}c_{11}\left(\xi_{11}^2 + \xi_{22}^2 + \xi_{33}^2\right) + c_{12}\left(\xi_{11}\xi_{22} + \xi_{22}\xi_{33} + \xi_{33}\xi_{11}\right) + \frac{1}{2}c_{44}\left(\xi_{12}^2 + \xi_{23}^2 + \xi_{13}^2\right). \quad (1.14)$$

For going over to the model of an isotropic medium, the elastic energy density (1.14) can be put as:

$$U = \frac{1}{2}c_{11}\left(\sum_i \xi_{ii}\right)^2 + \frac{1}{2}(c_{11} - c_{12} - 2c_{44})\sum_i \xi_{ii}^2 + \frac{1}{4}c_{44}\sum_{i,k}(\xi_{ik} + \xi_{ki})^2. \quad (1.15)$$

Equality

$$c_{11} - c_{12} - 2c_{44} = 0 \quad (1.16)$$

provides a transition to the isotropic medium model. Obviously, in this case, the elastic energy density $U(\xi_{in})$ is determined by two second-order elastic moduli.

To construct the equations of motion of the theory of elasticity, we use the variational principle. For this purpose, we write down the density of the Lagrangian as [41]:

$$L(\dot{u}_i, \xi_{in}) = T(\dot{u}_i) - U(\xi_{in}), \quad T(\dot{u}_i) = \sum_i \frac{\rho}{2}\dot{u}_i^2, \quad (1.17)$$

where $T(\dot{u}_i)$ is the kinetic energy density. The elastic energy density $U(\xi_{in})$ is computed from the expression (1.14) or (1.15). The variational principle claims that the integral $\int L(\dot{u}_i, \xi_{in})\, drdt$ must have an extremum for variations that disappear at the integration boundaries. Consequently, we get the Euler equation for this problem [41]:

$$-\frac{\partial}{\partial t}\frac{\partial L(\dot{u}_i, \xi_{in})}{\partial \dot{u}_i} - \sum_n \frac{\partial}{\partial x_n}\frac{\partial L(\dot{u}_i, \xi_{in})}{\partial \xi_{in}} = 0. \quad (1.18)$$

Plugging the expressions (1.17) and (1.14) into eq. (1.18) yields a set of equations of motion for the components of the displacement vector $\mathbf{u}(\mathbf{r}, t)$ in the form [42, 43, 47]:

$$p\ddot{u}_1 = c_{11}\frac{\partial^2 u_1}{\partial x_1^2} + (c_{12} + c_{44})\left(\frac{\partial^2 u_2}{\partial x_1 \partial x_2} + \frac{\partial^2 u_3}{\partial x_1 \partial x_3}\right) + c_{44}\left(\frac{\partial^2 u_1}{\partial x_2^2} + \frac{\partial^2 u_1}{\partial x_3^2}\right),$$

$$p\ddot{u}_2 = c_{11}\frac{\partial^2 u_2}{\partial x_2^2} + (c_{12} + c_{44})\left(\frac{\partial^2 u_1}{\partial x_1 \partial x_2} + \frac{\partial^2 u_3}{\partial x_2 \partial x_3}\right) + c_{44}\left(\frac{\partial^2 u_2}{\partial x_1^2} + \frac{\partial^2 u_2}{\partial x_3^2}\right), \quad (1.19)$$

$$p\ddot{u}_3 = c_{11}\frac{\partial^2 u_3}{\partial x_3^2} + (c_{12} + c_{44})\left(\frac{\partial^2 u_1}{\partial x_1 \partial x_3} + \frac{\partial^2 u_2}{\partial x_2 \partial x_3}\right) + c_{44}\left(\frac{\partial^2 u_3}{\partial x_2^2} + \frac{\partial^2 u_3}{\partial x_1^2}\right).$$

Using the expression (1.15) allows one to simplify the set of eq. (1.19) [43]

$$\rho\ddot{u}_i = (c_{11} - c_{12} - 2c_{44})\frac{\partial^2 u_i}{\partial x_i^2} + (c_{12} + c_{44})\nabla_i(\nabla \cdot \mathbf{u}) + c_{44}\Delta u_i, \tag{1.20}$$

where ∇ is the gradient operator and Δ is the Laplace operator. When $c_{11} - c_{12} - 2c_{44} = 0$, the first term in the expression (1.19) vanishes. Then, for an elastically isotropic medium, we obtain the known result [43, 46]

$$\rho\ddot{\mathbf{u}} = (c_{11} - c_{44})\nabla(\nabla \cdot \mathbf{u}) + c_{44}\Delta\mathbf{u}, \tag{1.21}$$

This equation has an important property: it is invariant with respect to the rotation of the axes because each term of the equation is an invariant. Thus, the relation (1.16) involves a condition that the crystal is elastically isotropic. In such a crystal, the phase velocities are isotropic and coincide in direction with the group velocities, the transverse modes are degenerate, and the longitudinal and transverse wave velocities do not coincide [43, 46]. The monograph [43] defines the anisotropy factor of cubic crystals by the following ratio:

$$A = \frac{2c_{44}}{c_{11} - c_{12}}. \tag{1.22}$$

For elastically isotropic materials, A = 1 and, accordingly, $2c_{44} = c_{11} - c_{12}$.

The solutions of the set of eq. (1.19) can be represented in the form of plane monochromatic waves for which the displacement vector appears as:

$$\mathbf{u} = A\mathbf{e}(\mathbf{q})\exp[i(\omega t - \mathbf{qr})], \tag{1.23}$$

where $q = (q_1, q_2, q_3)$ is the wave vector, $\mathbf{e}(\mathbf{q})$ is the polarization vector that indicates the direction in which the particles of matter oscillate. If the vector $\mathbf{e}(\mathbf{q})$ is parallel to the wave vector $\mathbf{q}$, the wave is called *longitudinal*. If the vector $\mathbf{e}(\mathbf{q})$ is perpendicular to the wave vector $\mathbf{q}$, the wave is referred to as *transverse*. As we will see later, in cubic crystal, elastic waves are neither purely longitudinal nor purely transverse due to the anisotropy of elastic energy. A plane wave defined by the expression (1.23) can be a solution to the equations of motion (1.19) conditional upon the existence of a definite relationship between the frequency ω and the wave number q, which is called a dispersion ratio. An elastically anisotropic continuous medium (anisotropic continuum models) suggests that the wave vector $\mathbf{q}$ of a phonon is much smaller than the Debye wave vector q_D, and the phonon spectrum with polarization λ is given by

$$\omega_q^\lambda = S_0^\lambda(\theta, \varphi)\, q. \tag{1.24}$$

Substituting the plane wave solution (1.23) into the equations of motion (1.19) gives the Christofel set of equations [42, 47] for determining the polarization vectors $e\,(q)$

and the phonon spectrum in cubic crystals. In the coordinate system along the edges of the cube, the set of equations can be reduced to [44]:

$$\sum_j e_j\{(n_i n_j - \varepsilon\delta_{ij}) + (k-1)n_i n_j(1-\delta_{ij})\} = 0, \quad \varepsilon = \frac{S_0(\theta,\varphi)^2\rho - c_{44}}{c_{11}-c_{44}}. \tag{1.25}$$

Here $n_j = q_j/q$ are the projections of the unit phonon vector $\mathbf{n} = (\sin\theta\cos\varphi, \sin\theta\sin\varphi, \cos\theta)$ on the corresponding coordinate axes, δ_{ij} is the Kronecker symbol. As can be seen from eq. (1.25), the dimensionless parameter k-1 is the only one that characterizes the influence of elastic anisotropy on the dynamic characteristics of elastic waves in cubic crystals:

$$k-1 = \frac{c_{12}+c_{44}}{c_{11}-c_{44}} - 1 = \frac{c_{12}+2c_{44}-c_{11}}{c_{11}-c_{44}}. \tag{1.26}$$

The condition for the existence of a non-trivial solution to the set of homogeneous eq. (1.25) delivers the cubic equation for the phonon spectrum:

$$\varepsilon^3 - \varepsilon^2 - (k^2-1)\,\varepsilon\xi - (1-k)^2\,(2k+1)\eta = 0, \tag{1.27}$$

where $\xi = n_1^2 n_2^2 + n_1^2 n_3^2 + n_2^2 n_3^2$ and $\eta = n_1^2 n_2^2 n_3^2$ are cubic harmonics.

Equation (1.27) has three solutions that correspond to three acoustic modes: longitudinal (L) one and two transverse ones (t_1, t_2):

$$\begin{aligned} S_0^\lambda(\theta,\varphi) &= \sqrt{\frac{c_{44}}{\rho}}\left(1+\frac{c_{11}-c_{44}}{c_{44}}\varepsilon^\lambda\right)^{1/2}, \quad \varepsilon^\lambda = \frac{1}{3} + z^\lambda, \\ z^L &= \frac{2}{3}r\cos\frac{Q}{3}, \quad z^{t_1,t_2} = \frac{2}{3}r\cos\left(\frac{Q}{3} \mp \frac{2\pi}{3}\right), \\ r &= \sqrt{1+3(k^2-1)\xi}, \quad \cos Q = \frac{1}{r^3}(1+4.5(k^2-1)\xi + 13.5\eta(1-3k^2+2k^3)). \end{aligned} \tag{1.28}$$

The phonon polarization indices t_1 and t_2 correspond to the "fast" (upper) and "slow" (lower) transverse vibrational modes. Further, it will be shown that the polarization of transverse modes needs to be accounted for their classification and the dividing of transverse modes into fast and slow ones in some cases is physically incorrect.

After plugging the solutions (1.28) into the set of eq. (1.25), we can arrive at the polarization phonon vectors of various vibrational branches [44]:

$$e_j^\lambda = \frac{1}{A^\lambda}\left\{\frac{n_j}{\psi_j^\lambda}\right\}, \quad A^\lambda = \pm\sqrt{\sum_j \frac{n_j}{\left(\psi_j^\lambda\right)^2}}, \quad (\mathbf{e}^\lambda\mathbf{n}) = \frac{1}{A^\lambda}\sum_j \frac{n_j^2}{\psi_j^\lambda}, \quad \psi_j^\lambda = \varepsilon^\lambda + (k-1)n_j^2. \tag{1.29}$$

It is easy to verify that the following relations are satisfied for the polarization vectors:

$$\left(\mathbf{e}^\lambda, \mathbf{e}^{\lambda'}\right) = \delta_{\lambda,\lambda'}$$

It should be noted that although the formulas (1.29) are simple; however, they are difficult to use in calculating the phonon relaxation frequencies due to the 0/0 type uncertainties arising for certain symmetric directions (see below). To evaluate the values of e_j in these directions, we have to return to the original set of eq. (1.25).

For an isotropic medium, the parameter $k-1=0$. The spectrum anisotropy disappears, and eq. (1.27) is simplified to:

$$\varepsilon^2(\varepsilon-1)=0. \tag{1.30}$$

One of its solutions gives the phase velocity of longitudinal phonons $S^L=\sqrt{c_{11}/\rho}$.

Two other solutions coincide and give the phase velocities of the transverse phonons $S^{t_1}=S^{t_2}=\sqrt{c_{44}/\rho}$. At the same time, for the polarization vectors, we have $\mathbf{e}^L=\mathbf{n}$ and $(\mathbf{e}^t\mathbf{n})=0$. Thus, in an isotropic medium, purely longitudinal and purely transverse waves with fixed velocities travel. The latter are controlled by two second-order elastic moduli c_{11} and c_{44}, and the crystal density ρ. Unlike isotropic media, in cubic crystals, quasi-longitudinal or quasi-transverse vibrations propagate.

For each direction of the wave vector in the crystal, there are three independent waves with their phase velocities $S_0^\lambda(\theta,\varphi)$ and mutually perpendicular displacements. In the general case, none of these displacements coincides either with the normal to the wave front or with the perpendicular direction to the normal: i.e. the waves are neither purely longitudinal, nor purely transverse [42, 47]. However, in the symmetric directions, such as [100], [110], and [111], pure modes propagate, and in directions of the [100] and [111] type, the transverse modes are degenerate (see Figure 1.1). For elastically isotropic media, the parameter k is equal to unity, and the transverse modes are degenerate for all the directions. As can be seen from Figure 1.1, for cubic crystals, the speed of sound and, accordingly, the phonon spectra differ significantly from those in isotropic media. It should be noted that the spectrum anisotropy and the degeneracy points in the vibrational modes of transverse phonons cause essential differences in the phonon relaxation frequencies in anharmonic scattering processes in cubic crystals from those in isotropic media [38, 49].

An important role in the anisotropy of the spectrum and angular dependencies of the phonon polarization vectors is played by the absolute value and sign of the parameter $k-1=(c_{12}+2c_{44}-c_{11})/(c_{11}-c_{44})$. The sign of the parameter $k-1$ indicates dividing all cubic crystals into crystals with positive $(k-1>0)$ and negative $(k-1<0)$ anisotropy of the second-order elastic moduli (see Table 1.2). The first type includes crystals such as Ge, Si, diamond, GaSb, etc. (see Table 1.2). For them, the velocity of longitudinal phonons is minimum in the [100] directions and that of transverse phonons is maximum, whereas the velocity of the longitudinal phonons is maximum in the [111] directions and that of the transverse phonons is minimum in the [110] directions (see Figure 1.1a, c):

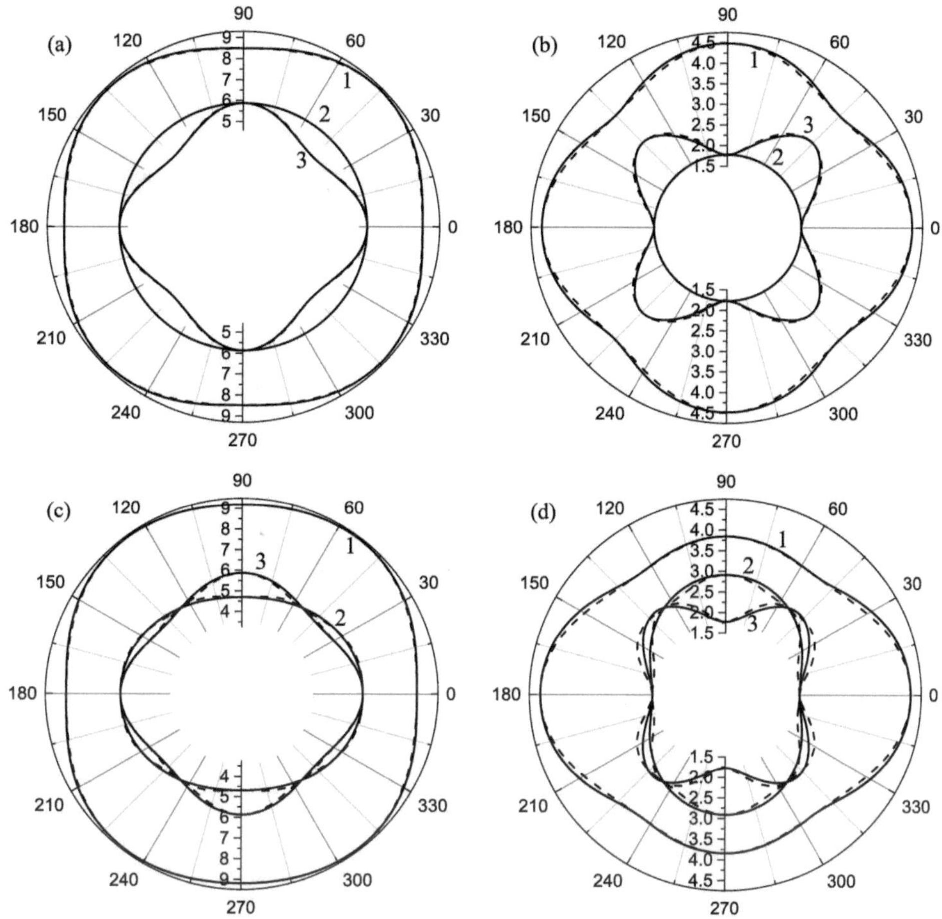

Figure 1.1: Angular dependencies of the sound velocity $S_0^\lambda(\theta,\varphi)$ in Si (a,c) and KCl (b,d) crystals for wave vectors lying in the cube face plane ($\varphi = 0$), (a,b) or in the cube diagonal plane ($\varphi = \pi/4$), (c,d). Solid curves 1 refer to quasi-longitudinal waves, solid curves 2 and 3 relate to t_1 and t_2 quasi-transverse modes, respectively. Dashed lines are the approximation (1.32).

$$S^L_{[100]} = \sqrt{\frac{c_{11}}{\rho}},\ S^L_{[111]} = S^L_{[100]}\left(1+\frac{2\Delta C}{3c_{11}}\right)^{1/2},\ S^L_{[110]} = S^L_{[100]}\left(1+\frac{\Delta C}{2c_{11}}\right)^{1/2},\ \Delta = (c_{12}+2c_{44}-c_{11}),$$
$$S^t_{[100]} = \sqrt{\frac{c_{44}}{\rho}},\ S^t_{[111]} = S^t_{[100]}\left(1-\frac{\Delta C}{3c_{44}}\right)^{1/2},\ S^{t1}_{[110]} = S^t_{[100]},\ S^{t2}_{[110]} = S^t_{[100]}\left(1-\frac{\Delta C}{2c_{44}}\right)^{1/2}. \tag{1.31}$$

The second type of cubic crystals ($k-1<0$) includes crystals of KCl, NaCl, etc. (see Table 1.2). As can be seen from a comparison of Figures 1.1a–d, the form of the spectra of the vibrational branches for crystals of the first and second type is qualitatively

Table 1.2: Second-order elastic moduli (10^{12} dyne/cm^2) and classification of cubic crystals in accordance with the anisotropy parameter $k-1$ (density ρ (g/cm^3)). The data are taken from [46, 50–59].

Type	Crystal	c_{11}	c_{12}	c_{44}	ρ	$k-1$	The degree of ionicity of the bond
I	Ge	1.289	0.483	0.671	5.32	0.87	0
	Si	1.677	0.65	0.804	2.3301	0.67	0
	Diamond	10.76	1.25	5.76	3.512	0.40	0
	GaAs	1.1904	0.5384	0.5952	5.317	0.90	0.12
	HgSe	0.69	0.51	0.23	8.26	0.61	0.15
	GaN	2.93	1.59	1.55	6.15	1.275	0.43
	GaSb	0.885	0.404	0.433	5.62	0.85	0.26
	InSb	0.672	0.367	0.302	5.76	0.81	0.32
	LiF	1.246	0.424	0.649	2.646	0.78	0.92
	MgO	2.86	0.87	1.48	3.58	0.703	0.84
	YAG	3.281	1.064	1.137	4.55	0.03	
	Na	0.0615	0.0469	0.0592	1.01	45.13	Metal
	Li	0.148	0.125	0.108	0.55	4.825	≫
	K	0.0457	0.0374	0.0263	0.91	2.284	≫
	Cu	1.684	1.214	0.754	8.94	1.116	≫
	Ag	1.22	0.915	0.448	10.49	0.766	≫
	Pb	0.466	0.392	0.144	11.34	0.655	≫
	Ni	2.465	0.473	1.247	8.9	0.412	≫
	Au	1.86	1.57	0.42	19.32	0.382	≫
	Al	1.069	0.626	0.285	2.7	0.162	≫
II	KCl	0.398	0.062	0.0625	1.98	−0.63	0.95
	NaCl	0.575	0.099	0.133	2.214	−0.48	0.94
	PbS	1.27	0.298	0.248	7.5	−0.466	0.20
	CaF_2	1.74	0.56	0.359	3.211	−0.33	0.89
	SrF_2	1.24	0.43	0.31	2.44	−0.204	0.89
	YIG	2.69	1.077	0.764	5.17	−0.04	
	Mo	4.6	1.76	1.1	10.19	−0.183	Metal
	W	5.01	1.98	1.514	19.2	−0.001	Metal

different, whereas the phonon spectra differ only quantitatively within one type of crystals. For these crystals, in contrast crystals of the first type, the velocity of longitudinal phonons is maximum in the [100] direction and minimum in the [111] direction, the velocity of transverse phonons is maximum in the [110] direction and minimum in the [100] direction (see Figure 1.1b, d). Although Ge, Si, and diamond falling into the first group of cubic crystals are covalent, and KCl, NaCl, and CaF_2 are ionic crystals, dividing cubic crystals by their elastic properties into two types is not related to the nature of their chemical bonding. It is worth emphasizing that the first group contains semiconductor crystals of GaSb and InSb, as well as ionic compounds LiF and MgO. Besides, metals of cubic symmetry also belong to both the first and second groups.

Apparently, there is no other physical criterion except for the analysis of the above Christofel equations that allows one to divide cubic crystals into two types (see also [44]). The deviation of sound velocities in cubic crystals, such as Ge, against elastically isotropic media, is relatively small; it does not exceed 23%. This means that, according to perturbation theory, the anisotropy of the spectrum can be taken into account. Bearing the cubic anisotropy in mind, we can also simplify the analytical calculations of the phonon relaxation frequencies. For this, the expressions for sound velocities (1.28) can be linearly approximated in cubic harmonics $\xi(\theta,\varphi)$ and $\eta(\theta,\varphi)$ with good accuracy. This is done in the following way:

$$S_0^\lambda(\theta,\varphi) = S_{[100]}^\lambda\{1 + a_1^\lambda \xi(\theta,\varphi) + a_2^\lambda \eta(\theta,\varphi)\}. \tag{1.32}$$

Coefficients a_1^λ and a_2^λ are found from the condition that, in all crystallographic directions of the type [100], [110] and [111], the expressions (1.32) coincide with the exact solution:

$$\begin{aligned} &a_1^L = 4\left[\sqrt{1+\frac{\Delta C}{2c_{11}}} - 1\right], \quad a_2^L = 27\left[\sqrt{1+\frac{2\Delta C}{3c_{11}}} + \frac{1}{3} - \frac{4}{3}\sqrt{1+\frac{\Delta C}{2c_{11}}}\right],\\ &a_1^{t_1} = 0, \quad a_2^{t_1} = 27\left[\sqrt{1-\frac{\Delta C}{3c_{44}}} - 1\right],\\ &a_1^{t_2} = 4\left[\sqrt{1-\frac{\Delta C}{2c_{44}}} - 1\right], \quad a_2^{t_1} = 27\left[\sqrt{1-\frac{\Delta C}{3c_{44}}} + \frac{1}{3} - \frac{4}{3}\sqrt{1-\frac{\Delta C}{2c_{44}}}\right] \end{aligned} \tag{1.33}$$

The analysis performed has shown that the approximation defined by formulas (1.32) and (1.33) (see Figure 1.1, dashed lines) reproduces well all the features of the vibrational branches of phonons in cubic crystals for an arbitrary direction. Approximation accuracy for wave vectors lying in the plane of the cube face amounts to about 1% for Si and 2% for KCl, while the maximum error increases to 3% for Si and 7% for KCl for the diagonal cross-section.

As for the phonon polarization vectors, quasi-longitudinal or quasi-transverse vibrations propagate in cubic crystals in an arbitrary direction, which does not coincide with any of the symmetric directions. In this case, the contribution of the transverse component to quasi-longitudinal vibrations in cubic crystals of the first and second types is small, and it can be neglected. Estimates show (see Figure 1.2, curves 1' and 2') that the quantity $(\mathbf{e}^L\mathbf{n}) = 1 - \delta^L$ differs little from unity, with the parameter taking the following values: $\delta^L < 0.02$ for crystals of the type of Ge, Si, diamond, GaSb, and $\delta^L < 0.03$ for KCl type crystals. The situation for the polarization vectors of quasi-transverse oscillations is generally more complicated, and the longitudinal component for quasi-transverse modes may be significant. In the literature, this issue is at present far from being studied well enough.

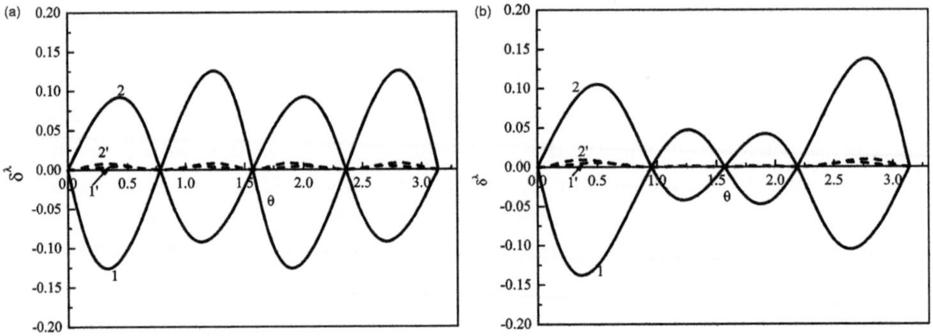

Figure 1.2: Angular dependencies of the quantities δ^L and δ^{t_2}, which characterize deviations of polarization vectors from pure modes for wave vectors lying in the plane of a cube face (a) and in a diagonal plane (b). Curves 1.1' mark Si crystals, curves 2.2' are for NaCl. Curves 1, 2 are intended for quasi-transverse modes and curves 1', 2' refer to for quasi-longitudinal modes.

Therefore, we will further examine the angular dependencies of the polarization vectors for the two most relevant cases, namely: (1) for wave vectors lying in the planes of the faces of the cube, (2) for wave vectors lying in the diagonal planes. For the phonon wave vector located in the XZ-plane, we have the quantity $\mathbf{n} = (\sin\theta, 0, \cos\theta)$, functions $\xi = n_1^2 n_2^2 = \sin^2\theta \, \cos^2\theta, \eta = 0$, and the cubic equation can be written as:

$$\varepsilon\left(\varepsilon^2 - \varepsilon - (k^2 - 1)\xi\right) = 0. \tag{1.34}$$

Its solution for longitudinal and transverse modes has the form:

$$\begin{aligned} &\varepsilon^{t_1} = 0, \quad \varepsilon^{L,\, t_2} = 1/2 \pm \sqrt{1/4 + (k^2 - 1)\xi}, \\ &S_0^{t_1} = S_{[100]}^t, \quad S_0^{L,\, t_2} = S_{[100]}^t \sqrt{1 + \frac{C_{11} - C_{44}}{C_{44}} \varepsilon^{L,\, t_2}}. \end{aligned} \tag{1.35}$$

Substituting these solutions into (1.25) (or using formulas (1.29)) yields an expression for the polarization vectors:

$$\begin{aligned} &\mathbf{e}^{t_1}(\theta) = \{0, 1, 0\}, \quad (\mathbf{e}^{t_1}\mathbf{n}_1) = 0, \quad e_x^{t_2} = \frac{n_x \psi_z \, sign(n_z \psi_z)}{\psi_2}, \\ &e_y^{t_2} = 0, \quad \psi_{x,z} = \varepsilon^{t_2}(k-1) n_{x,z}^2, \\ &e_z^{t_2} = \frac{n_z \psi_x sign(n_z \psi_z)}{\psi_2} = -\sqrt{1 - \left(e_x^{t_2}\right)^2}, \\ &\psi_2 = \sqrt{\varepsilon^{t_2}(14(k-1)\xi)2k(k-1)\xi}, \\ &(\mathbf{e}^{t_2}\mathbf{n}) = \delta^{t_2} = \frac{sign(n_z \psi_z)}{\psi_2}\left(\varepsilon^{t_2} + 2(k-1)\xi\right) = \cos(\theta_e - \theta), \end{aligned} \tag{1.36}$$

where θ_e is the angle between the polarization vector $\mathbf{e}^{t_2}$ and the Z-axis. So, the mode ε^{t_1} is purely transverse relative to the polarization vector directed perpendicular to the XZ-plane at hand. It is a fast mode for cubic crystals with positive anisotropy of the second-order elastic moduli $(k-1>0)$ and a slow one for crystals with negative anisotropy $(k-1<0)$. The mode ε^{t_2} is generally a mixed transverse-longitudinal mode. For it, the polarization vector lies in the XZ-plane, with, in type-I crystals (Ge, Si), it corresponds to the slow mode, and in type-II cubic crystals (KCl, NaCl), it being a fast transverse mode (see Figure 1.1). The quantity δ^{t_2} is responsible for the deviation of the polarization vector $\mathbf{e}^{t_2}$ from the pure mode. The maximum values of the longitudinal components of this mode δ^{t_2} reach 15.5% for Ge crystals and 24% for KCl. The absolute values of $|k-1|$, which characterize the anisotropy of the elastic properties of cubic crystals, decrease upon transition from crystals Ge to Si, diamond, and YAG crystals. Therefore, the maximum values of δ^{t_2} diminish to 12.6% for Si crystals, 8.4% for diamond, 9.1% for NaCl, and to vanishingly small values for YAG and YIG.

As can be seen from Figure 1.3a, the values of the x-components of the polarization vectors e^{t_2} of the quasi-transverse modes in Si and NaCl crystals (curves 1 and 2) and the corresponding pure mode components $\mathbf{e}_0^{t_2}=(\cos\theta, 0, -\sin\theta)$ for which $(\mathbf{e}_0^{t_2}, \mathbf{n})=0$ for both types of crystals differ more significantly than that of δ^{t_2}. Let us look at the ranges of angles $0<\theta<\pi/4$ and $\pi/4<\theta<\pi/2$ for type-I crystals (Ge, Si) (see Figures 1.2 and 1.3). The angle θ_e between the wave vector and the polarization vector $\mathbf{e}^{t_2}$ exceeds the value of $\pi/2$ and $\delta^{t_2}<0$ and is less than $\pi/2$ and $\delta^{t_2}>0$, respectively. For type-II crystals (KCl, NaCl), the situation is the opposite (see Figures 1.2 and 1.3, curves 2).

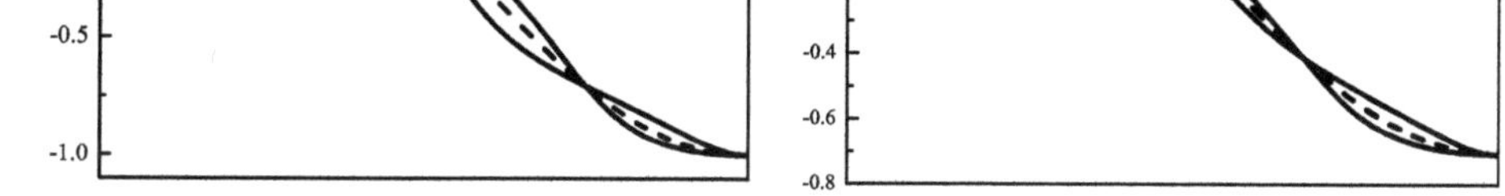

Figure 1.3: Angular dependencies of the x-component of the polarization vectors $\mathbf{e}^{t_2}$ of quasi-transverse modes in Si (curves 1) and NaCl (curves 2) crystals for phonon wave vectors lying in the plane of a cube face (a) or in the diagonal plane (b). The dashed curves 3 mark the corresponding pure-mode components.

More interesting results come from analysing the spectrum and polarization vectors for the diagonal cross-section: $\varphi = \pi/4$, $\mathbf{n} = (\sin\theta/\sqrt{2}, \sin\theta/\sqrt{2}, \cos\theta)$, $\xi = 2n_x^2(1 - 3/2n_x^2)$, $\eta = n_x^4(1 - 2n_x^2)$. In this case, the cubic eq. (1.27) can be factorized for the function $\psi_x = \varepsilon + (k-1)n_x^2$:

$$\psi_x \cdot \left[\psi_x^2 - \psi_x\left(1 + 3(k-1)n_x^2\right) + 2k(k-1)n_x^2\left(3n_x^2 - 1\right)\right] = 0. \tag{1.37}$$

One of the solutions of this equation, $\psi_x^{t_1} = 0$, gives:

$$\varepsilon^{t_1} = -\frac{1}{2}(k-1)\sin^2\theta, \quad S_0^{t_1}\left(\theta, \frac{\pi}{4}\right) = S_{[100]}^t\sqrt{1 - \frac{\Delta C}{2c_{44}}\sin^2\theta}. \tag{1.38}$$

The analysis showed that the polarization vector for this mode is perpendicular to the diagonal plane, and this solution is unique:

$$\mathbf{e}^{t_1} = \left(-1/\sqrt{2}, 1/\sqrt{2}, 0\right), \quad \left(\mathbf{e}^{t_1}, \mathbf{n}\right) = 0. \tag{1.39}$$

Thus, this mode is purely transverse despite the fact that the phonon spectrum for it is anisotropic (see Figure 1.1c, d). According to the concepts based on the model of an isotropic medium, this finding may seem unexpected at first glance. This is because the pure modes should correspond to the isotropic spectra of the vibrational branches, and the anisotropy of the spectrum of the transverse vibrational modes should lead to a deviation from the mutual perpendicularity of the phonon wave vector and the polarization vector. However, the symmetry analysis of elastic waves in crystals [42] indicates that for wave vectors lying in the plane of symmetry or in the plane perpendicular to the axis of symmetry of even order, one of the three normal waves is purely transverse, with its displacement being perpendicular to the plane at hand. The diagonal plane ($\varphi = \pi/4$) is just the plane of symmetry of the cubic crystal. Therefore, the result obtained above is consistent with the analysis performed in [42]. Note that the mode ε^{t_1} is a fast one for type-I cubic crystals and slow for type-II crystals in the ranges of angles $-\theta_{[111]} < \theta < \theta_{[111]}$ and $\pi - \theta_{[111]} < \theta < \pi + \theta_{[111]}$, where θ is the angle between the Z-axis and the [111] direction. In the angle range $\theta_{[111]} < \theta < \pi - \theta_{[111]}$ and $\pi + \theta_{[111]} < \theta < 2\pi - \theta_{[111]}$, it is a slow mode for type-I cubic crystals and fast for type-II crystals.

Two other solutions of eq. (1.37) correspond to the longitudinal and second transverse modes:

$$\begin{aligned} \psi_x^{L,t_2} &= 1/2\left(1 + 3(k-1)n_x^2\right) \pm \sqrt{1/4\left(1 + 3(k-1)n_x^2\right)^2 - 2k(k-1)n_x^2\left(3n_x^2 - 1\right)}, \\ S_0^{L,t_2} &= S_{[100]}^t\sqrt{1 + \frac{c_{11} - c_{44}}{c_{44}}\varepsilon^{L,t_2}}, \quad \varepsilon^{L,t_2} = \psi_x^{L,t_2} - (k-1)n_x^2. \end{aligned} \tag{1.40}$$

The ε^{t_2} mode is generally a mixed transverse-longitudinal mode. Note that it is a slow one for type-I cubic crystals and fast for type-II crystals in the range of angles

$-\theta_{[111]} < \theta < \theta_{[111]}$ and $\pi - \theta_{[111]} < \theta < \pi + \theta_{[111]}$. In the range of angles $\theta_{[111]} < \theta < \pi - \theta_{[111]}$ and $\pi + \theta_{[111]} < \theta < 2\pi - \theta_{[111]}$, it is fast for type-I cubic crystals and slow for type-II crystals (see Figure 1.1). Thus, the transverse modes intersect themselves in the [111] direction instead of touching the upper and lower transverse branches. So, dividing the transverse modes into fast and slow ones is not physically correct in this case. Therefore, the polarization vectors must be taken into account to classify the transverse modes. The vector polarization for the ε^{t_2} mode lies in the diagonal plane and can be described by:

$$e_x^{t_2} = e_y^{t_2} \frac{n_x \psi_z sign\,(n_z \psi_z)}{\psi_2}, \quad e_z^{t_2} = \frac{n_z \psi_x sign\,(n_z \psi_z)}{\psi_2} = -\sqrt{1 - 2\left(e_x^{t_2}\right)^2},$$
$$\psi_x = \psi_y = \varepsilon^{t_2} + (k-1)n_x^2, \quad \psi_z = \varepsilon^{t_2} + (k-1)n_z^2, \quad \psi_2 = \sqrt{2n_x^2\psi_z^2 + n_z^2\psi_x^2}, \tag{1.41}$$
$$(\mathbf{e}^{t_2}\mathbf{n}) = \delta^{t_2} = \frac{(2n_x^2\psi_z + n_z^2\psi_x)}{\psi_2} sign(n_z\psi_z) = \cos\,(\theta_e - \theta)\,.$$

The maximum value of the longitudinal component of the ε^{t_2} mode amounts to 16.5% for Ge crystals and reaches 27% for a KCl crystal (see Figure 1.2). Since the anisotropy of the elastic properties of cubic crystals decreases during the transition from Ge crystals to Si, diamond, and YAG crystals, the maximum values of δ^{t_2} drop up to 13.8% for Si crystals, 9.3% for diamond, 10.4% for NaCl and 0.6% for YAG.

Figure 1.3b illustrates that the values of the x-components of the polarization vectors e^{t_2} of the quasi-transverse modes in Ge and NaCl crystals (curves 1 and 2) and the corresponding components for the pure mode $\mathbf{e}_0^{t_2} = (\cos\theta/\sqrt{2}, \cos\theta/\sqrt{2}, -\sin\theta)$ for which $(\mathbf{e}_0^{t_2}\mathbf{n}) = 0$ for both types of crystals differ more significantly than those of δ^{t_2}. The expressions (1.41) imply that the z-components of the polarization vectors $\mathbf{e}^{t_2}$ can be uniquely determined through the x-components. Therefore, we may omit them. As can be seen from Figures 1.2 and 1.3 (curves 1), the angle θ^{t_2} is larger than $\pi/2$ in the range of angles $0 < \theta < \pi/4$ and $\delta^{t_2} < 0$ for type-I crystals (Ge, Si). At the same time, the angle θ^{t_2} is less than $\pi/2$ in the range of angles $\pi/4 < \theta < \pi/2$ and $\delta^{t_2} > 0$. For type-II crystals (KCl, NaCl), the situation with signs of the quantity δ^{t_2} is the opposite (see Figures 1.2 and 1.3, curves 2). So, the difference in the behavior of the polarization vectors for type-I and type-II crystals looks essential.

Having conducted the analysis, we can infer that both quasi-transverse modes are mixed transverse-longitudinal ones with a point of degeneration in the direction [001]. This holds true for the wave vectors lying in planes with the angles $\varphi \neq 0$, $\pi/4$, $3\pi/4$, $5\pi/4$, and $7\pi/4$. They can be classified as "fast" and "slow" modes, since $S^{t_1}(\theta, \varphi) \geq S^{t_2}(\theta, \varphi)$. Their polarization vectors $\mathbf{e}^{t_1}$ and $\mathbf{e}^{t_2}$ are significantly different from the polarization vectors of pure modes, $\mathbf{e}_0^{t_1} = (-\sin\varphi, \cos\varphi, 0)$ and $\mathbf{e}_0^{t_2} = (\cos\theta\cos\varphi, \cos\theta\sin\varphi, -\sin\theta)$, respectively. The former is perpendicular to the plane φ = const, the latter lies in the plane φ = const. As can be seen from Figure 1.4, $\theta \to 0$, the polarization vector $\mathbf{e}^{t_1}$ of the fast quasi-transverse mode tends to the vector

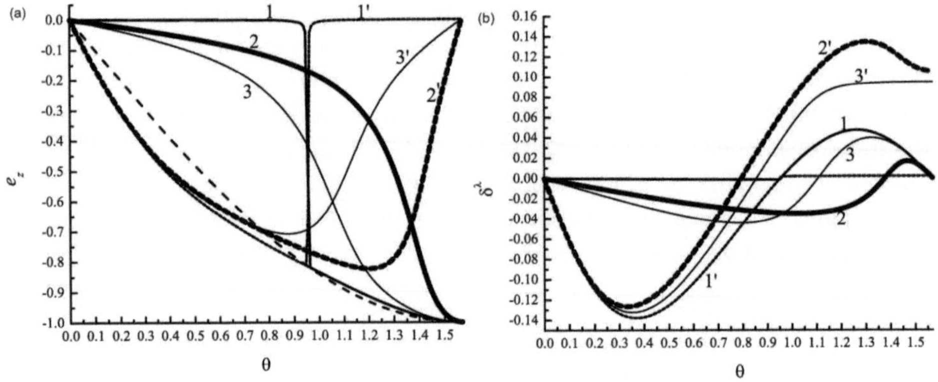

Figure 1.4: Angular dependencies of the z-components of the polarization vectors $\mathbf{e}^{t_1}$ and $\mathbf{e}^{t_2}$ (a) and the quantities δ^{t_1} and δ^{t_2} for quasi-transverse modes (b) in Si crystals. Curves 1–3 represent the fast mode ε^{t_1}, curves 1'–3' refer to the slow mode ε^{t_2}. Pair curves (1,1'), (2,2'), and (3,3') correspond to the angles $\varphi = \pi/4$; $\varphi = \pi/16$, and $\varphi = \pi/6$, respectively. Dashed curves relate to purely transverse modes.

$\mathbf{e}_0^{t_1}$, i.e. to the direction perpendicular to the plane φ = const. As the angle θ increases, the vector e^{t_1} deviates from the vector $\mathbf{e}_0^{t_1}$ and tends to the vector $\mathbf{e}_0^{t_2}$ as $\theta \to \pi/2$, i.e. it goes over into the plane φ = const (see Figure 1.4a). On the other hand, as $\theta \to 0$, the polarization vector e^{t_2} of the slow mode tends to the vector $\mathbf{e}_0^{t_2}$ lying in the plane φ = const, and when the angle θ increases, it leaves this plane and when $\theta \to \pi/2$ tends to the vector $\mathbf{e}_0^{t_1}$, i.e to the direction perpendicular to the plane φ = const.

The closer the angle φ is to $\pi/4$, the more sharply the angular dependencies of the components of the polarization vectors $\mathbf{e}^{t_1}$ and $\mathbf{e}^{t_2}$ change in the vicinity of the angle $\theta = \theta_{[111]}$ (see Figure 1.4a). The maximum values of the longitudinal components of the quasi-transverse modes δ^{t_1} and δ^{t_2} do not go beyond 17% for Ge crystals and 27% for KCl. However, the values and angular dependencies of the components of the polarization vectors $\mathbf{e}^{t_1}$ and $\mathbf{e}^{t_2}$ of the quasi-transverse modes and those of $\mathbf{e}_0^{t_1}$ and $\mathbf{e}_0^{t_2}$ of the corresponding pure modes differ more significantly. It is worth pointing out that the contribution of the transverse component to quasi-longitudinal vibrations in this case is also small, and it can be ignored. Here, we can emphasize one curious consequence of the outcomes obtained. Since the electron–phonon interaction through the deformation potential in metals and semiconductors is proportional to the scalar product $(\mathbf{e}^{\lambda}\mathbf{q})$, the electrons can "relax" their momentum due to the longitudinal component of quasi-transverse vibrations. On the other hand, quasi-transverse modes can transfer their momentum to electrons and enhance the effect of electron–phonon drag in such semiconductor crystals as Ge, Si, GaSb, HgSe, and InSb.

So, the main conclusions of the analysis of the dynamic characteristics of phonons in cubic crystals in the anisotropic continuum model can be formulated as follows [44]:

- All cubic crystals can be divided into two types: crystals with positive and negative anisotropy of second-order elastic moduli. The shape of the spectrum and the behavior of the polarization vectors of vibrational modes for crystals of the first and second type are qualitatively different.
- The contribution of the transverse component to quasi-longitudinal vibrations for the first- and second-type cubic crystals is small, and it can be neglected.
- Polarization of transverse modes needs to be accounted for their classifying. Dividing of the transverse modes into fast and slow modes adopted in [46, 47] is sometimes not physically correct.
- For wave vectors lying in the plane of a cube face or a diagonal plane, transverse modes with polarization vectors perpendicular to these planes are pure modes, and vibrational modes with polarization vectors lying in these planes are mixed transverse-longitudinal ones. For quasi-transverse modes, the longitudinal component is not small and cannot be left aside when analysing the absorption of ultrasound and phonon transport in cubic crystals.

1.2 Approximation of the Spectrum of Thermal Phonons in Cubic Crystals

To date, experimental research concerning the phonon spectrum in cubic crystals has only focused on symmetric directions [60–62]. For analysing the temperature dependencies of thermal conductivity, it is required to know the spectrum of phonons over the entire Brillouin zone. The anisotropic continuum model allows one to resolve this issue, but it is valid only for wave vectors that are much smaller than the Debye wave vectorq_D. Based on inelastic neutron scattering data of [60.61], we approximate the phonon spectrum for Si and Ge crystals over the entire Brillouin zone. For this purpose, we represent it through [39, 63]:

$$\begin{aligned} \omega_q^\lambda &= S_0^\lambda(\theta,\varphi)q\{1-f^\lambda(x,\theta,\varphi)\} \\ f^\lambda(x,\theta,\varphi) &= A^\lambda(\theta,\varphi)x^2 - B^\lambda(\theta,\varphi)x^4 + C^\lambda(\theta,\varphi)x^6 - \dots \end{aligned} \tag{1.42}$$

Here, the function $f^\lambda(x,\theta,\varphi)$ describes the dispersion of thermal phonons, the coefficients $A^\lambda(\theta,\varphi)$, $B^\lambda(\theta,\varphi)$, and $C^\lambda(\theta,\varphi)$ characterize the angular dependencies of the phonon spectrum, $x=q/q_{\max}$ is the reduced wave vector, and $q_{\max}(\theta,\varphi)$ is a wave vector at the boundary of the Brillouin zone. For symmetrical directions in a face-centered cubic lattice (fcc), it appears as

$$q_{max}^{[001]} = 2\pi/a, \quad q_{max}^{[101]} = (2\pi/a)\left(3\sqrt{2}/4\right), \quad q_{max}^{[111]} = (2\pi/a)\left(\sqrt{3}/2\right), \tag{1.43}$$

where a is the lattice constant (for silicon a = 5.431 Å, for germanium a = 5.66 Å). The quantity $q_{max}(\theta, \varphi)$ can be given by the following relations:

$$\begin{gathered} q_1 = (\theta, \varphi) = \frac{q_{max}^{[001]}}{|n_1|}, \quad q_2 = (\theta, \varphi) = \frac{q_{max}^{[001]}}{|n_2|}, \\ q_3 = (\theta, \varphi) = \frac{q_{max}^{[001]}}{|n_3|}, \quad q_4 = (\theta, \varphi) = \frac{3/2 q_{max}^{[001]}}{|n_1| + |n_2| + |n_3|}, \\ q_{max}(\theta, \varphi) = \min\{q_1(\theta, \varphi), q_2(\theta, \varphi), q_3(\theta, \varphi), q_4(\theta, \varphi)\}, \end{gathered} \tag{1.44}$$

where n_1, n_2, n_3 are components of the vector $\mathbf{n}$. In the long-wavelength limit $x \ll 1$, we can set $f^\lambda(x, \theta, \varphi) = 0$. Then, approximation (1.42) turns into the anisotropic continuum model (1.28), in which the anisotropy of the spectrum is determined by the phase velocity $S_0^\lambda(\theta, \varphi)$.

The experimental spectrum is approximated in two stages. The first stage includes the approximation for symmetric directions using the expansion in powers of a reduced wave vector. In doing so, we fix the angles $\theta = \theta_{[I]}$ and $\varphi = \varphi_{[I]}$ that correspond to symmetric directions, where $[I]$ = [100], [110], or [111]. Further, we expand the experimental spectrum in powers of the reduced wave vector x and find the coefficients $A_{[I]}^\lambda$, $B_{[I]}^\lambda$, and $C_{[I]}^\lambda$, for each acoustic mode. The values of these coefficients are listed in Table 1.3. As can be understood from Figure 1.5, the approximation of the phonon spectrum in crystals in Si and Ge by a seventh-degree polynomial reproduces rather well the experimental data for all vibrational modes in symmetric directions. The average error of this approximation is 0.9% for longitudinal phonons and amounts to 1.3% for transverse phonons. As can be evidenced from Table 1.3 below, the influence of dispersion on the spectrum of longitudinal phonons is maximum in the [110] direction and minimum in the [100] direction: the coefficient $A_{[101]}^L$ that specifies the role of dispersion when $x \ll 1$ in the [110] direction exceeds the coefficients $A_{[001]}^L$ and $A_{[111]}^L$ by 3.4 and 2.6 times, respectively. Note that the anisotropy- and dispersion-related effects for longitudinal phonons in Si and Ge are much weaker than for transverse ones. For transverse modes when $0.5 \le x \le 1$, there are plane sections, and the transverse mode t_1 has a maximum for x = 0.9. These features provoke the appearance of appropriate peaks in the density of phonon states [60, 61].

The second step involves the extrapolation of the phonon spectrum measured for symmetric directions over the entire Brillouin zone. It can be done by using the expansion in cubic harmonics similarly to how it was proposed in [39, 44]. According to [44], the expressions for sound velocities deduced within the anisotropic continuum model can be approximated with good accuracy in the linear approximation in the cubic harmonics $\xi(\theta, \varphi)$ and $\eta(\theta, \varphi)$. The all-direction averaged error for Si is less than 1% (see [44], Figure 2). Similarly to [39, 44], we present the

Table 1.3: Parameters determining the approximation of the phonon spectrum in Si and Ge. The spectrum is obtained from data on inelastic neutron scattering over the entire Brillouin zone [39, 63].

Longitudinal mode						Transverse modes					
	Si	Ge		Si	Ge		Si	Ge		Si	Ge
$A^L_{[001]}$	0.27	0.20	a^L_1	9.60	13.5	$A^t_{[001]}$	1.35	1.45	a^{t1}_1	−1.91	−1.91
$B^L_{[001]}$	0.13	0.10	a^L_2	−77.67	−116.2	$B^t_{[001]}$	1.22	1.27	a^{t1}_2	17.25	16.71
$C^L_{[001]}$	0.064	0.073	b^L_1	18.21	25.9	$C^t_{[001]}$	0.45	0.44	b^{t1}_1	−1.63	−2.21
$A^L_{[101]}$	0.91	0.87	b^L_2	−129.44	−231.9	$A^{t1}_{[101]}$	0.71	0.76	b^{t1}_2	14.94	21.49
$B^L_{[101]}$	0.72	0.75	c^L_1	8.86	9.64	$B^{t1}_{[101]}$	0.72	0.57	c^{t1}_1	−0.26	−1.43
$C^L_{[101]}$	0.21	0.25	c^L_2	−30.51	−80.9	$C^{t1}_{[101]}$	0.43	0.28	c^{t1}_2	2.66	17.47
$A^L_{[111]}$	0.35	0.24				$A^{t2}_{[101]}$	0.90	0.98	a^{t2}_1	−1.34	−1.29
$B^L_{[111]}$	0.30	0.11				$B^{t2}_{[101]}$	0.56	0.69	a^{t2}_2	12.12	11.11
$C^L_{[111]}$	0.18	0.089				$C^{t2}_{[101]}$	0.16	0.23	b^{t2}_1	−2.18	−1.81
						$A^t_{[111]}$	1.35	1.42	b^{t2}_2	19.81	17.88
						$B^t_{[111]}$	1.23	1.34	c^{t2}_1	−2.56	−1.92
						$C^t_{[111]}$	0.46	0.51	c^{t2}_2	23.44	21.84

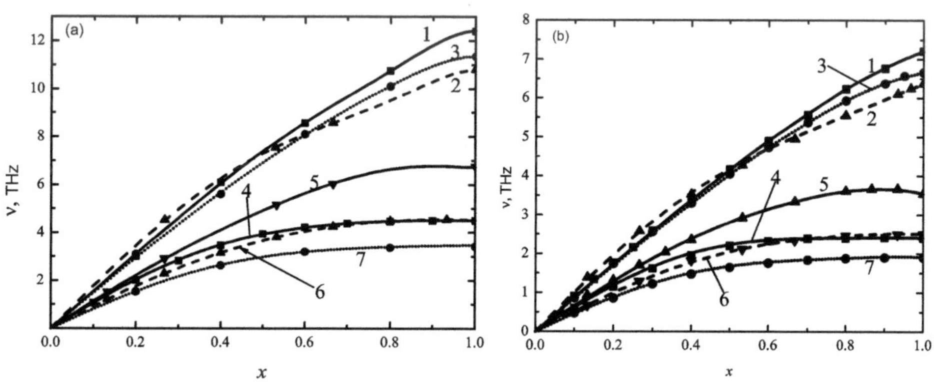

Figure 1.5: Dependencies of the spectra in Si (a) and Ge (b) for longitudinal (curves 1, 2, 3) and transverse phonons (curves 4, 5, 6, 7) on the reduced wave vector $x = q/q_{max}$ in the approximation of the experimental data [60, 61]. They are designated by symbols and a seventh-degree polynomial for different symmetry directions: curves 1, 4 refer to the [001] direction; curves 2, 5, and 6 denote the [110] direction, (5 is a fast transverse mode, 6 is a slow transverse mode); curves 3, 7 mark the [111] direction.

coefficients $A^\lambda(\theta,\varphi)$, $B^\lambda(\theta,\varphi)$, and $C^\lambda(\theta,\varphi)$ for the entire Brillouin zone in the linear approximation in the cubic harmonics $\xi(\theta,\varphi)$ and $\eta(\theta,\varphi)$ in such a way:

$$\begin{aligned}A^\lambda(\theta,\varphi)&=A^\lambda_{[001]}\left\{1+a_1^\lambda\xi(\theta,\varphi)+a_2^\lambda\eta(\theta,\varphi)\right\},\\ B^\lambda(\theta,\varphi)&=B^\lambda_{[001]}\left\{1+b_1^\lambda\xi(\theta,\varphi)+b_2^\lambda\eta(\theta,\varphi)\right\},\\ C^\lambda(\theta,\varphi)&=C^\lambda_{[001]}\left\{1+c_1^\lambda\xi(\theta,\varphi)+c_2^\lambda\eta(\theta,\varphi)\right\},\end{aligned}\tag{1.45}$$

Given that $\xi(0,0)=\eta(0,0)$ for the [001] symmetric directions, $\xi(\pi/4,0)=0.25$, $\eta(\pi/4,0)=0$, for the [101] directions, $\xi(\theta_{[111]},\pi/4)=1/3$, $\eta(\theta_{[111]},\pi/4)=1/27$ for the [111] directions, we find the coefficients a_i^λ, b_i^λ, and c_i^λ:

$$\begin{aligned}a_1^\lambda&=4\left(A^\lambda_{[110]}-A^\lambda_{[100]}\right)/A^\lambda_{[100]}, & a_2^\lambda&=9\left(3A^\lambda_{[111]}-4A^\lambda_{[110]}+A^\lambda_{[100]}\right)/A^\lambda_{[100]},\\ b_1^\lambda&=4\left(B^\lambda_{[110]}-B^\lambda_{[100]}\right)/B^\lambda_{[100]}, & b_2^\lambda&=9\left(3B^\lambda_{[111]}-4B^\lambda_{[110]}+B^\lambda_{[100]}\right)/B^\lambda_{[100]},\\ c_1^\lambda&=4\left(C^\lambda_{[110]}-C^\lambda_{[100]}\right)/C^\lambda_{[100]}, & c_2^\lambda&=9\left(3C^\lambda_{[111]}-4C^\lambda_{[110]}+C^\lambda_{[100]}\right)/C^\lambda_{[100]}.\end{aligned}\tag{1.46}$$

The values of these coefficients for Si and Ge crystals are given in Table 1.3. The approximation above preserves the cubic anisotropy of the spectrum; it allows one to analyse the change in the spectrum anisotropy during the transition from the case of long-wavelength phonons ($x<<1$) to the case of short-wavelength ones ($x\sim1$).

Comparison of the spectra calculated according to formulas (1.42)–(1.46) for Si and Ge and experimental data in [60, 61] showed that the anisotropic continuum model describes well the spectrum for all acoustic modes for $x\le2$. When $x\approx0.5$, the aforesaid approximation makes allowance for all the features of the spectrum of vibrational modes, with an error not exceeding 5%, but on approaching the boundary of the Brillouin zone, the approximation error rises (see Figure 1.5, curves 5). At the boundary of the Brillouin zone in symmetrical directions, the approximation error is small, it is less than 1.5%. However, in directions other than symmetrical, the maximum deviation from experimental data can reach 20%.

It should be noted that for Ge crystals, the dependencies of the phonon spectrum on the reduced wave vector in the [210]-type directions were measured in [60]. Therefore, the spectrum of acoustic modes can be refined for wave vectors in the plane of a cube face. Let us approximate the spectrum with a seventh-degree polynomial in the [210] direction as it was done earlier for symmetric directions. Further, we calculate the coefficients $A^\lambda_{[210]}$, $B^\lambda_{[210]}$, and $C^\lambda_{[210]}$ (see Table 1.4). Since $\eta=0$ in the plane of the face of the cube, we take $(\xi(\theta,\varphi))^2$ as the third harmonics. Thus, the expressions for the coefficients $A^\lambda(\theta,\varphi)$, $B^\lambda(\theta,\varphi)$, and $C^\lambda(\theta,\varphi)$ can be modified:

$$
\begin{aligned}
A^{\lambda}(\theta,\varphi) &= A^{\lambda}_{[001]}\{1+a_1^{\lambda}\xi(\theta,\varphi)+a_2^{\lambda}\eta(\theta,\varphi)+a_3^{\lambda}\xi^2(\theta,\varphi)\},\\
B^{\lambda}(\theta,\varphi) &= B^{\lambda}_{[001]}\{1+b_1^{\lambda}\xi(\theta,\varphi)+b_2^{\lambda}\eta(\theta,\varphi)+b_3^{\lambda}\xi^2(\theta,\varphi)\},\\
C^{\lambda}(\theta,\varphi) &= C^{\lambda}_{[001]}\{1+c_1^{\lambda}\xi(\theta,\varphi)+c_2^{\lambda}\eta(\theta,\varphi)+c_3^{\lambda}\xi(\theta,\varphi)\}.
\end{aligned}
\tag{1.47}
$$

It should be drawn attention that $\xi(\pi/2,\varphi_{[210]})=4/25, \eta(\pi/2,\varphi_{[210]})=0$ for the [210] direction.

Then, we write down the coefficients $a_i^{\lambda}, b_i^{\lambda}$, and c_i^{λ} that govern the phonon spectrum as follows:

$$
\begin{aligned}
a_1^{\lambda} &= \left(-369A^{\lambda}_{[100]}-256A^{\lambda}_{[110]}+625A^{\lambda}_{[210]}\right)\Big/-\left(36A^{\lambda}_{[100]}\right),\\
a_2^{\lambda} &= \left(-117A^{\lambda}_{[100]}-832A^{\lambda}_{[110]}+324A^{\lambda}_{[111]}+625A^{\lambda}_{[210]}\right)\Big/-\left(12A^{\lambda}_{[100]}\right),\\
a_3^{\lambda} &= 25\left(9A^{\lambda}_{[100]}+16A^{\lambda}_{[110]}-25A^{\lambda}_{[210]}\right)\Big/-\left(9A^{\lambda}_{[100]}\right),\\
b_1^{\lambda} &= \left(-369B^{\lambda}_{[100]}-256B^{\lambda}_{[110]}+625B^{\lambda}_{[210]}\right)\Big/-\left(36B^{\lambda}_{[100]}\right),\\
b_2^{\lambda} &= \left(-117B^{\lambda}_{[100]}-832B^{\lambda}_{[110]}+324B^{\lambda}_{[111]}+625B^{\lambda}_{[210]}\right)\Big/-\left(12B^{\lambda}_{[100]}\right),\\
b_3^{\lambda} &= 25\left(9B^{\lambda}_{[100]}+16A^{\lambda}_{B[110]}-25B^{\lambda}_{[210]}\right)\Big/-\left(9B^{\lambda}_{[100]}\right),\\
c_1^{\lambda} &= \left(-369C^{\lambda}_{[100]}-256C^{\lambda}_{[110]}+625C^{\lambda}_{[210]}\right)\Big/-\left(36C^{\lambda}_{[100]}\right),\\
c_2^{\lambda} &= \left(-117C^{\lambda}_{[100]}-832C^{\lambda}_{[110]}+324C^{\lambda}_{[111]}+625C^{\lambda}_{[210]}\right)\Big/-\left(12C^{\lambda}_{[100]}\right),\\
c_3^{\lambda} &= 25\left(9C^{\lambda}_{[100]}+16C^{\lambda}_{[110]}-25C^{\lambda}_{[210]}\right)\Big/-\left(9C^{\lambda}_{[100]}\right).
\end{aligned}
\tag{1.48}
$$

The values of these coefficients for Ge crystals are given in Table 1.4. The quadratic term of cubic harmonics in (1.47) causes a change in the approximation spectrum of phonon modes for wave vectors in the plane of the cube face. To illustrate the change, we compare the dependencies depicted in Figures 1.6 and 1.7. The figures display that, at the boundary of the Brillouin zone, the harmonics $(\xi(\theta,\varphi))^2$ involves significantly improving the agreement of the approximation results with the experimental data. In this case, the spectrum of longitudinal phonons in the plane of the face can be approximated with a maximum error of 4% instead of 17%. The error is equal to 8% and decreases up to 5% for the fast and slow modes, respectively. Thus, we have demonstrated that the error in approximating the spectrum can be dramatically declined by including higher-order terms of the cubic harmonics $\xi(\theta,\varphi)$ and $\eta(\theta,\varphi)$ into the expansion when collecting experimental data for asymmetric cross-sections.

To avoid complicating the problem of calculating the thermal conductivity any further, we restrict ourselves to the linear approximation of the spectrum in cubic harmonics according to the expressions (1.45). As an example, we choose the change

Table 1.4: Parameters that determine the approximation of the phonon spectrum in Ge crystals over the entire Brillouin zone taking into account the direction [210].

	Longitudinal mode	Mode t_1	Mode t_2
$A^{\lambda}_{[210]}$	1.05	1.17	1.10
$B^{\lambda}_{[210]}$	1.22	1.13	1.06
$C^{\lambda}_{[210]}$	0.59	0.48	0.39
a^{λ}_{1}	49.96	−0.68	−2.53
a^{λ}_{2}	−5.51	20.31	7.84
a^{λ}_{3}	−146.0	−5.55	4.16
b^{λ}_{1}	148.84	−0.52	−1.93
b^{λ}_{2}	136.82	25.03	17.03
b^{λ}_{3}	−491.6	−8.28	−1.4
c^{λ}_{1}	105.3	−0.63	−2.43
c^{λ}_{2}	206.1	19.59	18.04
c^{λ}_{3}	−382.6	−6.73	−0.83

of the phase velocity of phonons in Si upon transition from long-wavelength phonons to short-wavelength ones:

$$S^{\lambda}(x,\theta,\varphi)=S^{\lambda}_{0}(\theta,\varphi)\left\{1-f^{\lambda}(x,\theta,\varphi)\right\}. \tag{1.49}$$

Figure 1.8 outlines that the anisotropy of the phase velocity of phonons changes qualitatively when passing from the anisotropic continuum model $(x=0)$ (Figure 1.8, curves 1) to phase velocities at the boundary Brillouin zones (Figure 1.8, curves 5). The velocity of the longitudinal phonons is minimum in the [001] directions and is maximum in the [111] directions for $x=0$. However, for $x=1.0$, it increases significantly and remains unchanged in the [001] and [111] directions, respectively. Nevertheless, in the first case, its value is 6% less than the maximum. The minimum value is realized in the [101] directions (see Figure 1.8a).

The anisotropy of the phase velocities of the transverse phonons of both vibrational branches also changes qualitatively (see Figures 1.8a, b). For example, for wave vectors in the plane of a cube face, the fast mode t_1 is isotropic when $x=0.01$, but becomes anisotropic when $x=1.0$, with the velocity being maximum along the [101] direction. In contrast to this, the slow mode t_2 that is anisotropic in the long-wavelength limit $x<<1$ becomes almost isotropic when $x=1.0$. For the longitudinal phonons,

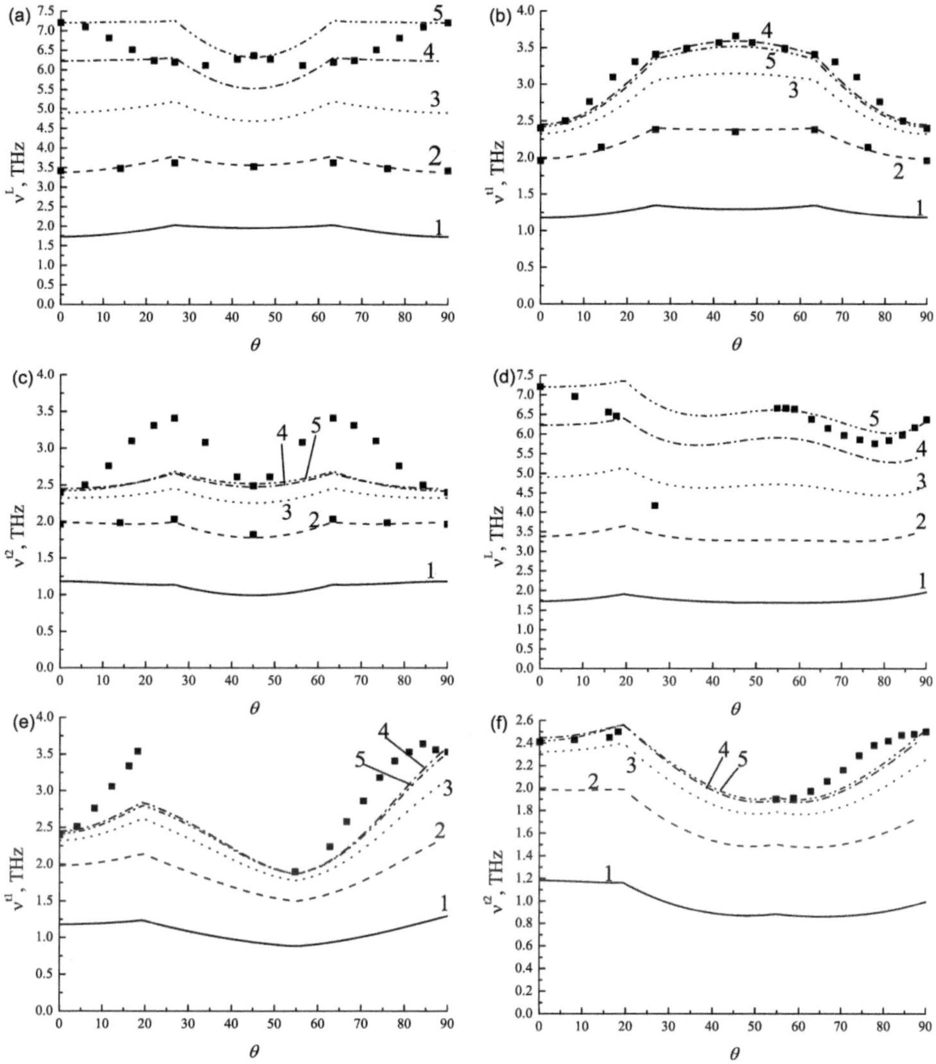

Figure 1.6: Phonon spectrum of Ge. Thermal phonon dispersion in the linear approximation in cubic harmonics is taken into account for $x = 0.2$ (curve 1), $x = 0.4$ (curve 2), $x = 0.6$ (curve 3), $x = 0.8$ (curve 4), and $x = 1$ (curve 5) in the plane of a cube face (a,b,c) and in a diagonal plane (d,e,f). A longitudinal mode is shown in the (a) and (d); (b) and (e) refer to t_1; (c) and (f) are for t_2. The experimental data of [60, 61] are denoted by symbols.

dispersion does not give rise to a qualitative change in the spectrum; whereas the dispersion has a stronger influence on the transverse phonons (see Figure 1.5). For all symmetrical directions, plane sections of the spectrum in the short-wavelength region $0.5 < x < 1$ emerge; the group velocity has anomalously small values (see Figure 1.5). As

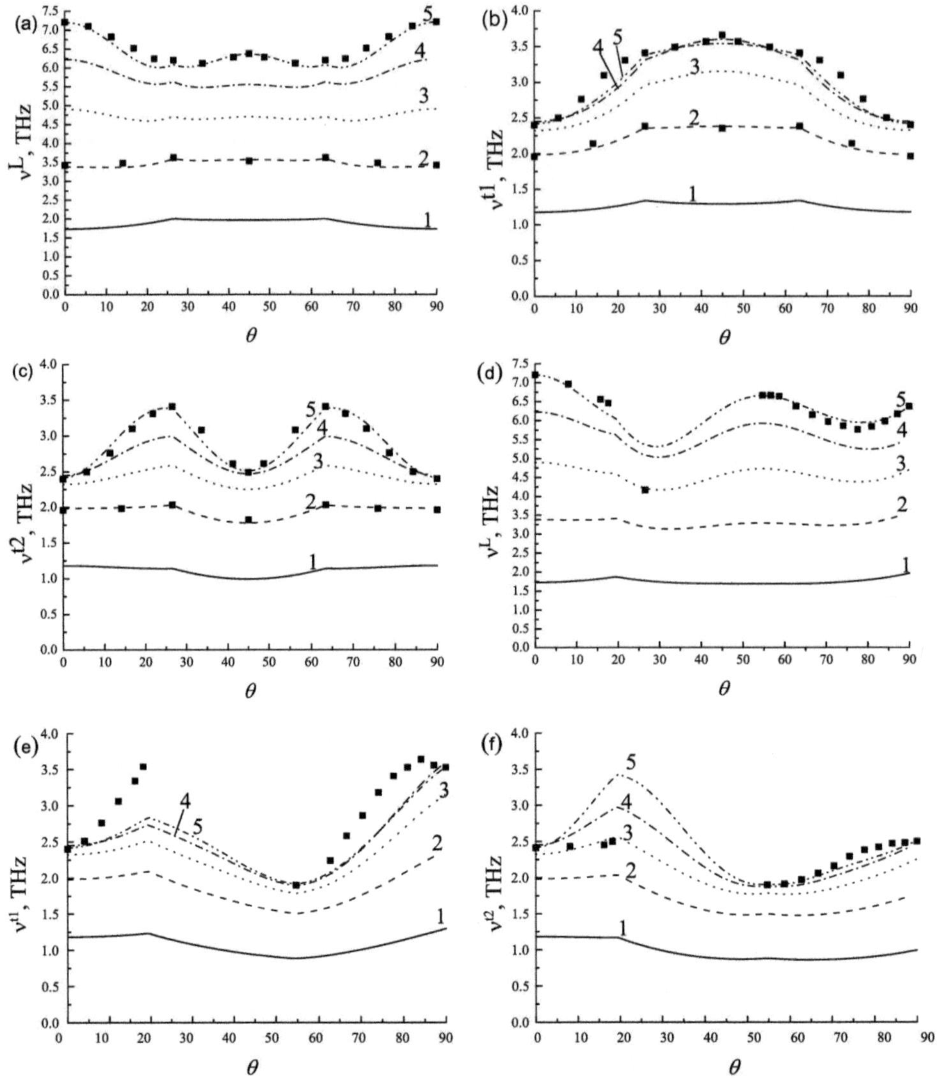

Figure 1.7: Phonon spectrum of Ge. Thermal phonon dispersion in the quadratic approximation in cubic harmonics is taken into account for $x = 0.2$ (curve 1), $x = 0.4$ (curve 2), $x = 0.6$ (curve 3), $x = 0.8$ (curve 4), and $x = 1$ (curve 5) in the plane of a cube face (a,b,c) and in a diagonal plane (d,e,f). A longitudinal mode is shown in the (a) and (d); (b) and (e) refer to t_1; (c) and (f) are for t_2. The experimental data of [60, 61] are denoted by symbols.

we will see later, a qualitative change in the anisotropy of the spectrum of thermal phonons leads to a change in the directions in which the phonons of various vibrational branches are focused.

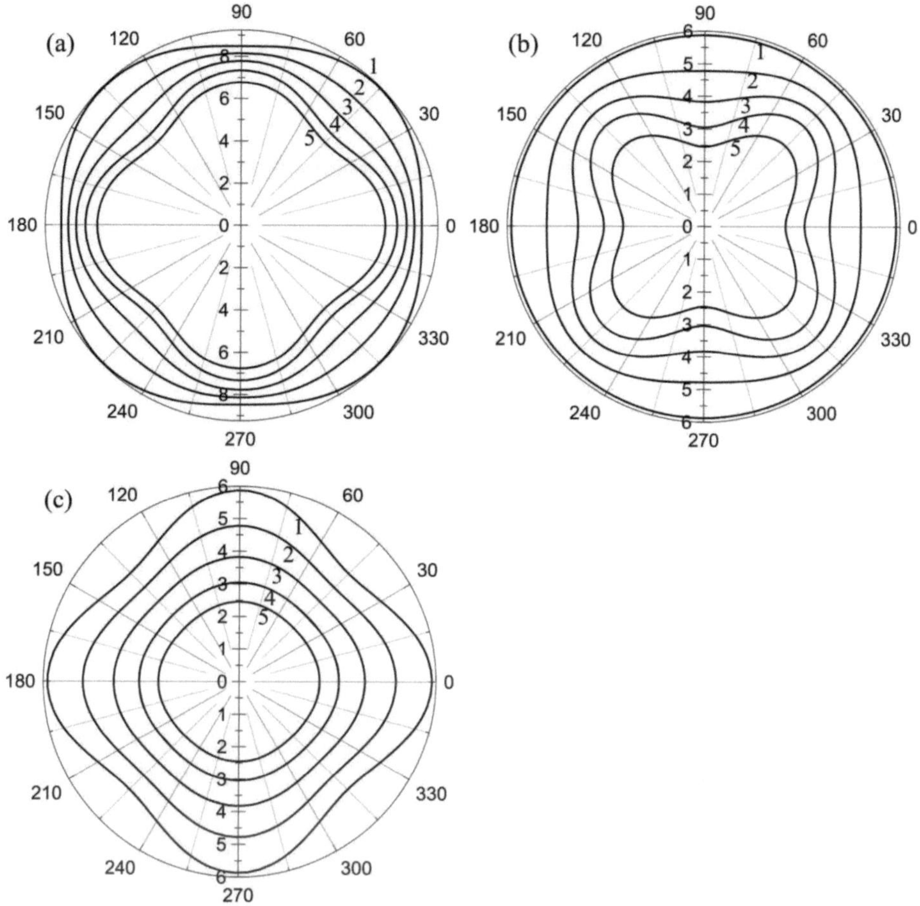

Figure 1.8: Angular dependencies of the sound velocity $S^{\lambda}(x, \theta, 0)$ normalized to 10^5cm/s on the angle q in Si crystals for wave vectors lying in the plane of a cube face: (a) for longitudinal phonons, (b), (c) for fast and slow transverse phonons, respectively, when the values of the reduced wave vector are: $x = 0$(curves 1), $x = 0.4$(curves 2), $x = 0.6$(curves 3), $x = 0.8$(curves 4), and $x = 1$(curves 5).

1.3 Calculation of Heat capacity of Si and Ge Crystals Using the Approximation Phonon Spectrum

To start with, we calculate the heat capacity of Si and Ge crystals using the anisotropic continuum model and the approximation spectrum of thermal phonons in accordance with formulas (1.28) and (1.42). Next, we compare the calculation results with the experimental data of [64, 65]. Also, we demonstrate how much the proposed approximation of the thermal phonon spectrum improves the agreement between the calculations of the specific heat of Si and Ge crystals with the results of [64, 65] compared to the anisotropic continuum model. The expression for molar heat capacity, according to [9, 41, 66], has the form:

$$C_V^\lambda = V_M C_V^\lambda = \frac{k_B \cdot V_M}{(2\pi)^3} \int d\mathbf{q} z_\lambda^2 \frac{\exp(z_\lambda)}{(\exp(z_\lambda) - 1)^2}, \quad z_\lambda = \frac{\hbar\omega_q^\lambda}{k_B T}, \tag{1.50}$$

where k_B is the Boltzmann constant, T is the temperature, ω_q^λ is the phonon frequency with polarization λ, C_V^λ is the heat capacity at constant volume, V_M is the molar volume, which is equal to 12.10 and 13.64 cm^3/mol, respectively, for Si and Ge crystals. The upper limit of integration over $\mathbf{q}$ in the expression (1.50) is specified by the Debye wave vector q_D[66]:

$$q_D = (6\pi^2 N/V)^{1/3} = (6\pi^2 N_A/V_M)^{1/3}, \tag{1.51}$$

where N is the number of unit cells, V is the crystal volume, $N_A \approx 6.022 \cdot 10^{23}$ cm^3/mol is the Avogadro number. The expression (1.50) implies that, at temperatures much lower than the Debye temperature T_D, the heat capacity obeys the *Debye law* and dependence T^3. In the case of high temperatures ($T \gg T_D$), the well-known Dulong and Petit law comes from the formula (1.50). It claims that the molar heat capacity of all solids at high temperatures tends to the limit

$$C_V = 3k_B N_A = 3R. \tag{1.52}$$

The experiments in [64, 65] took measurements of the heat capacity under constant pressure C_P for Si and Ge crystals. It was shown in [65] that the $C_P - C_V$ difference for the crystals tested is small, amounts to of the order of $(1-23) \cdot 10^{-3} J/\text{mol K}$ and can be neglected. Consequently, we accept that $C_P \approx C_V$.

Let us compare the results of calculating the heat capacity with the experimental data. Figure 1.9 sketches that the long-wavelength phonons make the main contribution to the heat capacity at temperatures below 6 and 10 K, respectively, for Ge and Si crystals. Therefore, the dashed curves plotted within the anisotropic continuum model are in good agreement with the data of [64, 65]. At higher temperatures, the anisotropic continuum model gives overestimated values for the energy of phonon modes and, appropriately, underestimated values of the heat capacity for Ge and Si crystals. Moreover, a maximum deviation from experiment takes place at T = 15–30 K and at T = 25–55 K for Ge and Si crystals, respectively, and goes up to 55–65%. For these intervals, the specific heat capacity calculated within the anisotropic continuum model is 2.7 times less than the experimental data of [64, 65] (see Figure 1.9b, curve 1a). The deviation is mainly due to neglecting by the strong dispersion of the transverse phonons (see Figure 1.5). As can be seen from Figure 1.5, for $q > q_D/2$, the spectrum of the slow mode reaches a plane section, which leads to a significant increase in the contribution of this mode to the heat capacity.

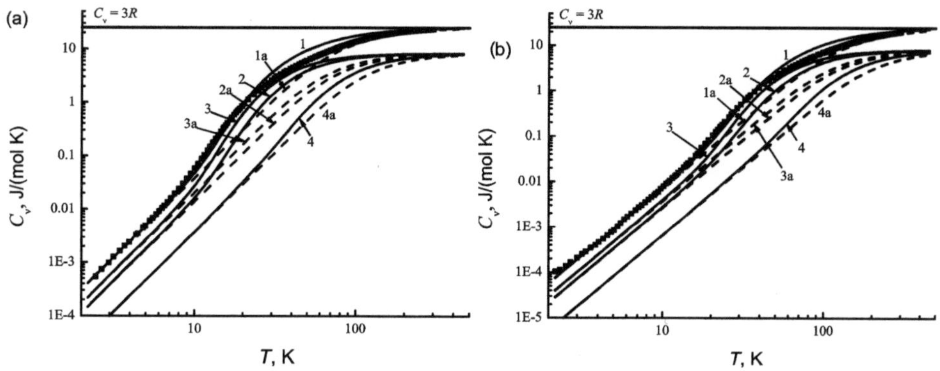

Figure 1.9: Temperature dependencies of the molar heat capacity of Ge (a) and Si (b) crystals in the anisotropic continuum model (curves 1a, 2a, 3a, 4a) and for the model taking the dispersion of thermal phonons into account (curves 1, 2, 3, 4). Curves 1 refer to the total heat capacity, curves 2 correspond to a fast transverse mode, curves 3 relate to a slow transverse mode, and curves 4 are for a longitudinal mode. The straight line $C_V = 3R = 24.9 J/\text{mol K}$ reflects the classic Dulong and Petit law (1.52). The experimental data of [64, 65] are designated by symbols.

The best way to understand the difference between the calculated outcomes is matching the curves 3 and 3a. The former represents the anisotropic-continuum-model-based temperature dependencies. The latter makes allowance for the dispersion using the approximation spectrum. It is seen that in the indicated interval the contribution to the heat capacity from the t_2 mode 2.5 times increases. This significantly reduces the discrepancy between the calculation results and the experimental data. Moreover, the heat capacity values obtained using the approximation spectrum of thermal phonons for Ge crystals at temperatures lower than 20 K, and less than 37 K for Si, are consistent with the experimental data with an error not exceeding 5%. With increasing the temperature, the role of short-wavelength phonons rises and the discrepancy between the calculation results and the findings observable enlarges. In this range, the calculation results lie above the experimental data. This points to the fact that the approximation designed somewhat overestimates the role of dispersion for short-wavelength phonons in directions different from symmetric ones. A maximum deviation for Ge and Si crystals is achieved at T = 30–40 K and at T = 55–70 K, respectively, and approaches 40–50% (see Figure 1.9). At $T > 300$ K, the molar heat capacity for both models, as well the experimental data tends to 3 R, with the contributions to the heat capacity from different branches of the phonon spectrum turning out to be equal to each other (see Figure 1.9). The slow transverse mode (curves 3, 3a) contributes dominantly to the heat capacity over the entire temperature range. For example, for Si crystals at T = 3 K, its contribution is 53%, and at T = 40–50 is 61%. The contribution of longitudinal phonons is much smaller than the contributions of transverse phonons. It does not exceed 9% at T = 3 K and 4% at T = 40–50 K.

So, we have proved that the proposed approximation of the spectrum of thermal phonons satisfactorily describes the experimental data concerning the heat capacity of Ge and Si crystals and can be applied for calculating the lattice thermal conductivity of cubic crystals.

1.4 Group Velocity and Peculiarities of the Phonon Propagation in Cubic Crystals with Various Types of Elastic Energy Anisotropy

In elastically anisotropic crystals, due to non-collinearity of phase and group velocities, the phonon flux emitted by a point heat source is focused along certain directions in the crystal lattice [16–18]. Figure 1.10 helps explain this effect by exemplifying three cases of an isoenergetic surface or a surface of constant frequency. In an isotropic medium, the directions of the group and phase velocities of phonons and the wave vector coincide. There is no focusing effect: the density of states is isotropic; it is constant for all directions (see Figure 1.10a). In cubic crystals, the anisotropy of the elastic properties poses the phonon focusing effect. That is to say, directions emerge for propagating dominantly phonons of a given vibrational mode [16–18]. To illustrate this effect, some cases are given in Figures 1.10b where the cross-section of the constant frequency surface has the shape of a square with sides perpendicular to the [100] and [101] directions. The group velocity being perpendicular to the constant frequency surface, a significant part of phonons propagates in the [100] directions in the first case and in the [101] directions in the second one. Obviously, the first case shows that the phonons are focused in the [100] directions and defocused in the [101] directions. The second case represents an opposite situation: the phonons are focused in the [101] directions and defocused in the directions [100].

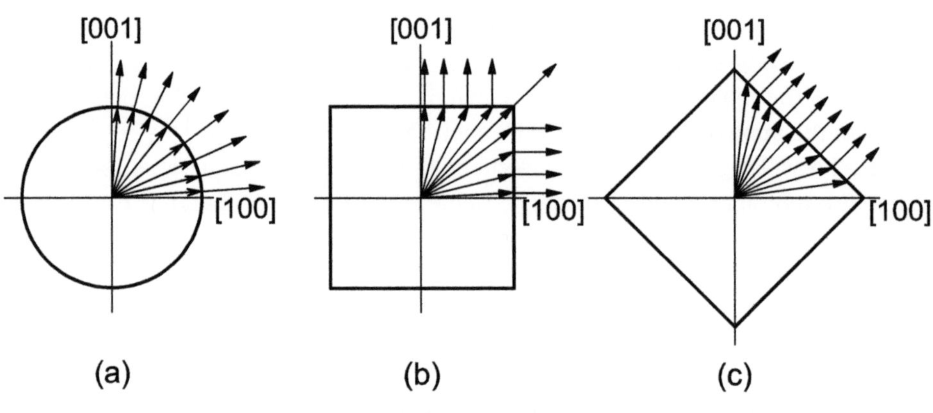

Figure 1.10: Scheme of cross-sections of isoenergy surfaces. It illustrate phonon focusing: (a) in the case of an isotropic medium, (b) focusing in the [100] direction, and (c) focusing in the [101] direction.

Let us look at phonon focusing in cubic crystals with positive (LiF, GaAs, InSb Ge, Si, MgO, diamond, YAG) and negative (CaF_2, NaCl, YIG) anisotropy of second-order elastic moduli within the anisotropic continuum model. Section 1.1 deals with the angular dependencies of phase velocities for all acoustic modes in cubic crystals of the same type. It is shown that they are qualitatively similar: the maximum and minimum values of phase velocities for all crystals are achieved in the same directions. The latter differ only in greater or lesser anisotropy of the phase velocities (see Figure 1.1a). The angular dependencies of the phase velocities for all acoustic modes in cubic crystals of different types are qualitatively different: the maximum values of phase velocities in type-I crystals correspond to the minimum values for type-II crystals (see Section 1.1).

As an example of type-I and type-II crystals, we regard Si and Ge, and CaF_2, respectively. In elastically anisotropic crystals, the phonon transport and focusing of phonons may be characterized by their group velocity. For an arbitrary direction of the wave vector, it can be represented in the form [40]:

$$\mathbf{V}_g^{\lambda}(\theta,\varphi) = S^{\lambda}(\theta,\varphi)\tilde{\mathbf{V}}_g^{\lambda}(\theta,\varphi), \quad \tilde{\mathbf{V}}_g^{\lambda}(\theta,\varphi) = \mathbf{n} + S_{\theta}^{\lambda}\mathbf{e}_{\theta} + S_{\varphi}^{\lambda}\mathbf{e}_{\varphi},$$
$$S_{\theta}^{\lambda}(\theta,\varphi) = \left(\frac{1}{S^{\lambda}}\right)\frac{\partial S^{\lambda}}{\partial\theta}, \quad S_{\varphi}^{\lambda}(\theta,\varphi) = \frac{1}{\sin\theta}\left(\frac{1}{S^{\lambda}}\right)\frac{\partial S^{\lambda}}{\partial\varphi}. \tag{1.53}$$

Here $\mathbf{e}_{\theta} = (\cos\theta\cos\varphi, \cos\theta\sin\varphi, -\sin\theta)$, $\mathbf{e}_{\varphi} = (-\sin\varphi, \cos\varphi, 0)$, and $\mathbf{n}$ is the unit wave vector of a phonon. The n, $\mathbf{e}_{\theta}$, and $\mathbf{e}_{\varphi}$ vectors form a mutually orthogonal triple of unit vectors.

To examine the influence of phonon focusing on the propagation of phonon modes in Si and CaF_2 crystals, we resort to the simplest case when wave vectors lie in the {100} plane of a cube face ($\varphi = 0$). In this case, $S_{\varphi}^{\lambda}(\theta,0) = 0$, and only the angular component S_{θ}^{λ} is responsible for the deviation of the group velocity vector from the wave vector direction. In the angular dependencies of the phase velocities $S^{\lambda}(\theta,0)$, we mark the directions of the angular component S_{θ}^{λ} of the group velocity vector with small arrows (see Figure 1.11a, b). The greater the anisotropy of the spectrum, the greater the angular components S_{θ}^{λ} and S_{φ}^{λ} of the group velocity are, and the greater the angle the group velocity vector deviates from the direction of the wave vector. It should be emphasized that the group velocity vector deviates towards the maximum values of the phase velocity. So, for example, $S_{\theta}^{t_2} < 0$ for a slow transverse mode in the angle range of $0 \le \theta \le \pi/4$ for Si. The quantity $S_{\theta}^{t_2}$ becomes positive for the angles $\pi/4 \le \theta \le \pi/2$, therefore, the group velocity vector for the mode t_2 deviates towards the [001] and [010] directions (see Figure 1.10a, c). In these directions, the phonons of the slow transverse mode are focused. For the longitudinal phonons, $S_{\theta}^{L} > 0$ in the range of angles $0 \le \theta \le \pi/4$, but $S_{\theta}^{L} < 0$ when $\pi/4 \le \theta \le \pi/2$. Therefore, the group velocity vectors deviate from $\mathbf{n}$ to the [101] focusing direction (see Figure 1.11a, b). In Figure 1.11a, these directions are marked with bold radial arrows. For the fast transverse mode in Si in the {100} plane, the spectrum is isotropic and there is no focusing

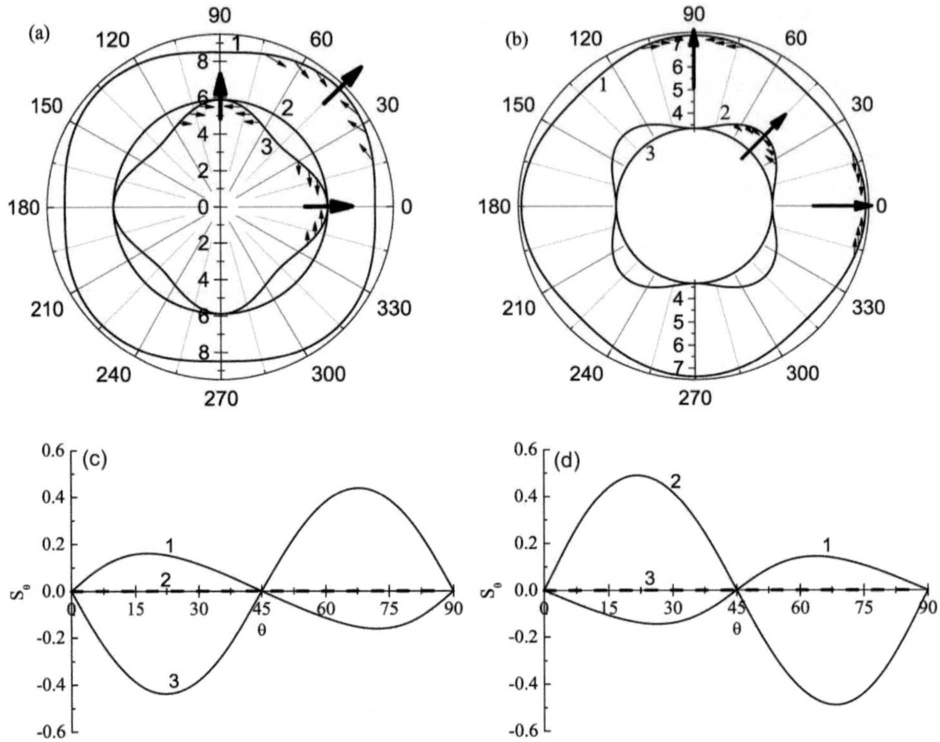

Figure 1.11: Angular dependencies of the sound velocities (a,b) and dimensionless group velocity components $S_\theta^\lambda(\theta,0)$ (c,d) in Si (a,c) and CaF_2 (b,d) crystals for wave vectors lying in the plane of a cube face. Curves 1 are longitudinal phonons, curves 2, 3 refer to fast and slow transverse phonons, respectively. Focusing directions are designated by bold radial arrows.

effect. Recall that $S_\theta^\lambda = S_\varphi^\lambda = 0$ in isotropic media, the directions of the group and phase velocities coincide, and there is no focusing effect in them. For wave vectors in the {100} planes, this equality holds for fast transverse modes in type-I crystals and for slow transverse modes in type-II crystals, therefore, there is no focusing effect for them. As can be reflected from Figures 1.11c, d (curves 2), the phonons of the fast transverse mode in CaF2 crystals deviate from the [001] directions to the [101] focusing directions. For longitudinal phonons, on the contrary: the group velocity vectors deviate from the [101] directions to the focusing directions [001]. The focusing directions in Figure 1.11 are denoted by bold radial arrows. In one and the same crystal, they are different for different modes.

For crystals of one type, the directions of focusing and defocusing for the vibrational modes coincide, whereas the directions of focusing in type-I crystals turn into the directions of defocusing in type-II crystals. For example, in Si crystals, longitudinal phonons are focused in the [101] directions, and defocused in the [001] directions. In CaF_2 crystals, longitudinal phonons are focused in the [001] directions, and defocused in the [111] directions. In Si, the slow mode with the polarization vector in the plane of a cube face is focused in the [001] directions and defocused in the [101] directions, whereas the focusing and defocusing directions are opposite for the fast transverse mode with the polarization vector in the plane of the face of the cube in CaF_2. Thus, there is a deep physical basis for classifying cubic crystals, as is proposed in [44].

As is evident from the foregoing, a simple way of finding the phonon focusing directions can be suggested avoiding complex group-velocity structures. Figure 1.11 displays that the phonon-focusing directions coincide with the directions of reaching maxima of the phonon phase velocities. Consequently, the plotted angular dependencies $S_0^\lambda(\theta, \varphi)$ are sufficient to determine the focusing directions and the maximums of the phase velocities.

So, we have shown that for type-I and type-II crystals, not only the spectrum and polarization vectors but also the phonon focusing directions are qualitatively different.

1.5 The Influence of Focusing on the Density of Phonon States in Cubic Crystals

The method proposed above for analysing component directions of group velocities allows determining only the direction of focusing and defocusing of phonons. In order to quantitatively assess the influence of anisotropy of elastic energy on the propagation of acoustic modes in cubic crystals, as well the influence of phonon focusing on the density of phonon states, it is necessary to construct an isoenergetic surface or a surface of constant frequency [67.68]

$$\mathbf{q} = \mathbf{n} \cdot \omega^\lambda / S^\lambda(\theta, \varphi). \tag{1.54}$$

Then, the definition (1.54) implies that the phonon group velocity vector $\mathbf{V}_g^\lambda$ is orthogonal to the isoenergetic surface. This means that the phonon wave vectors can be unambiguously connected with the corresponding group velocity vectors for all acoustic modes and the angles between the directions of the group and phase velocities can be computed. The theory of elastic waves also uses other definitions that are equivalent to the constant frequency surface: a surface of slowness [18], a back surface [68] or a refraction surface [42]. These surfaces can be obtained from a surface of constant frequency by dividing the expression (1.54) by the frequency ω [18, 42, 68]. The surface of slowness is a geometric place of the ends of the vector ξ^λ, measuring from one point:

$$\xi^\lambda = \mathbf{q}/\omega^\lambda = \mathbf{n}/S^\lambda(\theta, \varphi)\,. \tag{1.55}$$

The monograph [42] defines this vector as a vector of refraction. In various literature, we often encounter the terms of an isoenergetic surface or a constant frequency surface [67]. Therefore, we will adhere to this terminology. With a view to quantifying the features of the propagation of phonon modes in nanostructures with various types of anisotropy of elastic energy, we should make a graphic image of the isoenergetic surface.

Below, we resort to the findings of [69–71] to investigate the influence of focusing on the propagation of phonons and the density of phonon states in elastically anisotropic crystals. Figure 1.1 sketches the maximum anisotropy having the phase velocities $S^{\lambda}(\theta,\varphi)$ for slow and fast transverse modes in the {100} and {110} planes. We illustrate the phonon focusing by the example of the t_1 and t_2 modes in Si and CaF_2 crystals for wave vectors lying in the {100} and {110} planes. In doing so, we construct isoenergetic surfaces for them (see Figure 1.12). The phonon group velocity determining the direction of the energy transfer is perpendicular to this surface. As can be seen from Figure 1.12, the phonon focusing is characterized by the following angles: $\theta_1, \theta_2, \theta_3,$ and θ_4.

The angle θ_1 governs the direction of the wave vector for which the group velocity vector $\mathbf{V}_g^{t_2}(\theta_g)$ is parallel to the focusing direction. For type-I crystals and the wave vectors in the {100} plane, the group velocity vector is parallel to the [001] direction, therefore,$\theta_g^{\lambda}=n\pi/2$, where n is an integer (see Figure 1.12). The lines of zero curvature are distinguished peculiarities on the isoenergetic surface (slowness surface). They reflect a transition from the convex to concave regions (see Figure 1.12 (a)). In the three-dimensional case, many of these points form lines of zero curvature. For the chosen section, the angles $\pm\theta_2$ specify the directions of the wave vectors towards the zero curvature points. The latter are counted from the focusing directions (see Figure 1.12). The angles $\pm\theta_3$ define the directions of group velocities at points of zero curvature on the isoenergetic surface. At these points, the phonon group velocity vectors have a maximum convergence to the [100] direction. We denote this angle as $\theta_3=\theta_g^{\lambda}(\theta_2)$ (see Figure 1.12). Obviously, the size of the sector of the wave vectors corresponding to the focusing region is determined by the group velocities $\mathbf{V}_g^{t_2}(\pm\theta_4)$ that are collinear to the vectors $\mathbf{V}_g^{t_2}$ at the points of zero curvature, namely at $-\theta_4\le\theta\le\theta_4$ (see Figure 1.12).

To find the characteristic angles $\theta_i^{t_2\{100\}}$, we determine the direction of the group velocity $\theta_g^{\lambda}(\theta,\varphi)$ through the θ and φ angular coordinates of the wave vector. The condition $(\mathbf{V}_\mathbf{g}^{\lambda}\mathbf{q})=V_g^{\lambda}q\cos(\alpha^{\lambda}(\theta,\varphi))$ yields an angle between the group velocity and the wave vector

$$\alpha^{\lambda}(\theta,\varphi)=\pm\ \arccos\left(1+\left(S_{\theta}^{\lambda}\right)^2+\left(S_{\varphi}^{\lambda}\right)^2\right)^{-0.5}. \tag{1.56}$$

Consider the cases of wave vectors lying in the {100} and {110} planes for which the angles are $\varphi=0$ and $\pi/4$, and the group velocity component is $S_{\varphi}^{\lambda}=0$. Then

$$\alpha^{\lambda}(\theta)=\text{arctg}\ S_{\theta}^{\lambda}(\theta). \tag{1.57}$$

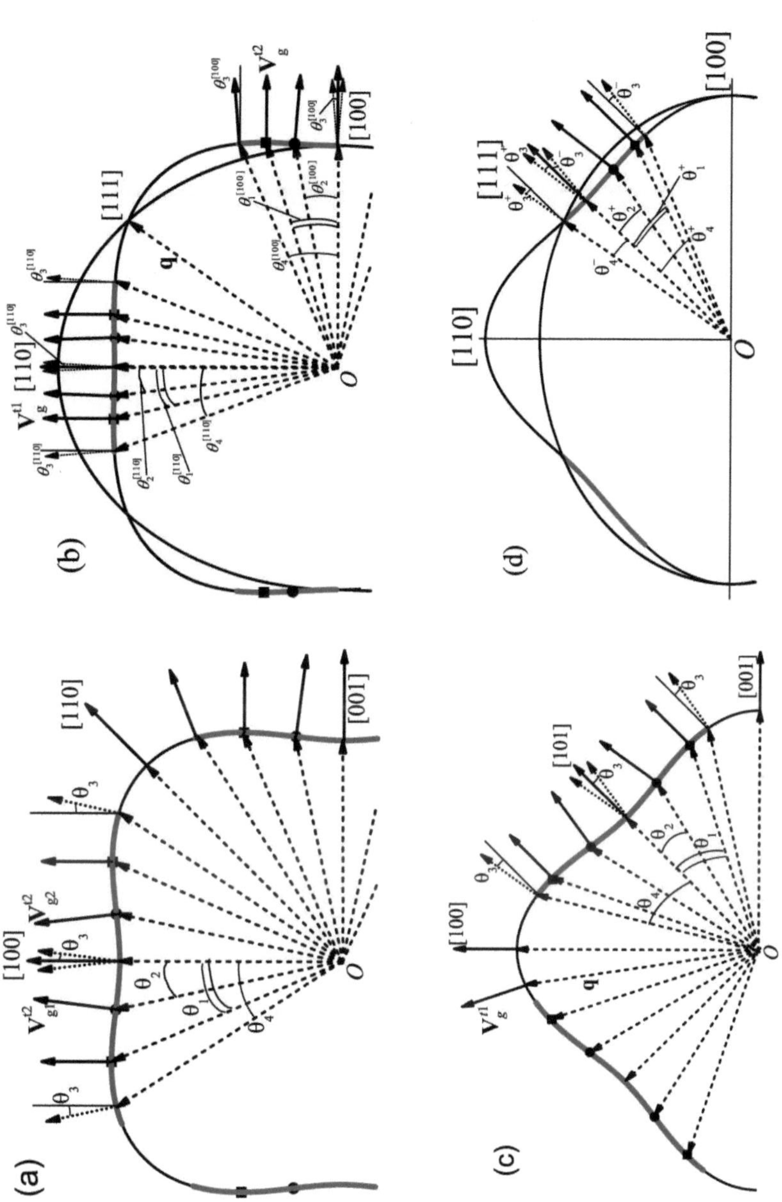

Figure 1.12: Scheme that illustrates focusing of slow and fast transverse modes in Si (a,b) and CaF_2 (c, d) crystals for cross-sections of an isoenergetic surface; (a, c) The plane XZ intersects the isoenergetic surface (for the mode t_2); and (b, d) A diagonal plane intersects the isoenergetic surface (for the t_1 and t_2 modes). Arrows stand for wave vectors inside the surface and for the corresponding group velocities of phonons outside this surface.

In the coordinate system along the edges of the cube, the angle between the Z-axis and the group velocity direction is equal to:

$$\theta_g^\lambda = \theta + \alpha^\lambda = \theta + \text{arctg } S_\theta^\lambda(\theta). \tag{1.58}$$

It is readily apparent that the sign of the component of S_θ^λ is responsible for in which direction the group velocity vector deviates relative to the phonon wave vector. Since, in an isotropic medium, the directions of the phonon phase and group velocities coincide with the direction of the phonon wave vector, the model of an isotropic medium can be treated as a comparison system when analysing the influence of phonon focusing on the change in the density of phonon states in elastically anisotropic crystals. Figure 1.13 shows a diagram for calculating the characteristic angles $\theta_i^{t_2\{100\}}$. The angles $\theta_i^{t_2\{100\}}$ come from the condition $\theta_g^\lambda = 0$:

$$\theta_1 + \text{arctg } S_\theta^\lambda(\theta_1) = 0. \tag{1.59}$$

For searching the θ_2 and θ_3 angles, it is necessary to construct the function $\theta_g^{t_2}(\theta, 0)$ (curve 3):

$$\theta_g^\lambda(\theta) = \theta + \text{arctg } S_\theta^\lambda(\theta) \tag{1.60}$$

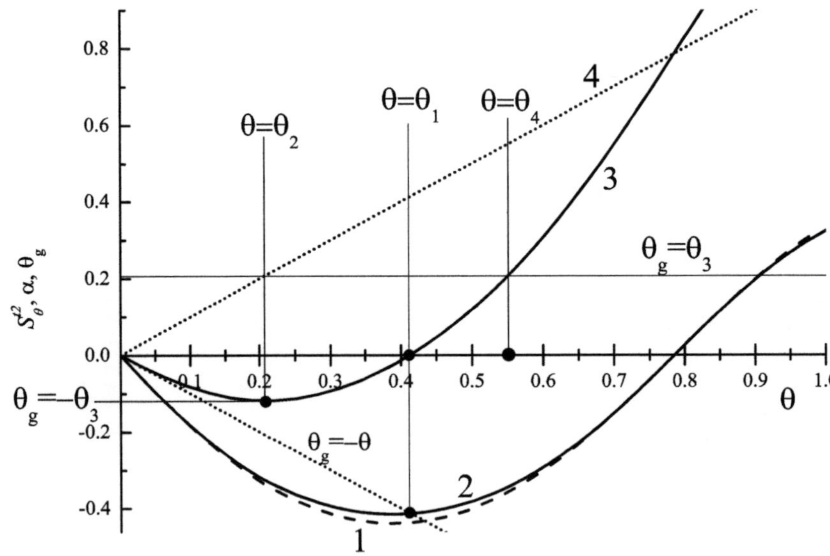

Figure 1.13: Diagram for calculating the characteristic angles θ_1, θ_2, θ_3, and θ_4 for the mode t_2 in silicon: the curve 1 is the function $S_\theta^{t_2}(\theta, 0)$; the curve 2 is the function $\alpha^{t_2}(\theta, 0)$; the curve 3 is the function $\theta_g^{t_2}(\theta, 0)$; the curve 4 is the function $\theta_g^{t_2} = \theta$.

and find its minimum. The position of the minimum $d\theta_g^{t2}(\theta,0)/d\theta|_{\theta=\theta_2}=0$ determines the point of zero curvature with the θ_2 angle, and the value of the function gives the magnitude of the θ_3 angle (see Figure 1.13).

$$\theta_3 = \theta_2 + \text{arctg } S_\theta^\lambda(\theta_2). \tag{1.61}$$

After performing the above operations, we construct the vectors of the group velocities $V_g^{t2}(\pm\theta_2^{t2})$ and determine the θ_3 angles of "convergence" of the group velocities. The $2\theta_3$ angle assigns the region of phonon focusing. Obviously, the group velocities $V_g^{t2}(\pm\theta_4^{t2})$ that are collinear to the directions of V_g^{t2} at the points of zero curvature, namely at $-\theta_4^{t2}\le\theta\le\theta_4^{t2}$, ascertain the magnitude of the sector of the wave vectors corresponding to the phonon focusing (see Figure 1.12).

An equation for calculating the θ_4^λ angle has the form:

$$\theta_3^\lambda = \theta_g^\lambda(\theta_4^\lambda) = \theta_4^\lambda + \text{arctg } S_\theta^\lambda(\theta_4^\lambda). \tag{1.62}$$

The θ_4^λ angle can be graphically found by drawing a horizontal straight line at a height of θ_3 until it intersects the curve 3 (see Figure 1.13). The intersection point $\theta_g^\lambda(\theta_4^\lambda)=\theta_3^\lambda$ gives us the desired angle θ_4^λ that defines the focusing sector of $-\theta_4^\lambda\le\theta\le\theta_4^\lambda$ in the space of wave vectors. Thus, the diverging sector of $-\theta_4^\lambda\le\theta\le\theta_4^\lambda$ of wave vectors in the {100} plane for the mode t_2 turns into the sector of $-\theta_3^\lambda\le\theta_g\le\theta_3^\lambda$ of group velocities, which converge to the [001] direction.

In an isotropic medium, the directions of propagation of phonons and their wave vectors coincide. Consequently, phonons travelling in an isotropic medium in the sector of $-\theta_4^\lambda\le\theta\le\theta_4^\lambda$ will move in elastically anisotropic crystals in the substantially smaller sector of $-\theta_3^\lambda\le\theta_g\le\theta_3^\lambda$ (see Figure 1.12). Thus, the density of states for this mode in the focusing directions close to [100] significantly increases in comparison with an isotropic medium. The method proposed allows one to quantitatively estimate changes in angular density distributions of phonon states for slow quasi-transverse modes in the crystals tested. For this purpose, we introduce the average density of phonon states of the t_2 mode for wave vectors in the {100} plane per unit angle for the focusing regions of $-\theta_3^{t2}\le\theta_g\le\theta_3^{t2}$ $N_F^{t2\{100\}}$ and an isotropic medium N_{iso}. The diverging sector of wave vectors for an isotropic medium in the {100} plane with an $2\theta_4^{t2\{100\}}$ angle (for example, for Ge and Si, the $2\theta_4$ angles are equal to 62.9° and 57.3°, respectively) turning into a diverging sector of the phonon group velocity vectors with an $2\theta_3^{t2\{100\}}$ angle (for example, for Ge and Si, the $2\theta_3^{t2\{100\}}$ angles amount to 18.8° and 13.6°, respectively), the average density of states in the phonon focusing regions is greater than that in an isotropic medium. This can be expressed through the ratio:

$$n_{FI}^{t2\{100\}} = N_F^{t2\{100\}}/N_{Iso} = \theta_4^{t2\{100\}}/\theta_3^{t2\{100\}}. \tag{1.63}$$

For type-I crystals, the parameter $n_{FI}^{t_2\{100\}} = 1.98$ is minimum for GaN and $n_{FI}^{t_2\{100\}} = 4.98$ is maximum for MgO (see Table 1.5). From the foregoing, it follows that no direct correlation exists between the value of the $k-1$ anisotropy parameter and the value of the $n_{FI}^{t_2\{100\}}$ coefficient. The inverse relationship appears to take place: the larger the $k-1$ parameter, the smaller the $n_{FI}^{t_2\{100\}}$ coefficient. For GaN crystals, the k-1 parameter is equal to 1.275 and the $n_{FI}^{t_2\{100\}}$ coefficient is minimum, whereas, for MgO, the parameter $k-1 = 0.69$ has one of the minimum values in Table 1.2, and the $n_{FI}^{t_2\{100\}}$ coefficient is maximum. The same situation occurs in type-II crystals for the t_1 mode (see Table 1.5). In SrF_2 crystals, the parameter $k-1 = -0.2$. It has the minimum absolute value among those listed in the table, and the value of $n_{FI}^{t_1\{100\}} = 52$ is maximum. At the same time, the parameter $k-1 = -0.466$ has the maximum value in absolute magnitude for PbS, and the coefficient $n_{FI}^{t_1\{100\}} = 2.51$ is minimum for type-II crystals (see Table 1.5). For the defocusing region, the ratio

Table 1.5: Calculated values of the angles $\theta_i^{\lambda\{100\}}$ and the average density of states for focusing and defocusing regions in the {100} plane for the mode t_2 in type-I crystals and for the mode t_1 in type-II crystals.

Type-I	***k*-1**	$\theta_1^{t2\{100\}}$	$\theta_2^{t2\{100\}}$	$\theta_3^{t2\{100\}}$	$\theta_4^{t2\{100\}}$	$n_{FI}^{t2\{100\}}$	$n_{DI}^{t2\{100\}}$	$n_{FD}^{t2\{100\}}$
GaN	1.275	33.0°	14.3°	19.9°	39.4°	1.98	0.22	8.9
GaAs	0.90	28.6°	13.8°	12.0°	34.4°	2.86	0.32	8.9
Ge	0.87	25.7	12.2	9.4	31.4	3.33	0.38	8.7
GaSb	0.85	28.2°	13.5°	11.9°	34.2°	2.88	0.33	8.8
InSb	0.81	30.9°	14.3°	14.9°	36.7°	2.46	0.28	8.8
LiF	0.78	23.8°	11.7°	7.5°	29.1°	3.90	0.42	9.2
MgO	0.69	21.2°	10.9°	5.2°	25.7°	4.98	0.49	10.2
Si	0.67	23.6°	11.9°	6.8°	28.6°	4.22	0.43	9.9
Type-II	***k*-1**	$\theta_1^{t1\{100\}}$	$\theta_2^{t1\{100\}}$	$\theta_3^{t1\{100\}}$	$\theta_4^{t1\{100\}}$	$n_{FI}^{t1\{100\}}$	$n_{DI}^{t1\{100\}}$	$n_{FD}^{t1\{100\}}$
SrF_2	−0.20	7.8°	4.5°	0.17°	9.0°	51.99	0.80	64.7
CaF_2	−0.33	25.9°	13.4°	7.8°	31.0°	3.97	0.38	10.5
PbS	−0.466	30.3°	14.2°	14.5°	36.4°	2.51	0.28	8.9
NaCl	−0.48	27.5	13.2	11.8	34.0	2.87	0.33	8.6

$$n_{DI}^{\lambda\{100\}} = N_D^{\lambda\{100\}}/N_{Iso} = \left(\pi - 4\theta_4^{\lambda\{100\}}\right)/\left(\pi - 4\theta_3^{\lambda\{100\}}\right) \tag{1.64}$$

implies that the average density of states $n_{DI}^{\lambda\{100\}}$ for transverse modes is less than N_{iso}.

The expression (1.62) yields $n_{DI}^{t_2\{100\}} = 0.38$ and $n_{DI}^{t_2\{100\}} = 0.43$ for Ge and Si, respectively. The ratios between the average density of states for focusing and defocusing regions can be deduced from formulas (1.61) and (1.62):

$$n_{FD}^{\lambda\{100\}} = N_F^{\lambda\{100\}}/N_D^{\lambda\{100\}} = \theta_4^{\lambda\{100\}}\left(\pi - 4\theta_3^{\lambda\{100\}}\right)/\left[\left(\pi - 4\theta_4^{\lambda\{100\}}\right)\theta_3^{\lambda\{100\}}\right] \tag{1.65}$$

These ratios characterize the anisotropy of density of states.

Estimates claim that the maximum and minimum anisotropy of density of phonon states for the t_1 mode is observed for Si ($n_{FD}^{t_2\{100\}} = 9.9$) and for Ge ($n_{FD}^{t_2\{100\}} = 8.7$), respectively. In this case, there is also no direct correlation between the value of the anisotropy parameter k-1 and the value of $n_{FD}^{t_2\{100\}}$.

However, as can be seen from the table, the maximum values of $n_{FD}^{t_2\{100\}}$ are due to smallness of the $\theta_3^{t_2\{100\}}$ angles that characterize the phonon focusing region. This effect is inherent to type-II crystals (see Table 1.5). So, for example, for SrF_2 crystals, the angle is minimum ($\theta_3^{t_1\{100\}} = 0.17^\circ$), and the $n_{FD}^{t_1\{100\}}$ coefficient is abnormally large ($n_{FD}^{t_1\{100\}} = 64.7$). It should be pointed out that the total number of states concentrated in the focusing region ($2\theta_3^{t_1\{100\}}$) for SrF_2 is relatively not great compared to other compounds. Therefore, as we will see later, this abnormal makes no essential influence on the relaxation characteristics of phonons in the regime of the Knudsen phonon gas flow. For the rest of type-II crystals, the $n_{FD}^{t_1\{100\}}$ coefficient varies from 8.9 for PbS to 10.5 for CaF_2.

Diamond should be excluded from Table 1.5 because its anisotropy parameter turns out to be too small to provide sections of negative curvature in the isoenergetic surfaces for quasi-transverse acoustic modes. It is this case (the presence of negative curvature sections in the isoenergetic surfaces of acoustic modes) that meets certain difficulties in analysing the phonon focusing using the "enhancement factor" (see, for example, [18, 72–75]). This is due to the fact that, firstly, the "enhancement factor" goes into infinity at points of zero curvature, and secondly, the dependencies of group velocities on the direction of the wave vector become an ambiguous function of angles in the negative curvature regions in the isoenergetic surface (see Figure 1.12). Unlike this approach, we give below a simple and intuitive method that allows one to estimate the average density of states for the phonon focusing and defocusing regions.

The situation concerning the focusing of transverse phonons, whose wave vectors lie in the diagonal plane, is more complicated than in the plane of the cube face. Indeed, in the case under consideration, the dividing of transverse modes into fast and slow ones is physically incorrect (see [44], as well as Section 1.1). This can be explained by the fact that one of the transverse modes with a polarization vector lying in the diagonal plane is a mixed transverse-longitudinal mode, and another mode with a polarization vector perpendicular to this plane is pure. The former is a slow mode in the angle range of $\theta_{[111]} < \theta < \pi - \theta_{[111]}$ and $\pi + \theta_{[111]} < \theta < 2\pi - \theta_{[111]}$ and a fast mode in the angle region of $-\theta_{[111]} < \theta < \theta_{[111]}$ and $\pi - \theta_{[111]} < \theta < \pi + \theta_{[111]}$. The isoenergetic surface has appropriate negative curvature regions for both transverse modes, as marked bold line in Figure 1.14.

However, for planes close to {110} when $\varphi = \pi/4 \pm \delta$ (where $\delta << 1$), there occurs separation of the fast and slow branches instead of their intersection, with the polarization vectors sharply changing in the vicinity of this transition (see Section 1.1, Figure 1.4). For $\delta = \pm 0.1$, it can be noticed that the degeneracy of the branches is removed in the spectrum of transverse phonons for directions in the [111] neighborhood and they are divided into fast and slow transverse modes. Since, in calculating the kinetic characteristics of phonons, integration is carried out over all directions of the phonon wave vectors, the singularity at an isolated point ($\theta_{[111]}$ and $\varphi = \pi/4$) makes no contributions to the relaxation characteristics of the phonon gas. Therefore, we will further assume that the dividing of the transverse modes into fast and slow ones for all cross-sections is valid.

In the plane in the isoenergetic surfaces of the slow transverse mode in the vicinity of the [111] direction, there are regions with negative curvature for crystals of both types. With increasing the parameter δ, the surface region with a negative curvature first shrinks. Then, as the values of δ rises, the isoenergetic surface becomes completely convex (see Figure 1.16). So, for example, a CaF_2 type-II crystal demonstrates a similar scenario for $\delta = 3\pi/16$. For $\delta = \pi/4$, we go over to the plane of the face of the cube, with the spectrum of the slow transverse mode becoming isotropic (see Figure 1.16). The same situation occurs for type-I crystals such as Si, MgO, GaN, etc. (see Figure 1.14b, d). For them, the vicinity of the [111] direction has regions with negative curvature in the isoenergetic surface. This results in appearing local focusing maxima in the vicinity of the [111] directions for the t_2 mode (see Figure 1.14b, d).

In type-I crystals for wave vectors in the diagonal plane, the fast transverse mode focuses in the [110] direction, and the slow transverse mode focuses in the [100] and [111] directions (see Figure 1.14). Figure 1.14 makes clear that the focusing of the phonons of the t_1 mode in type-II crystals for wave vectors in the {110} plane occurs differently than in type-I crystals. First, the minimum in the isoenergetic surface and the focusing direction of the phonons of the t_1 fast mode do not coincide with the symmetric direction. Secondly, the negative curvature region in the isoenergetic surface becomes asymmetric relative to the minimum point (see Table 1.7 and Figure 1.14c). So, for example, for CaF_2, the minimum point is located at $\theta_F = 47.6^{\circ}$.

Let us calculate the average values of the density of states in type-I crystals for the focusing and defocusing regions of the t_2 mode. It should be paid attention that

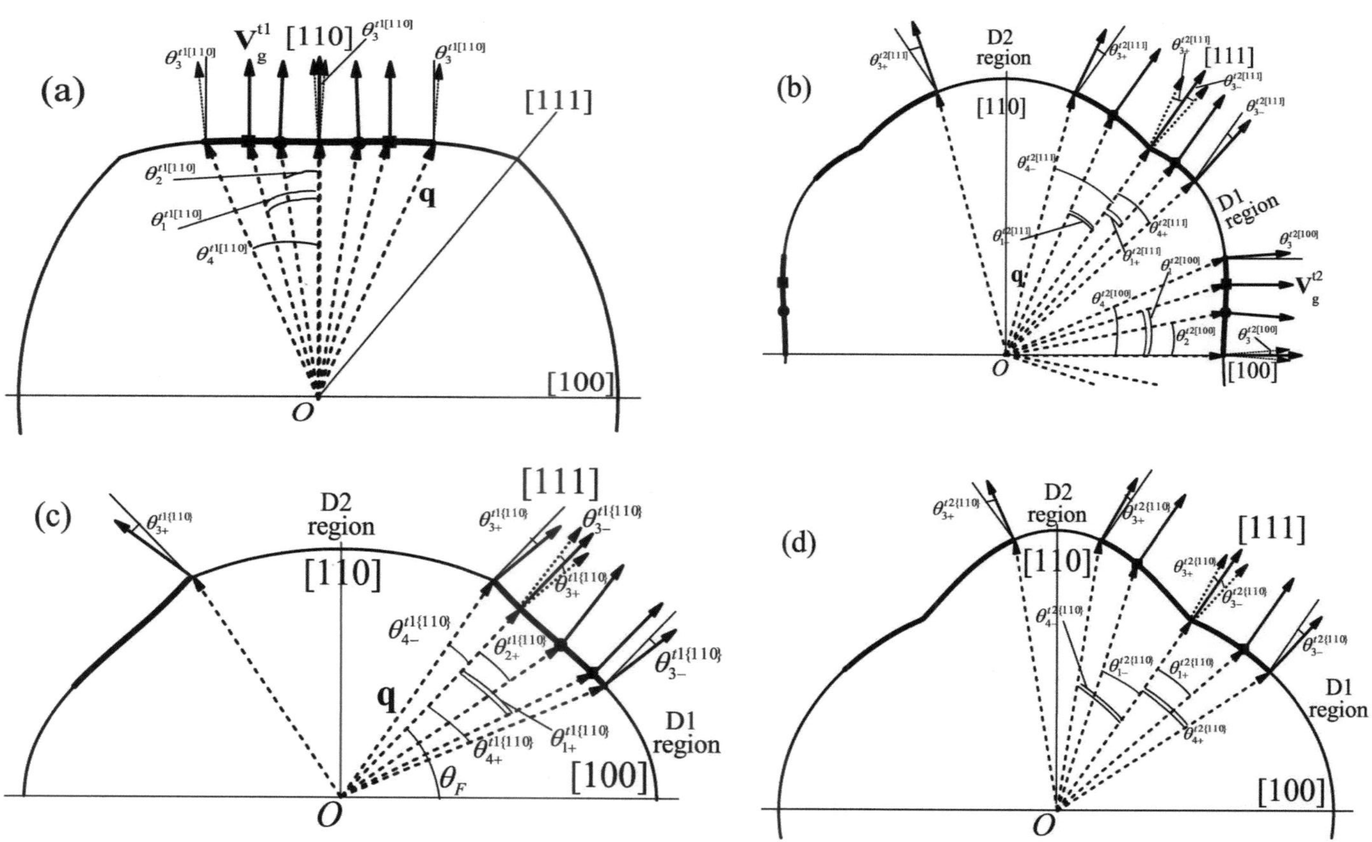

Figure 1.14: A scheme that illustrates focusing of fast (a,c) and slow (b,d) transverse modes in Si (a,b) and CaF_2 (c,d) crystals for intersecting the isoenergetic surface by a plane close to the diagonal plane ($\varphi = \pi/4 \pm \delta$, where $\delta \to 0$). Arrows stand for wave vectors inside the surface and the corresponding group velocities of phonons outside this surface. D1 and D2 are the regions of defocusing of phonons.

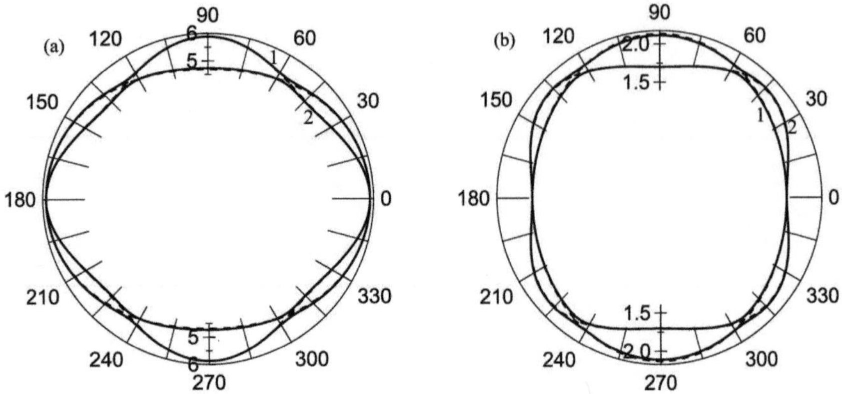

Figure 1.15: Angular dependencies of the phase velocities normalized to 10^5 cm/s (a) and isoenergetic surfaces (b) in Si crystals for wave vectors lying near the diagonal plane: $\varphi = \pi/4 + 0.1$ is solid curves and $\varphi = \pi/4 + 0.01$ is dashed curves. Curves 1 and 2 are for quasi-transverse modes t_1 and t_2, respectively.

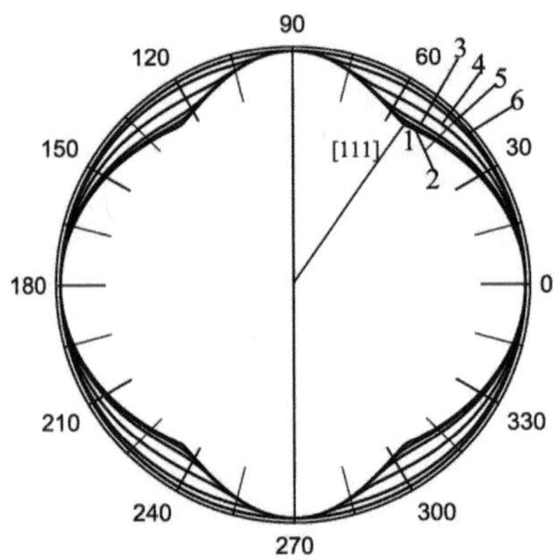

Figure 1.16: Angular dependencies of isoenergetic surfaces in CaF_2 crystals for a slow transverse mode in the planes $\varphi = \pi/4 + \delta$: $\delta = 0.01$ is for curve 1, $\delta = 0.1$ is for curve 2, $\varphi = \pi/16$ is for curve 3, $\varphi = 2\pi/16$ is for curve 4, $\varphi = 3\pi/16$ is for curve 5, and $\varphi = \pi/4$ is for curve 6.

the t_2 mode has two phonon focusing and defocusing regions in the diagonal plane, against the plane of the cube face (see Figure 1.14b). The $n_{FI[100]}^{t_2\{110\}}$ coefficient can be derived from the expression (1.63) for the first focusing region, and for the second region, it is given by:

$$n_{FI[111]}^{t2\{110\}} = \left(\theta_{4[111]+}^{t2\{110\}} + \theta_{4[111]-}^{t2\{110\}}\right) \Big/ \left(\theta_{3[111]+}^{t2\{110\}} + \theta_{3[111]-}^{t2\{110\}}\right). \tag{1.66}$$

For Si crystals, according to the data from the Table 1.6, we have: (1) $n_{FI\{100\}}^{t_2\{100\}} = 6.33$ and (2) $n_{FI[111]}^{t_2\{110\}} = 1.28$. The defocusing regions D1 and D2 can be viewed from Figures 1.14b: (1) $\theta_3^{t_2[100]} < \theta_g < \theta_{[111]} - \theta_{3-}^{t_2[111]}$ and (2) $\pi - (\theta_{[111]} + \theta_{3+}^{t_2[111]}) < \theta_g < \theta_{[111]} + \theta_{3+}^{t_2[111]}$. Then, the average densities of states for both regions relative isotropic medium are:

Table 1.6: Angles $\theta_i^{\lambda\{110\}}$ that characterize the focusing of fast and slow transverse modes for wave vectors lying in the diagonal plane ($\varphi = \pi/4$).

Crystal (mode, focusing direction)	$\theta_1^{\lambda\{110\}}$		$\theta_2^{\lambda\{110\}}$		$\theta_3^{\lambda\{110\}}$		$\theta_4^{\lambda\{110\}}$		$n_{FI}^{\lambda\{110\}}$	$n_{DI}^{\lambda\{110\}}$
GaN (t_1, [110])	28.0°		14.4°		8.1°		33.3°		4.09	0.69
GaN (t_2, [100])	27.5°		12.3°		14.7°		33.8°		2.30	0.37 (1)
Si (t_1, [110])	13.4°		7.5°		0.86°		15.7°		18.2	0.83
Si (t_2, [100])	18.1°		9.6°		3.4°		21.7°		6.33	0.40 (1)
MgO (t_1, [110])	7.9°		4.5°		0.1°		8.75°		88.2	0.90
MgO (t_2, [100])	16.0°		8.6°		2.3°		18.7°		8.15	0.46 (1)
	$\theta_{1+}^{t2\{110\}}$	$\theta_{1-}^{t2\{110\}}$	$\theta_{2+}^{t2\{110\}}$	$\theta_{2-}^{t2\{110\}}$	$\theta_{3+}^{t2\{110\}}$	$\theta_{3-}^{t2\{110\}}$	$\theta_{4+}^{t2\{110\}}$	$\theta_{4-}^{t2\{110\}}$	$n_{FI}^{t2\{110\}}$	$n_{DI(2)}^{t2\{110\}}$
GaN (t_2, [111])	8.8°	18.2°	0°	0°	23.3°	23.3°	14.8°	30.0°	0.96	0.44
Si (t_2, [111])	6.4°	10.9°	0°	0°	12.6°	12.6°	12.0°	20.4°	1.28	0.66
MgO (t_2, [111])	5.9°	9.8°	0°	0°	11.1°	11.1°	11.3°	18.4°	1.34	0.70

$$n_{DI(1)}^{t2\{110\}} = N_{DI(1)}^{t2\{110\}} / N_{Iso} = \left(\theta_{[111]} - \theta_{4+}^{t2[111]} - \theta_{4}^{t2[100]}\right) \Big/ \left(\theta_{[111]} - \theta_{3-}^{t2[111]} - \theta_{3}^{t2[100]}\right), \qquad (1.67)$$

$$n_{DI(2)}^{\{110\}t2} = N_{DI(2)}^{\{110\}t2} / N_{Iso} = \left(\pi - 2\left(\theta_{[111]} + \theta_{4-}^{t2[111]}\right)\right) \Big/ \left(\pi - 2\left(\theta_{[111]} + \theta_{3+}^{t2[111]}\right)\right). \qquad (1.68)$$

Let us look into Table 1.6 and find the values of the $n_{DI(1)}^{t_2\{110\}}$ coefficient for Si: (1) $n_{DI(1)}^{t_2\{110\}} = 0.40$ and (2)$n_{DI(1)}^{t_2\{110\}} = 0.66$. As a result, the ratio between the average densities of states for the phonon focusing and defocusing regions yields: (1) $n_{FD[100]}^{t_2\{110\}} = 15.8$ and (2) $n_{FD[111]}^{t_2\{110\}} = 1.9$.

Here, it is worth giving the estimates of the average densities of states for the focusing and defocusing regions for the t_2 mode by the example of CaF_2 (a type-II crystal). The t_2 mode has one focusing region in the vicinity of the [111] direction and two phonon defocusing regions (see D1 and D2 regions in Figure 1.14d). The relative density of states $n_{FI[111]}^{t_2\{110\}}$ in the focusing region for a CaF_2 crystal can be evaluated through the formula (1.66). For the defocusing regions, it is easy to see that

$$n_{DI(1)}^{t2\{110\}} = N_{DI(1)}^{t2\{110\}} / N_{Iso} = \left(\theta_{[111]} - \theta_{4+}^{t2[111]}\right) / \left(\theta_{[111]} - \theta_{3-}^{t2[111]}\right), \qquad (1.69)$$

$$n_{DI(2)}^{t2\{110\}} = N_{DI(2)}^{t2\{110\}} / N_{Iso} = \left(\pi - 2\left(\theta_{[111]} + \theta_{4-}^{t2[111]}\right)\right) / \left(\pi - 2\left(\theta_{[111]} + \theta_{3+}^{t2[111]}\right)\right). \qquad (1.70)$$

Let us consult Table 1.7 and find for CaF_2 the following values: $n_{DI(1)}^{t_2\{110\}} = 0.77$, and $n_{DI(2)}^{t_2\{110\}} = 0.39$. As a result, the ratio between the average densities of states for the phonon focusing and defocusing regions yields: (1) $n_{FD(1)}^{t_2\{110\}} = 2.6$ and (2)$n_{FD(2)}^{t_2\{110\}} = 5.1$. Now, we should examine the angular dependencies of the average density of states for the fast transverse mode t_2 in crystals of the second group for wave vectors in the diagonal plane. In this plane, the t_2 mode focuses in the directions close to [111] at $\theta = \theta_F$. It has two defocusing regions: D1 in the range of angles $-\left(\theta_F - \theta_{3-}^{t_1}\right) < \theta_g < \theta_F - \theta_{3-}^{t_1}$ and D2 in the range of angles $\theta_F + \theta_{3+}^{t_1} < \theta_g < \pi - \theta_F - \theta_{3+}^{t_1}$, as depicted in Figure 1.14c. For CaF_2, in the focusing region for the t_2 mode, we have:

Table 1.7: Angles $\theta_i^{\lambda\{110\}}$ that characterize the focusing of fast and slow transverse modes for wave vectors lying in the diagonal plane for CaF_2 crystal. $n_{DI(1),(2)}^{\lambda\{110\}}$ is the density of states in the [100] (1) and [110] (2) directions.

Mode/ focusing direction	$\theta_{1+}^{\lambda\{110\}}$	$\theta_{1-}^{\lambda\{110\}}$	$\theta_{2+}^{\lambda\{110\}}$	$\theta_{2-}^{\lambda\{110\}}$	$\theta_{3+}^{\lambda\{110\}}$	$\theta_{3-}^{\lambda\{110\}}$	$\theta_{4+}^{\lambda\{110\}}$	$\theta_{4-}^{\lambda\{110\}}$	$n_{FI}^{\lambda\{110\}}$	$n_{DI(1)}^{\lambda\{110\}}$	$n_{DI(2)}^{\lambda\{110\}}$
t_1, $\theta_F = 47.6°$	20.5°	8.0°	11.7°	0°	3.3°	3.7°	23.4°	8.2°	4.67	0.55	0.87
t_2, [111]	14.1°	20.2°	0°	0°	12.0°	12.0°	21.7°	26.1°	1.99	0.77	0.39

$$n_{FI}^{t1\{110\}} = \left(\theta_{4+}^{t1\{110\}} + \theta_{4-}^{t1\{110\}}\right) / \left(\theta_{3+}^{t1\{110\}} + \theta_{3-}^{t1\{110\}}\right) = 4.67. \tag{1.71}$$

For the defocusing regions, it is not hard to make sure that:

$$n_{DI(1)}^{t1\{110\}} = N_{DI(1)}^{t1\{110\}} / N_{Iso} = \left(\theta_F - \theta_{4+}^{t1\{110\}}\right) / \left(\theta_F - \theta_{3-}^{t1\{110\}}\right), \tag{1.72}$$

$$n_{DI(2)}^{t1\{110\}} = N_{DI(2)}^{t1\{110\}} / N_{Iso} = \left(\pi - 2\left(\theta_F + \theta_{4-}^{t1\{110\}}\right)\right) / \left(\pi - 2\left(\theta_F + \theta_{3+}^{t1\{110\}}\right)\right). \tag{1.73}$$

For CaF_2 crystals, formulas (1.72) and (1.73) bring to $n_{DI(1)}^{t_1\{110\}} = 0.55$ and $n_{DI(2)}^{t_1\{110\}} = 0.87$. As a result, for the ratio of the average densities of states in the phonon focusing and defocusing regions we come up with: (1) $n_{FD(1)}^{t_1\{110\}} = 8.5$ and (2) $n_{FD(2)}^{t_1\{110\}} = 5.4$.

Tables 1.6 and 1.7 indicate that the following equality $\theta_3^{t_2[111]} \equiv \theta_{3-}^{t_2[111]} = \theta_{3+}^{t_2[111]}$ is fulfilled for the slow transverse mode in the [111] direction; it holds true for both types of the crystals. This is not surprising since there occurs internal conical refraction of transverse phonons in the [111] direction (see [76–78]). It has been shown in [76] that, in cubic crystals, phonons of a slow transverse mode propagate in a cone with an opening angle of $2\theta_3^{t_2[111]}$ in the vicinity of the [111] direction. According to [76], the $\theta_3^{t_2[111]}$ angle can be expressed in terms of elastic second-order moduli as follows:

$$\theta_3^{t2[111]} = arctg\left|\frac{1}{\sqrt{2}}\frac{(k-1)(c_{11} - c_{44})}{(c_{11} - c_{12} + c_{44})}\right|. \tag{1.74}$$

For wave vectors lying in planes close to the diagonal plane $\varphi = \pi/4 \pm \delta$ (where $\delta << 1$), the curvature of the isoenergetic surface, as well as the characteristic angles $\theta_i^\lambda(\delta)$ are functions of the angle δ (see Figures 1.14–1.16). A scheme of determination of the angles $\theta_i^{t_2\{110\}}$ for $\delta << 1$ in the diagonal plane is shown in Figure 1.17, and their values are presented in Table 1.8. In the case of asymmetric negative-curvature regions, the relative density of states for the t_1 and t_2 modes in the phonon focusing region is governed by the angles $\theta_{4\pm}^\lambda(\delta)$ and $\theta_{3\pm}^\lambda(\delta)$ and the expression:

$$n_{FI}^{\lambda\{110\}} = N_F^{\lambda\{110\}} / N_{Iso} = \left(\theta_{4+}^\lambda(\delta) + \theta_{4-}^\lambda(\delta)\right) / \left(\theta_{3+}^\lambda(\delta) + \theta_{3-}^\lambda(\delta)\right). \tag{1.75}$$

In order to estimate changes in the density of phonon states for the wave vectors in the planes close to the diagonal plane, we establish the dependencies of the angles $\theta_{3\pm}^{t_2[111]}(\delta)$ and $\theta_{3\pm}^{t_2[111]}(\delta)$ at small values of δ (see Table 1.8). Since the functions $\theta_g^\lambda(\theta, \pi/4 \pm \delta)$ are even functions of the parameter δ, the dependencies $\theta_{3\pm}^{t_2[111]}(\delta)$ and $\theta_{4\pm}^{t_2[111]}(\delta)$ should be approximated by biquadratic functions of the type:

$$\theta_{3\pm}^{t2[111]}(\delta) = a^\pm(\delta)^4 + b^\pm(\delta)^2 + c^\pm. \tag{1.76}$$

Let us compute the angles $\theta_{3\pm}^{t_2[111]}(\delta)$ for the values of the parameters $\delta = 0^+$, 0.05, and 0.1 and find unknown coefficients $a^\pm, b^\pm$, and $c^\pm$, for example, for CaF_2:

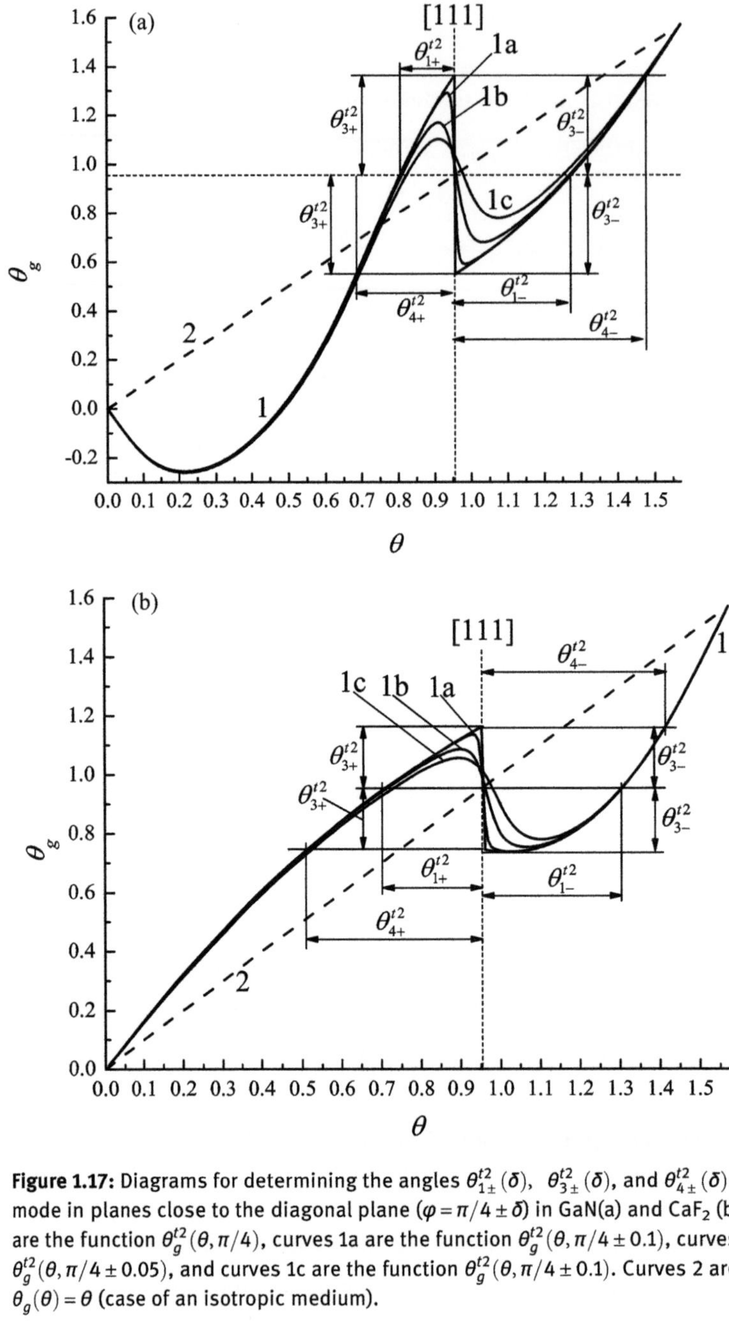

Figure 1.17: Diagrams for determining the angles $\theta_{1\pm}^{t2}(\delta)$, $\theta_{3\pm}^{t2}(\delta)$, and $\theta_{4\pm}^{t2}(\delta)$ for a slow transverse mode in planes close to the diagonal plane ($\varphi = \pi/4 \pm \delta$) in GaN(a) and CaF_2 (b) crystals: curves 1 are the function $\theta_g^{t2}(\theta, \pi/4)$, curves 1a are the function $\theta_g^{t2}(\theta, \pi/4 \pm 0.1)$, curves 1b are the function $\theta_g^{t2}(\theta, \pi/4 \pm 0.05)$, and curves 1c are the function $\theta_g^{t2}(\theta, \pi/4 \pm 0.1)$. Curves 2 are the function $\theta_g(\theta) = \theta$ (case of an isotropic medium).

Table 1.8: The angles $\theta_{3\pm}^{t2[111]}(\delta)$ and $\theta_{4\pm}^{t2[111]}(\delta)$ for the parameters $\delta = 0.05$ and 0.1, which characterize the focusing of slow transverse modes in the vicinity of the [111] direction.

Crystal	δ	$\theta_{3+}^{t2[111]}(\delta)$	$\theta_{3-}^{t2[111]}(\delta)$	$\theta_{4+}^{t2[111]}(\delta)$	$\theta_{4-}^{t2[111]}(\delta)$
GaN	0.05	12.2°	15.8°	12.3°	26.7°
	0.1	7.3°	11.2°	10.9°	23.1°
Si	0.05	5.1°	6.3°	8.3°	15.7°
	0.1	2.1°	3.4°	6.7°	12.2°
MgO	0.05	5.1°	4.2°	7.8°	13.8°
	0.1	3.4°	2.6°	6.8°	11.3°
CaF_2	0.05	7.5°	11.6°	20.9°	26.2°
	0.1	4.9°	10.9°	17.9°	25.4°

1) For the angel $\theta_{3\pm}^{t_2[111]}$: $a^+ = 2537,\ b^+ = -37.76,\ c^+ = 0.21$;
2) For $\theta_{3-}^{t_2[111]}$: $a^+ = 116.4,\ b^+ = -3.08,\ c^+ = 0.21$.

Next, we average the function $\theta_3^{t_2[111]}(\delta)$ over the angle δ in the interval $[0, \Delta]$ in the following way:

$$\left\langle \theta_3^{t_2[111]}(\delta) \right\rangle = \frac{1}{\Delta}\int_0^{\Delta} d\delta\theta_3^{t_2[111]}(\delta). \tag{1.77}$$

We take the averaging region $\Delta = 0.1\,\text{rad}\,(5.7°)$ and obtain: $\langle \theta_{3+}^{t_2[111]} \rangle = 7.7°$ and $\langle \theta_{3-}^{t_2[111]} \rangle = 11.5°$. Similarly, we carry out the averaging of the angles $\theta_{4\pm}^{t_2[111]}$. The average values of the angles $\langle \theta_i^{t_2[110]} \rangle$ are given in Table 1.9.

Let us further explore and compare the angular dependencies of the average densities of phonon states in type-I (Ge, Si, GaN) and type-II (CaF_2, PbS) crystals for wave vectors in the plane of the cube face.

The regions of the maxima of the density of states for the type-I crystals (Ge, Si, GaN) correspond to the minima of that for the type-II crystals (CaF_2, PbS), as in Figure 1.18a.

This can be explained by the fact that the focusing directions in the type-I crystals correspond to the defocusing directions in the type-II crystals (see Figure 1.18a). For the type-I crystals, the values of the density of states reach their maximum for Si crystals, with latter demonstrating the minimum value of the anisotropy parameter and the $2\theta_3^{t_2\{100\}}$ angle that is responsible for the phonon focusing region. The values of the density of states reach their minimum for GaN crystals, which the maximum value of the anisotropy parameter and the $2\theta_3^{t_2\{100\}}$ angle are inherent to.

Table 1.9: The average values of the angles $\left\langle \theta_{3,4}^{t2\{110\}} \right\rangle$, which characterize the focusing of slow transverse modes in the vicinity of the [111] direction, as well as the values of the densities of states in the phonon focusing $\left\langle n_{FI}^{t2\{110\}} \right\rangle$ and defocusing $\left\langle n_{DI(1),(2)}^{t2\{110\}} \right\rangle$ regions and the ratios of the average densities of states $\left\langle n_{FD(1),(2)}^{t2\{110\}} \right\rangle$.

Crystal	$\left\langle \theta_{3+}^{t2[111]} \right\rangle$	$\left\langle \theta_{3-}^{t2[111]} \right\rangle$	$\left\langle \theta_{4+}^{t2[111]} \right\rangle$	$\left\langle \theta_{4-}^{t2[111]} \right\rangle$	$\left\langle n_{FI}^{t2\{110\}} \right\rangle$	$\left\langle n_{DI(1)}^{t2\{110\}} \right\rangle$	$\left\langle n_{DI(2)}^{t2\{110\}} \right\rangle$	$\left\langle n_{FD(1)}^{t2\{110\}} \right\rangle$	$\left\langle n_{FD(2)}^{t2\{110\}} \right\rangle$
GaN	12.9°	16.1°	12.5°	26.6°	1.35	0.15	0.39	8.95	3.5
Si	5.6°	6.7°	8.5°	15.8°	1.97	0.43	0.66	4.63	3.0
MgO	5.6°	4.9°	8.1°	14.0°	2.11	0.47	0.72	4.49	2.9
CaF_2	7.7°	11.5°	20.5°	26.1°	2.42	0.79	0.33	3.06	7.3

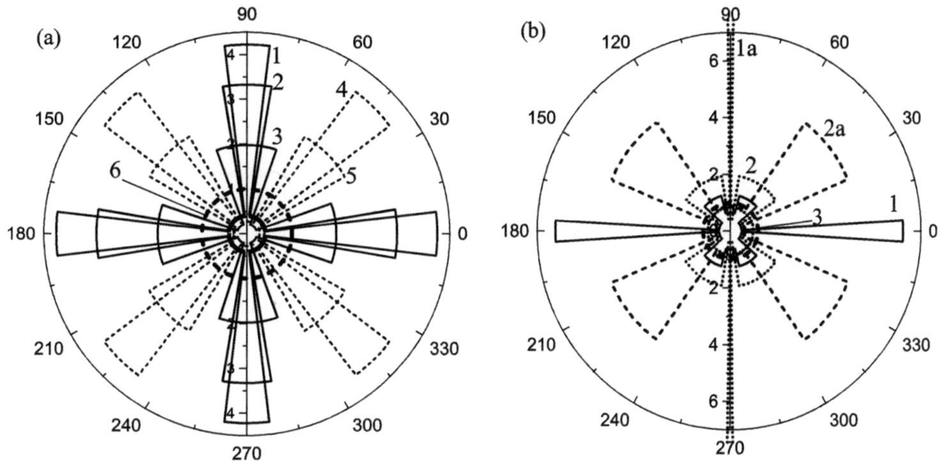

Figure 1.18: (a) Angular dependencies of the average densities of phonon states in the {100} plane for the t_2 mode in Si (1), Ge (2), GaN (3) crystals and the fast t_1 mode in CaF_2 (4) and PbS (5) crystals. Curve 6 is the density of phonon states in the model of an isotropic medium. (b) Angular dependencies of the average densities of phonon states in the {110} plane ($\varphi = \pi/4$) in Si (1.1a) and CaF_2 (2.2a) crystals for slow (curves 1 and 2) and fast (curves 1a and 2a) transverse phonons. Curve 3 is the density of phonon states for an isotropic medium model.

The angular dependencies of the densities of phonon states for wave vectors in the diagonal plane ($\varphi = \pi/4$) being too involved, we present them only for Si and CaF_2 crystals (see Figure 1.18b). For Si crystals, in the focusing region of the fast transverse mode in the [110] direction, the angle $\theta_3^{t_1}$ is larger than the angle $\theta_4^{t_1}$ (see Table 1.6). Therefore, the maximum density of states for this mode significantly exceeds the density of states in an isotropic medium ($n_{FI}^{t_1\{119\}} = 18.2$) and the maxima of the density of states for the slow transverse mode that focuses in the [100] and [111] directions (see Table 1.6 and Figure 1.18b, curves 1 and 1a). Moreover, the maximum density of states for the t_2 mode in the [001] direction exceeds that in the [111] direction by almost 5 times. In CaF_2 crystals, phonons of the fast and slow transverse mode are focused in the $\theta = \theta_F = 47.6°$ and [111] directions, respectively. The maximum density of states for the mode t_2 in the [111] direction turns out to be two times higher than that corresponding to the fast transverse mode (see Figure 1.18b, curves 2 and 2a). Since the focusing regions of these modes overlap, it can be expected that the mode-averaged free path may markedly increase in the vicinity of [111]. Thus, the analysis of the anisotropy of the phonon-state densities confirms the justification of the dividing of cubic crystals into two types in accordance with the sign of the anisotropy parameter. As we will see later, the thermal conductivity and phonon mean free paths of the crystals become affected by a sharp change in the density of phonon states upon transition from the focusing to defocusing regions.

For the t_2 mode in Si and CaF_2 crystals, let us compare the angular dependencies of the average densities of phonon states in the diagonal plane $\varphi = \pi/4$ and phonon-state densities averaged over wave vectors lying in planes close to the diagonal plane $\varphi = \pi/4 \pm \delta$ (where $\delta \le 0.1$). As can be understood from Figures 1.18b and 1.19a, for Si crystals, the density states for the t_2 mode in the [111] direction at $\varphi \equiv \pi/4$ is $n_{FI[111]}^{t_2\{110\}} = 1.28$. However, after averaging over planes close to the diagonal plane, it increases 1.5 times: $\langle n_{FI[111]}^{t_2\{110\}} \rangle = 1.97$ (see Figure 1.19a). A weaker effect is observed for type-II CaF_2 crystals: $n_{FI[111]}^{t_2\{110\}} = 1.95$ at $\varphi \equiv \pi/4$. After averaging over the angle δ, the maximum density of states increases 1.2 times: $\langle n_{FI[111]}^{t_2\{110\}} \rangle = 2.42$ (see Figure 1.19b).

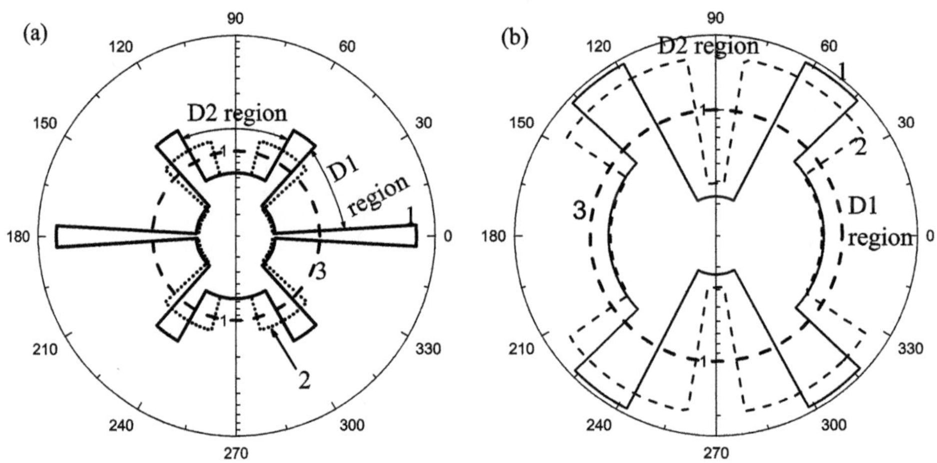

Figure 1.19: Angular dependencies of the phonon density of states of slow transverse phonons in Si (a) and CaF_2 (b) crystals for wave vectors in the diagonal plane. Curves 1 are the density of states of slow transverse phonons for averaged angles $\left\langle \theta_{3\pm}^{t2[111]} \right\rangle$ and $\left\langle \theta_{4\pm}^{t2[111]} \right\rangle$ (see Table 1.9). Curves 2 are the density of states of slow transverse phonons for angles $\theta_{3\pm}^{t2[111]}$ and $\theta_{4\pm}^{t2[111]}$, determined at $\varphi = \pi/4$ (see Tables 1.6 and 1.7). Curves 3 are the density of phonon states for an isotropic medium model. The (D1) and (D2) phonon defocusing regions are indicated corresponding to Figure 1.14.

This effect can be explained as follows. With an increase in the deviation of δ from the diagonal plane, the curvature of the concave section of the isoenergetic surface in the vicinity of the [111] direction diminishes. As a result, it becomes flatter, with the angles $\theta_{3\pm}^{t_2[111]}(\delta)$ that govern the phonon focusing regions decreasing. After averaging, the angle $\langle \theta_{3\pm}^{t_2[111]}(\delta) \rangle$ becomes less than $\theta_{3\pm}^{t_2[111]}(\pi/4)$ (see Figure 1.16 and Tables 1.6–1.8). Since the angle $\theta_{3\pm}^{t_2[111]}(\delta)$ changes slightly with increasing δ, the averaged values $\langle n_{FI[111]}^{t_2\{110\}}(\delta) \rangle$ of the density of states turn out to be greater than those at $\varphi \equiv \pi/4$. The averaged values $\langle n_{FI[111]}^{t_2\{110\}} \rangle$ and $\langle \theta_{3\pm}^{t_2[111]} \rangle$ will be applied in Chapter 4 for analysing the anisotropy of the thermal conductivity and mean free paths in the diagonal plane.

As for the longitudinal phonons, in contrast to the quasi-transverse modes, the slowness surfaces for all the cubic crystals examined contain no negative curvature regions and are convex everywhere (see Figure 1.20). Their group velocity dependencies on the directions of the phonon wave vector are an unambiguous function of angles for crystals of both types. The focusing effect is less pronounced and occurs less dramatically than that for the quasi-transverse modes (see Figure 1.20). As can be seen from Figure 1.20, based on the construction of isoenergetic surfaces, the analysis of the influence of focusing of longitudinal phonons on the group velocity directions confirms the results of the qualitative analysis carried out in Section 1.4. The focusing directions of longitudinal phonons for the type-I crystals are converted to the defocusing directions for the type-II crystals and vice versa.

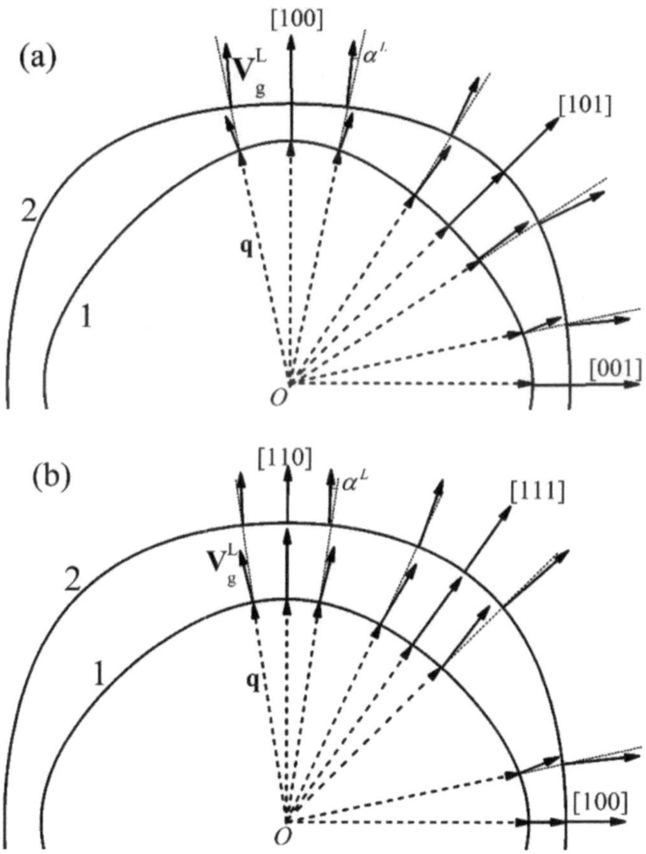

Figure 1.20: Focusing of longitudinal modes in Si and CaF_2 crystals. Sections of the slowness surface by (a) {100} and (b) {110} planes in Si crystals (curves 1) and CaF_2 (curves 2) in an anisotropic continuum model. The dashed arrows show the wave vectors and the solid arrows show the corresponding phonon group velocities.

1.6 Phonon Flux Enhancement Factor

The studies of the phonon pulse propagation in elastically anisotropic crystals have shown high anisotropy of the spatial distribution of the energy flux of acoustic oscillations of various polarizations (see [16–18, 45, 75, 79]). The research in [16] reports that the amplitudes of phonon pulses in LiF and KCl crystals strongly depend on their polarizations and propagation directions. For LiF crystals in the [100] direction, the flux intensity of transverse phonons appears to be 100 times higher than that for longitudinal phonons, whereas it is vice versa in KCl crystals: the pulse amplitude of transverse phonons is seven times lower than for longitudinal phonons. On the other hand, for LiF crystals in the [110] direction, the flux intensity of a fast transverse mode 10 and 20 times exceeds that for longitudinal and slow transverse modes, respectively. To quantitatively describe these effects, Maris [17] used the isotropic medium model as a reference system and introduced the notion of the phonon flux enhancement factor A^{λ}, i.e., the "enhancement factor", which is also known as the "amplification factor [18, 73, 80]." According to [17], the coefficient A^{λ} is defined as the ratio of the phonon flux of a given polarization λ for a chosen wave vector direction to the corresponding flux in an isotropic medium. At the first acquaintance with [17, 18, 45, 73–75, 79], an impression arises that the "enhancement factor" has the meaning of the enhancement factor of the phonon density of states for elastic anisotropic crystals relative to the isotropic medium. However, as we will see later, the differential nature of calculating the coefficient A^{λ} leads to a number of peculiarities that make it difficult to interpret experimental data on phonon pulse propagation in elastically anisotropic crystals. Going the coefficient A^{λ} into infinity at zero-curvature points on the isoenergetic surface prevents from making quantitative estimates of changes in the density of phonon states in the crystals due to the effect of phonon focusing.

Despite a big number of papers dedicated to analysing the enhancement factor [17, 18, 45, 73–75, 80–82, 84], some problems are currently far from being solved. First and foremost, an analytic expression for the coefficient A^{λ} in cubic crystals has not been derived. To date, we have made out with the above issue: a new method of calculating the enhancement factor and a simple analytic expression for it are proposed. This offers an opportunity to qualitatively describe the differences between the enhancement factors A^{λ} for cubic crystals with a various type of anisotropy of elastic energy. Research [44] of the influence of the elastic energy anisotropy on the spectrum and polarization vectors of vibrational modes has shown that all cubic crystals may be divided into two types depending on the sign of the parameter k-1 ($k-1=(c_{12}+2c_{44}-c_{11})/(c_{11}-c_{44})$, c_{ij} are elastic second-order moduli): crystals with positive $k-1>0$ (Fe, Cu, MgO, InSb, GaAs, GaN, Ge, Si, diamond, and LiF) and negative $k-1<0$ (CaF_2, SrF_2, and PbS) anisotropy of elastic second-order moduli (see Table 1.2, as well as [44]). The goal of the present work is to investigate the physical aspects of the influence of focusing on propagation of phonon pulses and compute the enhancement factors for single-crystalline samples with a various type of elastic energy anisotropy.

1.6.1 Analytic Calculation of the Phonon Flux Enhancement Factor

To start with, we calculate the coefficient A^λ as it was done in [17]. Let a phonon source (heater) be on one of the sample faces, and a phonon detector is on the opposite face. They are connected by a vector **R**. In an isotropic medium, waves leaving the source will reach the detector if their wave vectors lie within the solid angle $\delta\Omega_q = \sin\theta d\theta d\varphi$, where θ, φ are angular variables of the vector **q** in the vicinity of the direction of **R** (see Figure 1.21a). In the case of an anisotropic medium, a wave with the wave vector **q** and polarization λ reaches the detector center if the group velocity directions are within the solid angle $\delta\Omega_V$ in the vicinity of the vector **R**, i.e. (see Figure 1.21b): $\mathbf{V}_g^\lambda \| \mathbf{R}$. As can be seen from Figures 1.22, the condition for the transverse waves is fulfilled for at least two or three waves in the directions of wave vectors corresponding to negative-curvature regions on the isoenergetic surface. In [17], it has been shown that the enhancement factor $A^\lambda(\theta, \varphi)$ that characterizes the difference between phonon fluxes in crystals in an isotropic medium can be determined as the ratio of solid angles for wave vectors and group velocities:

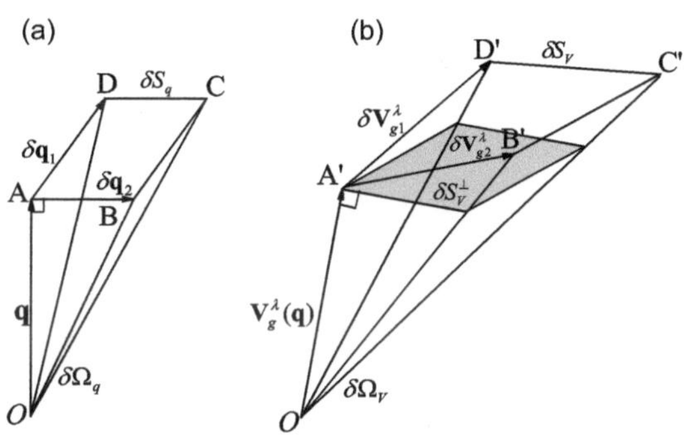

Figure 1.21: Schematic representation of solid angles: (a) $\delta\Omega_q$ in the **q**-space and (b) $\delta\Omega_V$ in the group velocity space.

$$A^\lambda(\theta, \varphi) = \delta\Omega_q / \delta\Omega_V^\lambda. \tag{1.78}$$

Since the distribution of wave vectors, in contrast to group velocities, is isotropic in the crystal, the phonon flux density increases or decreases by the magnitude of $A^\lambda(\theta, \varphi)$ with respect to the isotropic medium. Therefore, the problem is reduced to calculating the ratio of the solid angle in the q-space to the solid angle in group velocity space. In the **q**-space, a parallelogram *ABCD* is observed from point O under the angle $\delta\Omega_q$ (see Figure 1.21a). The vectors $\mathbf{q} = q\mathbf{n}$, $\delta\mathbf{q}_1 = \mathrm{q}\mathbf{n}_1$, and $\delta\mathbf{q}_2 = \mathrm{q}\mathbf{n}_2$ constitute a triad of mutually orthogonal vectors [40]. Then, for an isotropic medium, we have

$$\delta\Omega_q = (\delta q / q)^2. \tag{1.79}$$

The corresponding solid angle for the same variations of the $\delta\mathbf{q}_1$ and $\delta\mathbf{q}_2$ wave vectors in the group velocity space $\delta\Omega_V$ is the angle under which the parallelogram $A'B'C'D'$ is seen (see Figure 1.21b):

$$\begin{aligned} \delta V_{g1i}^{\lambda} &= V_{gi}^{\lambda}(\mathbf{q}+\delta\mathbf{q}_1) - V_{gi}^{\lambda}(\mathbf{q}) = \partial V_{gi}^{\lambda}/\partial\mathbf{q}\cdot\delta\mathbf{q}_1, \\ \delta V_{g2i}^{\lambda} &= V_{gi}^{\lambda}(\mathbf{q}+\delta\mathbf{q}_2) - V_{gi}^{\lambda}(\mathbf{q}) = \partial V_{gi}^{\lambda}/\partial\mathbf{q}\cdot\delta\mathbf{q}_2, \end{aligned} \tag{1.80}$$

where V_{gi}^{λ} is the i-th group velocity component in the orthogonal coordinate system.

Then, the solid angle in the group velocity space acquires the form [17]:

$$\delta\Omega_V = \delta S_V^{\perp}/\left(V_g^{\lambda}\right)^2 = \left|\left(\left[\delta\mathbf{V}_{g1}^{\lambda}\times\delta\mathbf{V}_{g2}^{\lambda}\right]\cdot\mathbf{V}_g^{\lambda}\right)/\left(V_g^{\lambda}\right)^3\right|. \tag{1.81}$$

The enhancement factor $A^{\lambda}(\theta,\varphi)$ is presented as in [17]:

$$\left(A^{\lambda}(\theta,\varphi)\right)^{-1} = \frac{\delta\Omega_V}{\delta\Omega_q} = \frac{q^2}{\left(V_g^{\lambda}\right)^3}\left|\left(\left[\left\{\mathbf{n}_1\cdot\frac{\partial\mathbf{V}_{g1}^{\lambda}}{\partial\mathbf{q}}\right\}\times\left\{\mathbf{n}_2\cdot\frac{\partial\mathbf{V}_{g2}^{\lambda}}{\partial\mathbf{q}}\right\}\right]\cdot\mathbf{V}_g^{\lambda}\right)\right|. \tag{1.82}$$

We further show that the spherical coordinate system $(n_1 = e_\theta,\ n_2 = e_\phi)$ should be used in calculating the factor $A^{\lambda}(\theta,\varphi)$. This makes it possible to arrive at an analytical expression for it. The group velocity vectors and the δV_{g1}^{λ} derivative in the spherical coordinate system can be found as follows:

$$\begin{aligned} &\mathbf{V}_g^{\lambda}(\theta,\varphi) = S^{\lambda}(\theta,\varphi)\tilde{\mathbf{V}}_g^{\lambda}(\theta,\varphi), \quad \tilde{\mathbf{V}}_g^{\lambda}(\theta,\varphi) = \mathbf{n} + S_{\theta}^{\lambda}\mathbf{e}_{\theta} + S_{\varphi}^{\lambda}\mathbf{e}_{\varphi}, \\ &S_{\theta}^{\lambda}(\theta,\varphi) = \left(\frac{1}{S^{\lambda}}\right)\frac{\partial S^{\lambda}}{\partial\theta}, \\ &S_{\varphi}^{\lambda}(\theta,\varphi) = \frac{1}{\sin\theta}\left(\frac{1}{S^{\lambda}}\right)\frac{\partial S^{\lambda}\partial V_{gi}^{\lambda}}{\partial\varphi\ \partial\mathbf{q}} = \frac{\partial V_{gi}^{\lambda}}{\partial q}\mathbf{n} + \frac{1}{q}\frac{\partial V_{gi}^{\lambda}}{\partial\theta}\mathbf{e}_{\theta} + \frac{1}{q\sin\theta}\frac{\partial V_{gi}^{\lambda}}{\partial\varphi}\mathbf{e}_{\varphi}. \end{aligned} \tag{1.83}$$

Here, the phase velocity $S^{\lambda}(\theta,\varphi)$ in the anisotropic continuum model has been determined in [44, 69], and the vectors n, e_θ, and e_φ form a mutually orthogonal triad of unit vectors: $\mathbf{n} = \mathbf{q}/\mathrm{q} = (\sin\theta\ \cos\varphi, \sin\theta\ \sin\varphi, \cos\theta)$ is the unit wave vector of a phonon, $e_\theta = (\cos\theta\ \cos\varphi, \cos\theta\ \sin\varphi, -\sin\theta)$, $e_\varphi = (-\sin\varphi, \cos\varphi, 0)$. From formulas (1.80)–(1.83) for the vectors δV_{g1}^{λ} and δV_{g2}^{λ}, we get:

$$\begin{aligned} \delta V_{g1i}^{\lambda} &= \left(\partial V_{gi}^{\lambda}/\partial\mathbf{q}\cdot\delta\mathbf{q}_1\right) = \left(\mathbf{e}_{\theta}\partial V_{gi}^{\lambda}/\partial\mathbf{q}\right)\cdot\delta q = (\delta q/q)\cdot\left(\partial V_{gi}^{\lambda}\partial\theta\right), \\ \delta V_{g2i}^{\lambda} &= \left(\partial V_{gi}^{\lambda}/\partial\mathbf{q}\cdot\delta\mathbf{q}_2\right) = \left(\mathbf{e}_{\varphi}\partial V_{gi}^{\lambda}/\partial\mathbf{q}\right)\cdot\delta q = (\delta q/q\sin\theta)\cdot\left(\partial V_{gi}^{\lambda}/\partial\varphi\right). \end{aligned} \tag{1.84}$$

and the expression for the area δS_V, $\delta S_V = [\delta\mathrm{V}_{g1}^{\lambda}\times\delta\mathrm{V}_{g2}^{\lambda}]$. Figure 1.21b shows that the area δS_V is not orthogonal to the group velocity vector V_g^{λ}. Therefore, we project it onto the surface perpendicular to the unit vector of the group velocity: $\delta S_V^{\perp} = (\delta S_V\cdot V_g^{\lambda})/V_g^{\lambda}$ (see the shaded area in Figure 1.21b). Consequently, the $A^{\lambda}(\theta,\varphi)$ factor can be written in the spherical coordinate system as:

$$\left(A^{\lambda}(\theta,\varphi)\right)^{-1} = \frac{\delta\Omega_V}{\delta\Omega_q} = \frac{1}{\left(V_g^{\lambda}\right)^3 \sin\theta} \times \left| \left(\left[\frac{\partial \mathbf{V}_g^{\lambda}}{\partial\theta} \times \frac{\partial \mathbf{V}_g^{\lambda}}{\partial\varphi} \right] \cdot \mathbf{V}_g^{\lambda} \right) \right|. \tag{1.85}$$

The use of the Cartesian coordinate system leads to the result:

$$\begin{aligned}\left(A^{\lambda}(\theta,\varphi)\right)^{-1} = & \frac{1}{\left(V_g^{\lambda}\right)^3 \sin\theta} \cdot \left| \left(\frac{\partial V_{gy}^{\lambda}}{\partial\theta}\frac{\partial V_{gz}^{\lambda}}{\partial\varphi} - \frac{\partial V_{gz}^{\lambda}}{\partial\theta}\frac{\partial V_{gy}}{\partial\varphi} \right) V_{gx}^{\lambda} \right. \\ & \left. + \left(\frac{\partial V_{gz}^{\lambda}}{\partial\theta}\frac{\partial V_{gx}^{\lambda}}{\partial\varphi} - \frac{\partial V_{gx}^{\lambda}}{\partial\theta}\frac{\partial V_{gz}^{\lambda}}{\partial\varphi} \right) V_{gy}^{\lambda} + \left(\frac{\partial V_{gx}^{\lambda}}{\partial\theta}\frac{\partial V_{gy}^{\lambda}}{\partial\varphi} - \frac{\partial V_{gy}^{\lambda}}{\partial\theta}\frac{\partial V_{gx}^{\lambda}}{\partial\varphi} \right) V_{gz}^{\lambda} \right|,\end{aligned} \tag{1.86}$$

where $V_{gx}^{\lambda}(\theta,\varphi)$, $V_{gy}^{\lambda}(\theta,\varphi)$, and $V_{gz}^{\lambda}(\theta,\varphi)$ are the phonon group velocity components with polarization λ in the Cartesian coordinate system:

$$\begin{aligned} V_{gx}^{\lambda}(\theta,\varphi) &= S^{\lambda}(\theta,\varphi)\left\{\sin\theta\cos\varphi + S_{\theta}^{\lambda}(\theta,\varphi)\cos\theta\cos\varphi - S_{\varphi}^{\lambda}(\theta,\varphi)\sin\varphi\right\}, \\ V_{gy}^{\lambda}(\theta,\varphi) &= S^{\lambda}(\theta,\varphi)\left\{\sin\theta\sin\varphi + S_{\theta}^{\lambda}(\theta,\varphi)\cos\theta\sin\varphi + S_{\varphi}^{\lambda}(\theta,\varphi)\cos\varphi\right\}, \\ V_{gz}^{\lambda}(\theta,\varphi) &= S^{\lambda}(\theta,\varphi)\left\{\cos\theta - S_{\theta}^{\lambda}(\theta,\varphi)\sin\theta\right\}. \end{aligned} \tag{1.87}$$

When using the Cartesian coordinate system, the expression (1.86) for the enhancement factor contains more than 300 terms and is too cumbersome for analysis.

The papers [81, 82] report on another method of calculating the enhancement factor in terms of derivatives of the group velocity angles θ_V^{λ} and φ_V^{λ}. The latter are responsible for the direction of the vector V_g^{λ}. According to definition (1.78), the coefficient $A^{\lambda}(\theta,\varphi)$ can be presented in the following way:

$$\left(A^{\lambda}(\theta,\varphi)\right)^{-1} = \frac{\delta\Omega_V^{\lambda}}{\delta\Omega_q^{\lambda}} = \frac{d\left(\cos\theta_V^{\lambda}\right)d\varphi_V^{\lambda}}{d(\cos\theta)d\varphi}. \tag{1.88}$$

Here $\delta\Omega_q = d(\cos\theta)\, d\varphi$ is the solid angle in **q**-space, under which the area $|\delta S_q| = |\delta q_1 \times \delta q_2|$ formed by the vectors δq_1 and δq_2 is seen. $\delta\Omega_V^{\lambda} = d(\cos\theta_V^{\lambda})\, d\varphi_V^{\lambda}$ is the corresponding solid angle in the group velocity space, under which the area $|\delta S_V \cdot V_g^{\lambda}/V_g^{\lambda}|$ formed by the vectors δV_{g1}^{λ} and δV_{g2}^{λ} is visible (see Figure 1.21). Hence, the enhancement factor can be rewritten through the Jacobian of the transition from the variables $(\theta_V^{\lambda}, \varphi_V^{\lambda})$ to the variables (θ,φ)[81, 82]:

$$\begin{aligned}\left(A^{\lambda}(\theta,\varphi)\right)^{-1} &= \begin{vmatrix} \partial\left(\cos\theta_V^{\lambda}\right)/\partial(\cos\theta) & \partial\left(\cos\theta_V^{\lambda}\right)/\partial\varphi \\ \partial\varphi_V^{\lambda}/\partial(\cos\theta) & \partial\varphi_V^{\lambda}/\partial\varphi \end{vmatrix} \\ &= \frac{1}{\sin\theta}\left| \frac{\partial\left(\cos\theta_V^{\lambda}\right)}{\partial\theta}\frac{\partial\varphi_V^{\lambda}}{\partial\varphi} - \frac{\partial\left(\cos\theta_V^{\lambda}\right)}{\partial\varphi}\frac{\partial\varphi_V^{\lambda}}{\partial\theta} \right|.\end{aligned} \tag{1.89}$$

To establish the dependencies $\theta_V^\lambda(\theta,\varphi)$ and $\varphi_V^\lambda(\theta,\varphi)$, the papers [81, 82] have again employed the expressions for the group velocity in the Cartesian coordinate system:

$$\mathbf{V}_g^\lambda = V_g^\lambda\left(\sin\theta_V^\lambda\cos\varphi_V^\lambda\mathbf{i} + \sin\theta_V^\lambda\sin\varphi_V^\lambda\mathbf{j} + \cos\theta_V^\lambda\mathbf{k}\right). \tag{1.90}$$

As a result, we have:

$$\cos\theta_V^\lambda(\theta,\varphi) = \frac{V_{gz}^\lambda(\theta,\varphi)}{V_g^\lambda(\theta,\varphi)} \quad \text{and} \quad \varphi_V^\lambda(\theta,\varphi) = \operatorname{arctg}\frac{V_{gy}^\lambda(\theta,\varphi)}{V_{gx}^\lambda(\theta,\varphi)}. \tag{1.91}$$

In spite of significantly simplifying the numerical analysis of the enhancement factor by using formulas (1.89)–(1.91) as compared with the expressions (1.86)–(1.87), calculating the enhancement factor in the Cartesian coordinate system remains rather cumbersome. Besides, formulas (1.89)–(1.91) do not yield an analytical expression for the $A^\lambda(\theta,\varphi)$ factor.

In contrast to this, our analysis showed that the spherical coordinate system for group velocity vectors and their derivatives makes it possible to deduce an accurate analytical solution for the $A^\lambda(\theta,\varphi)$ enhancement factor and comprehensively analyse its features. To this purpose, we resort to formulas (1.80)–(1.83) to calculate the scalar product $([\partial\mathbf{V}_g^\lambda/\partial\theta\times\partial\mathbf{V}_g^\lambda/\partial\varphi]\cdot\mathbf{V}_g^\lambda)$ in the spherical coordinate system. Finally, we come to the expression for $A^\lambda(\theta,\varphi)$ in terms of the group velocity angular-components S_θ^λ and S_φ^λ and their derivatives [84]:

$$A^\lambda(\theta,\varphi) = \left(\tilde{V}_g^\lambda\right)^3\left|\left(1+(S_\theta^\lambda)^2+\frac{\partial S_\theta^\lambda}{\partial\theta}\right)\left(1+S_\theta^\lambda\frac{\cos\theta}{\sin\theta}+\left(S_\varphi^\lambda\right)^2+\frac{1}{\sin\theta}\frac{\partial S_\varphi^\lambda}{\partial\varphi}\right)\right.$$
$$\left.-\left(\frac{1}{\sin\theta}\frac{\partial S_\theta^\lambda}{\partial\varphi}+S_\varphi^\lambda\left(S_\theta^\lambda-\frac{\cos\theta}{\sin\theta}\right)\right)^2\right|^{-1}. \tag{1.92}$$

It is obvious that the angular derivatives S_θ^λ and S_φ^λ for the isotropic medium model are zero, and the enhancement factor is equal to unity. The further analysis of the peculiarities of the phonon flux propagation in elastically anisotropic crystals requires using the analytical expression (1.92).

The elastic wave spectrum and group velocities of real crystals have rather complicated shapes. To visualize the solid angle $\delta\Omega_q$ in $\mathbf{q}$-space and the corresponding angle $\delta\Omega_V$ in group velocity space, Figure 1.22 should be looked at. It sketches cross-sections of the isoenergetic and wave surfaces in group velocity space by the cube face plane for a slow transverse mode in a Si crystal. The wave surface is a geometric place of group velocity vector ends for every possible wave vector directions in the crystal [17]. According to [17] (see eq. (24.13)), it can be defined by the expression

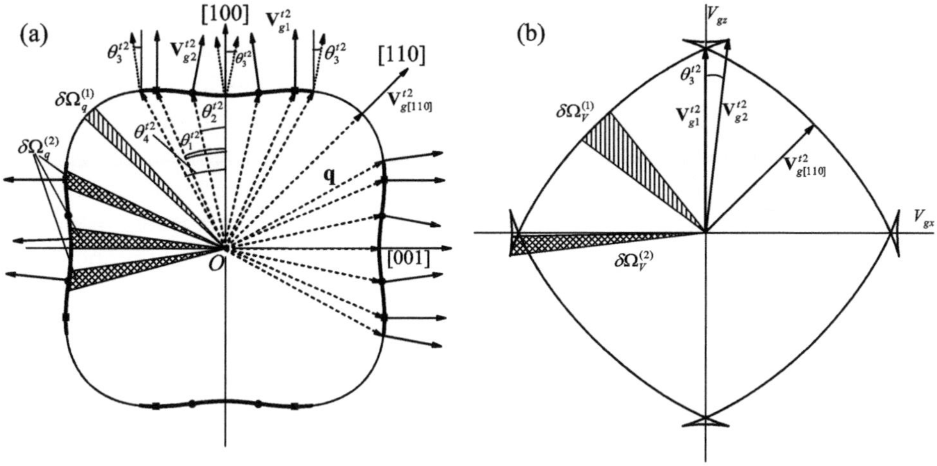

Figure 1.22: Cross-sections of the (a) isoenergetic and (b) wave surfaces in group velocity space by the cube face plane for a slow transverse mode in a Si crystal. The values of the angles θ_i^{t2} for the cube face plane are given in Table 1.6. The group velocity vectors $\mathbf{V}_{g1}^{t2}$, $\mathbf{V}_{g2}^{t2}$, and $\mathbf{V}_{g[110]}^{t2}$ in (a) correspond to the vectors $\mathbf{V}_{g1}^{t2}$, $\mathbf{V}_{g2}^{t2}$, and $\mathbf{V}_{g[110]}^{t2}$ in (b).

$$\left(\mathbf{n}\cdot \boldsymbol{V}_g^{\lambda}\right)/S^{\lambda}(\theta,\varphi)=1. \tag{1.93}$$

The wave surface can also be assigned parametrically, having taken the wave vector angles θ and φ as parameters. For the {010} cross-section, it is given by

$$\begin{cases} V_{gx}^{\lambda}=V_{gx}^{\lambda}(\theta,0), \\ V_{gy}^{\lambda}=0, \\ V_{gz}^{\lambda}=V_{gz}^{\lambda}(\theta,0). \end{cases} \tag{1.94}$$

Here the group velocity components $\mathrm{V}_{\mathrm{g}}^{\lambda}(\theta,\varphi)$ are controlled by formulas (1.87). In the vicinity of the [110] direction, the solid angle $\delta\Omega_V^{(1)}$ corresponds to the angle $\delta\Omega_q^{(1)}$ (see Figure 1.22). In this case, the direct calculation yields the factors $A_{[110]}^{t_2}<1$, i.e., $\delta\Omega_V^{(1)}>\delta\Omega_q^{(1)}$ and the slow transverse mode for this direction defocuses (see Figures 1.22).

In the vicinity of the [100] direction, the solid angle $\delta\Omega_V^{(2)}$ (see Figure 1.22b) corresponds to three different values of the solid angle $\delta\Omega_q^{(2)}$ (see Figure 1.22a). Therefore, the total ratio $\delta\Omega_q/\delta\Omega_V$ for the [100] direction is much more than unity, which indicates a high flux phonon enhancement in the vicinity of the focusing direction. Let us consider the isoenergetic surface for the t_2 mode in more detail. The work [71] declares that the function $\theta_g^{t_2}(\theta)<0$ in the vicinity of the [100] directions in the angular range $-\theta_1\le\theta\le\theta_1$ but it has a local minimum at $\theta_2=0.21$.

The angles $\pm\theta_2$ define the directions of the group velocity vectors $V_{g2}^{t_2}$ towards the wrinkle boundaries in the wave surface in the group velocity space, as in Figure 1.22b. The same angles define the wave vector directions towards zero-curvature points in the isoenergetic surface [71]. The convex-to-concave transition occurs at these points, and the curvature vanishes. The positions of the zero-curvature points are indicated by circles in Figure 1.22. In the three-dimensional case, these points are combined into zero-curvature lines in the isoenergetic surface. The phonon flux being orthogonal to this surface and inversely proportional to its curvature at the given point [17, 18, 45], the zero-curvature points mathematically match to infinite phonon flux emitted from a point heat source along an appropriate group velocity direction.

1.6.2 Analysis of the Angular Dependencies of the Enhancement Factor

To examine the angular dependencies of the enhancement factor in silicon crystals, we use eq. (1.92) (see Figure 1.23a). In the isotropic medium model, the coefficient is $A^{\lambda}(\theta, 0) = 1$. Therefore, the angular ranges in which the inequality $A^{\lambda}(\theta, 0) > 1$ is satisfied can be attributed to phonon focusing regions; the angular ranges in which the reverse inequality is met can be attributed to phonon defocusing regions. It is clear that the enhancement factor of the transverse mode t_2 becomes infinite at zero-curvature points ($\theta = \theta_2^{t_2}$) in the isoenergetic surface (see Figure 1.23a, curve 4). The coefficient $A^{\lambda}(\theta, 0)$ reaches its minimum in the [110] defocusing direction: $A_{[110]}^{t_2} = 0.23$ for Si. Longitudinal phonons are focused and defocused in the [111] and [100] directions, respectively. For these cases, $A_{[111]}^{L} = 2.20$ and $A_{[100]}^{L} = 0.27$. In the [110] direction, a local focusing maximum takes place for them, and $A_{[110]}^{L} = 1.42$. In CaF_2 type-II crystals, the enhancement factor of the fast transverse mode at zero-curvature points (at $\theta = \theta_2^{t_1}$) in the isoenergetic surface goes into infinity (see Figure 1.23b, curve 3). Longitudinal phonons are focused in the [100] direction, and the enhancement factor is $A_{[100]}^{L} = 3.21$; in the [110] direction, they are defocused, and $A_{[100]}^{L} = 0.78$.

The analysis of the enhancement factors for the fast transverse mode in type-I crystals and the slow transverse mode in type-II crystals has revealed an interesting feature. The fact is that the spectrum of both transverse modes in cubic crystals for wave vectors in the cube face plane is isotropic (see Figure 1.1). Consequently, the $A^{\lambda}(\theta, 0)$ factor is expected to be equal to unity, as found within the isotropic medium model. However, its magnitude turns out to be much more than unity for silicon crystals in directly calculating using the expressions (1.92) (see Figure 1.23a, curve 3); it is appreciably less than unity for the type-II CaF_2 crystal, (Figure 1.23c, curve 4).

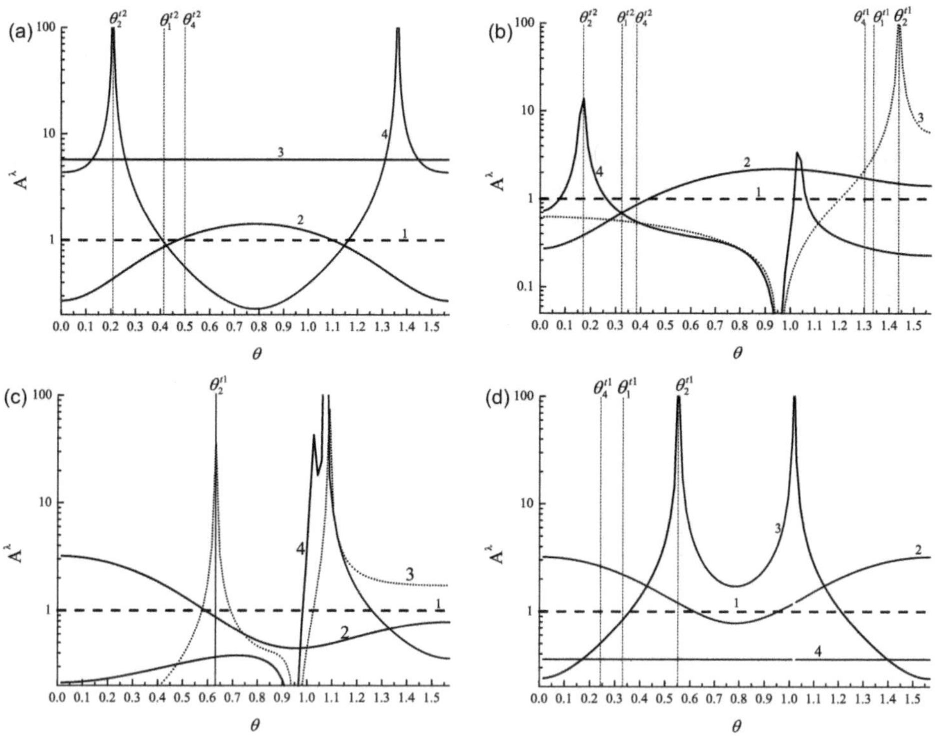

Figure 1.23: Angular dependencies of the enhancement factors $A^{\lambda}(\theta, \varphi)$ in Si (a, b) and CaF_2 (c,d) crystals for wave vectors in the cube face plane (a,c) and diagonal plane (b,d). Dashed lines (1) correspond to the isotropic medium model; curves 2, 3, and 4 correspond to longitudinal phonons, the fast transverse modet_1, and the slow transverse mode t_2, respectively.

Since this feature of the enhancement factor has been previously disregarded, we should treat it for crystals with different types of elastic energy anisotropy as thoroughly as possible. To this end, we expand the phase velocity of the fast transverse mode from eq. (1.28) for type-I crystals, near the cube face plane at $\varphi \ll 1$:

$$S^{t1}(\theta,\varphi) \approx \sqrt{\frac{c_{44}}{\rho}}\left(1-\frac{\delta}{2}\varphi^2\cdot\sin^2\theta\right), \delta = \left[\frac{c_{11}-c_{44}}{c_{44}}\right]\frac{(k-1)(2k+1)}{(k+1)}. \tag{1.95}$$

Equation (1.95) implies that $S^{t_1}(\theta, 0)$ is maximum for the type-I crystals. It follows from expression (1.92) that in the case $\varphi \to 0$ the contribution to the enhancement factor of the t_1 mode is made only by the second derivative $\partial S^{t_1}/\partial\varphi^2$ of the phase velocity:

$$A^{t1}(\theta,0) = \left|1+\frac{1}{\sin^2\theta}\frac{\partial^2 S^{t1}}{\partial\varphi^2}\sqrt{\frac{\rho}{c_{44}}}\right|^{-1} = \left|\frac{1}{1-\delta}\right|. \tag{1.96}$$

For most type-I semiconductor crystals, the values of the parameter δ fall into the range $0 \le \delta \le 2$ and the enhancement factor of a fast transverse mode can significantly exceed unity. However, for the type-I Na, K, Li, and Cu metal crystals, the parameter $\delta > 2$, and the enhancement factor of the fast transverse mode appears to be smaller than unity (see Table 1.10). As seen from Table 1.10, there is no correlation between the $A^{t_1}(\theta, 0)$ factor and the anisotropy parameter (k–1). So, for example, the anisotropy parameter for YAG crystals is small ($k-1 \approx 0.03$), and the enhancement factor is close to unity: $A^{t_1}(\theta, 0) = 1.08$. For HgSe crystals, the anisotropy parameter is 20 times higher; however, the enhancement factor is lower than that for YAG crystals. For GaN crystals ($A^{t_1}(\theta, 0) = 1.02$), the parameter k–1 is maximum. However, the enhancement factor is the same as for YAG crystals (see Table 1.10). Among the type-I crystals mentioned, the enhancement factor reaches a maximum value for MgO ($A^{t_1}(\theta, 0) = 16.11$) whose anisotropy parameter (k–1 = 0.7) differs slightly from that of

Table 1.10: Parameters k–1, δ and the $A^{t_1}(\theta, 0)$ enhancement actors for a fast transverse mode in type-II cubic crystals.

Crystal	k–1	δ	A^{t_1}
Fe	1.15	1.91	1.09
Cu	1.12	2.31	0.76
GaAs	0.9	1.49	2.04
InSb	0.81	1.63	1.58
MgO	0.7	1.06	16.11
GaSb	0.85	1.47	2.14
Ge	0.87	1.32	3.13
GaN	1.28	1.92	1.08
LiF	0.78	1.2	4.89
Si	0.67	1.17	5.74
HgSe	0.61	1.98	1.02
Diamond	0.4	0.55	2.22
YAG	0.03	0.076	1.08
Na	45.1	3.47	0.40
Li	4.83	3.31	0.43
K	2.28	2.98	0.51

HgSe (see Table 1.10). The enhancement factor becomes minimum ($A^{t_1}(\theta, 0) = 0.4$) for Na alkali metal whose anisotropy parameter of the elastic energy has a maximum magnitude ($k{-}1 = 45.1$).

In the type-II crystals ($k{-}1 < 0$), the phase velocity of the slow transverse mode t_2 takes minimum values for wave vectors in the cube face plane. The phonon spectrum for it at $\varphi \ll 1$ is defined by eq. (1.95), and the $A^{t_2}(\theta, 0)$ enhancement factor is identical to the expression (1.96). Since the parameter $\delta < 0$ for all type-II crystals, their enhancement factor appears to be less than unity (see Table 1.11 and Figure 1.23).

Table 1.11: Parameters $k{-}1$, δ and the A^{t2} enhancement factors for type-II crystals for a slow transverse mode and wave vectors lying in the cube face plane.

Crystal	KCl	NaCl	PbS	CaF_2	SrF_2	YIG
k-1	−0.63	−0.48	−0.466	−0.33	−0.204	−0.04
δ	−4.29	−2.12	−2.59	−1.80	−0.88	−0.17
A^{t2}	0.19	0.320	0.28	0.36	0.53	0.86

The data of Table 1.11 confirm that the enhancement factors for the type-II crystals are inversely proportional to the anisotropy parameter.

The maximum enhancement factor ($A^{t_2}_{\{100\}} \approx 0.86$) is inherent to YIG crystals whose anisotropy parameter minimum is $k-1 = -0.04$. The minimum enhancement factor ≈ 0.19 is characteristic of the most anisotropic KCl crystals with the anisotropy parameter $k-1 = -0.63$. Thus, phonon defocusing effects dominate in the enhancement factors of the slow transverse mode for all type-II crystals. Therefore, they are less than unity for wave vectors in the cube face plane.

As can be noticed from Figure 1.23, the enhancement factor of the slow transverse mode in the vicinity of the [100] direction as $\theta \to 0$ suffers a finite discontinuity. In this case, for Si crystals, this factor is different at $\varphi = 0$ and $\varphi = \pi/4$. At $\varphi = 0$, it appears to be much greater than unity ($A^{t_2}(0, 0) \approx 4.4$); at $\varphi = \pi/4$, it is less than unity ($A^{t_2}(0, 0) \approx 0.72$). This result has remained physically strange and not yet been explained. Figures 1.12 (see also [71]) indicate that the slow transverse mode is focused in the [100] direction as well as in the cube face plane ($\varphi = 0$) and in the diagonal plane ($\varphi = \pi/4$). Therefore, the enhancement factor in both cases should exceed unity. To explain this paradox, a detailed analysis of the behaviour of the $A^{t_2}(\theta, \varphi)$ factor at small angles θ is required. For this purpose, formulas (1.92) need to be drawn attention to. The point is that, in Section 1.5 (see also [71]), in analysing the phonon density of states and considering the isoenergetic surface cross-sections at $\varphi = 0$ and $\varphi = \pi/4$, we fixed the angle φ and accounted for only the phase velocity derivatives over θ, whereas in calculating the $A^{\lambda}(\theta, \varphi)$ factor, the derivatives over both angles were taken into account. The enhancement factor characterizes the isoenergetic surface and is inversely proportional to its Gaussian curvature, $A \sim 1/K$ [82]. Depending on its curvature, various kinds of peculiarities are possible to take place: (a) "a saddle" and (b) "a concave lobe" types [82]. The peculiarity type in the three-dimensional surface is defined by the Gaussian curvature K [83]. The latter can be represented as the product of the principal surface curvatures: $K = K_1K_2$. In [82], it has been shown that the surface has a saddle-type feature when $K < 0$ (K_1 and K_2 are of opposite sign). If both K_1 and K_2 are negative, the surface is concave ("a concave lobe") and the surface is convex if both K_1 and K_2 are positive.

To understand which kind of features arises in the vicinity of the [100] directions for various angles φ, we expand the expression for the phase velocity of the slow mode t_2 at $\theta \ll 1$. Then, we obtain [84]:

$$S^{t2}(\theta,\varphi) \approx \sqrt{\frac{C_{44}}{\rho}}\left(1-\theta^2\Delta_{t2}(\varphi)\right),\ \Delta_{t2}(\varphi) = \frac{C_{11}-C_{44}}{4C_{44}}(k-1)\left[\sqrt{(1+2k)\cos^2 2\varphi + k^2} + k + 1\right]. \tag{1.97}$$

For type-I crystals, the phase velocity of slow transverse phonons in the [100] direction has its absolute maximum. For the cube face plane, the second derivatives over θ and φ appear with different signs $\left(\partial^2 S^{t_2}/\partial\theta^2 < 0, \partial^2 S^{t_2}/\partial\varphi^2 > 0\right)$. Therefore, the saddle-point peculiarity emerges in the vicinity of the cube face at $\theta \ll 1$ in Si crystals. For the diagonal plane, the second derivative $\partial^2 S^{t_2}/\partial\theta^2$ remains negative but the second derivative $\partial^2 S^{t_2}/\partial\varphi^2$ becomes negative. Thus, the surface turns into a concave one ("a concave lobe") in the vicinity of the diagonal plane.

In the limiting case at $\theta \ll 1$, it follows from the expression (1.92) that the dependence of the enhancement factor of the slow transverse mode on the angle φ can be presented at $\varphi = 0$ and $\varphi = \pi/4$ as the product of two multipliers [84]:

$$A^{t2}(0^+,\varphi) = \frac{1}{\left|1-2\Delta_{t2}(\varphi)\right|} \cdot \frac{1}{\left|1-2\Delta_{t2}(\varphi)-\Delta''_{t2}(\varphi)\right|}, \tag{1.98}$$

where

$$\Delta''_{t2}(\varphi) = \frac{d^2\Delta_{t2}(\varphi)}{d\varphi^2} = -\frac{C_{11}-C_{44}}{4C_{44}}(k-1)(1+2k)\left\{\frac{4\cos 4\varphi}{\sqrt{(1+2k)\cos^2 2\varphi + k^2}} + \frac{(1+2k)\sin^2 4\varphi}{\left(\sqrt{(1+2k)\cos^2 2\varphi + k^2}\right)^3}\right\}.$$

For Si crystals at the angles $\varphi = 0$ and $\varphi = \pi/4$, we have:

$$\Delta_{t2}(0) = \frac{c_{11} - c_{44}}{2c_{44}}(k^2 - 1) = 0.96, \qquad (1 - 2\Delta_{t2}(0))^{-1} = -1.09,$$

$$\Delta''_{t2}(0) = -\frac{c_{11} - c_{44}}{c_{44}}\frac{(k-1)(1+2k)}{(k+1)} = -1.17, \; \left(1 - 2\Delta_{t2}(0) - \Delta''_{t2}(0)\right)^{-1} = 4,$$

$$\Delta_{t2}(\pi/4) = \frac{c_{11} - c_{44}}{4c_{44}}(k-1)(2k+1) = 0.78, \; (1 - 2\Delta_{t2}(\pi/4))^{-1} = -1.79, \tag{1.99}$$

$$\Delta''_{t2}(\pi/4) = \frac{c_{11} - c_{44}}{c_{44}}\frac{(k-1)(1+2k)}{k} = 1.88, \; \left(1 - 2\Delta_{t2}(\pi/4) - \Delta''_{t2}(\pi/4)\right)^{-1} = -0.41.$$

The values specified above yield the following enhancement factors: $A^{t_2}(0,0) = 4.36$ and $A^{t_2}(0, \pi/4) = 0.72$. As seen from the estimates (1.99), the major role in the fulfillment of inequalities $A^{t_2}(0,0) > 1$ and $A^{t_2}(0,\pi/4) < 1$ is played by their dependence on the derivative $\partial^2 S^{t_2}/\partial\varphi^2$. It is seen that the factors $A^{t_2}(0,0)$ and $A^{t_2}(0,\pi/4)$ contain the multipliers $\left(|1 - 2\Delta_{t_2}(0)| - |\Delta''_{t_2}(0)|\right)^{-1} = 4$ and $\left(|1 - 2\Delta_{t_2}(\pi/4)| - |\Delta''_{t_2}(\pi/4)|\right)^{-1} = 0.41$. The latter differ from each other by an order of magnitude due to changing the parameter $\Delta''_{t_2}(\varphi)$. In this case, the values of $|1 - 2\Delta_{t_2}(\varphi)|$ change to a lesser extent during the transition from the cube face plane to the diagonal plane. Using (1.98), we can arrive at $\varphi = 0$ and $\varphi = \pi/4$ at [84]:

$$K_1^{t2}(0,\varphi) = S^t_{[100]}\tilde{K}_1^{t2}(0,\varphi), \quad \tilde{K}_1^{t2}(0,\varphi) = 1 - 2\Delta_{t2}(\varphi),$$
$$K_2^{t2}(0,\varphi) = S^t_{[100]}\tilde{K}_2^{t2}(0,\varphi), \quad \tilde{K}_2^{t2}(0,\varphi) = 1 - 2\Delta_{t2}(\varphi) - \Delta''_{t2}(\varphi). \tag{1.100}$$

For type-I crystals with low elastic anisotropy (such as YAG), the isoenergetic surface of the mode t_2 is convex everywhere, and both principal surface curvatures K_1 and K_2 are positive (see Table 1.12). The conditions $K_1 > 0$, $K_2 > 0$ and expressions (1.99) lead to the inequalities under which the isoenergetic surface remains convex (see Figure 1.24):

$$1 < \frac{c_{11}}{c_{44}} < 1 + \frac{2k}{(k-1)(k+2)(2k+1)}. \tag{1.101}$$

Hence, we see that this region is bounded by curve 1:

$$c_{11}/c_{44} = 1 + 2k/\{(k-1)(k+2)(2k+1)\}. \tag{1.102}$$

It should be noted that K_1 and K_2 are positive for all the type-II crystals ($k-1 \le 0$). Therefore, the constant-energy surfaces of transverse phonons are convex at $\theta \ll 1$ (see Table 1.12).With increasing the anisotropy parameter $k-1$, we cross the curve 1 and enter the second region, for which the constant-energy surface of the slow transverse phonons is convex in the vicinity of the cube face plane at $\theta \ll 1$ ($K_1 > 0$ and $K_2 > 0$). In the vicinity of the diagonal plane, the saddle-point-type feature is implemented ($K_1 > 0$, $K_2 < 0$, see Table 1.12). For the second region, the system of the inequalities $K_1(0, 0) > 0$,

Table 1.12: Values of the enhancement factors $A^{t_2}(0^+,\varphi)$ and $A^{t_1}(0^+,\varphi)$ and parameters that characterize the isoenergetic surface curvatures for the first- and second-type crystals at $\theta \ll 1$.

Crystal	$\frac{c_{11}}{c_{44}}$	$1/\tilde{K}_1^{t2}(0,0)$	$1/\tilde{K}_2^{t2}(0,0)$	$1/\tilde{K}_1^{t2}(0,\frac{\pi}{4})$	$1/\tilde{K}_2^{t2}(0,\frac{\pi}{4})$	$A^{t2}(0,0)$	$A^{t2}(0,\frac{\pi}{4})$
Fe	1.99	−0.39	−1.49	0.50	−0.21	0.57	0.10
Cu	2.23	−0.3	−1.03	−0.38	−0.17	0.31	0.064
GaAs	2.00	−0.62	−8.31	−0.86	−0.29	5.16	0.25
InSb	2.23	−0.56	−6.37	−0.77	−0.26	3.56	0.20
MgO	1.94	−1.32	3.26	−2.33	−0.47	4.32	1.10
Ge	1.92	−0.77	50	−1.11	−0.34	38.5	0.38
LiF	1.92	−0.94	7.14	−1.47	−0.39	6.74	0.58
Na	1.04	−0.01	−0.01	−0.01	−0.01	0.0002	0.0002
Li	1.37	−0.09	−0.13	−0.1	−0.07	0.011	0.007
K	1.74	−0.16	−0.31	−0.19	−0.11	0.05	0.02
GaN	1.89	−0.37	−1.25	−0.47	−0.2	0.46	0.096
Si	2.09	−1.09	4.0	−1.79	−0.41	4.36	0.72
Diamond	1.87	6.25	1.41	2.94	−1.67	8.80	4.90
YaG	2.89	1.11	1.03	1.08	1.29	1.14	1.40
Crystal	$\frac{c_{11}}{c_{44}}$	$1/\tilde{K}_1^{t1}(0,0)$	$1/\tilde{K}_2^{t1}(0,0)$	$1/\tilde{K}_1^{t1}(0,\frac{\pi}{4})$	$1/\tilde{K}_2^{t1}(0,\frac{\pi}{4})$	$A^{t1}(0,0)$	$A^{t1}(0,\frac{\pi}{4})$
YIG	3.52	0.84	0.95	0.87	0.69	0.8	0.6
SrF_2	4.0	0.48	0.82	0.56	0.26	0.39	0.15
CaF_2	4.74	0.33	0.75	0.41	0.15	0.25	0.06
PbS	5.12	0.25	0.74	0.33	0.1	0.19	0.03
NaCl	4.33	0.29	0.78	0.38	0.11	0.23	0.04
KCl	6.37	0.18	0.75	0.25	0.05	0.13	0.01

$K_2(0,0) > 0$ at $\varphi = 0$ and $K_1(0,\pi/4) > 0$, $K_2(0,\pi/4) < 0$ at $\varphi = \pi/4$ gives the following expression:

$$1 + \frac{2k}{(k-1)(k+2)(2k+1)} < \frac{c_{11}}{c_{44}} < 1 + \frac{1}{k^2-1}. \tag{1.103}$$

This region is bounded by curves 1 and 2; the latter is defined by the expression $c_{11}/c_{44} = 1 + 1/(k^2-1)$. Among the crystals considered, diamond arrives at this parameter range (see Figure 1.24). In the third region, principal curvatures are opposite in sign in both planes: $K_1 < 0$, $K_2 > 0$ in the cube face plane and $K_1 > 0$, $K_2 < 0$ in the diagonal plane. Therefore, for both planes in the isoenergetic surface, the saddle point-type features are formed. Such a region is located between curves 2 and 3 and defined by the inequalities:

$$1+\frac{1}{k^2-1}<\frac{c_{11}}{c_{44}}<1+\frac{2}{(k-1)(2k+1)}. \tag{1.104}$$

However, none of crystals set in Table 1.12 arrived at this region. A further increase in the anisotropy parameter k–1 leads to arising the following features in the isoenergetic surface at $\theta \ll 1$, but namely, the saddle-point feature appears in the vicinity of the cube face plane for the mode t_2 ($K_1(0,0) < 0$, $K_2(0,0) > 0$); the isoenergetic surface becomes concave ($K_1(0, \pi/4) < 0$, $K_2(0, \pi/4) < 0$) in the vicinity of the diagonal plane. The above inequalities create the fourth region bounded by curves 3 and 4 (see Figure 1.24):

$$1+\frac{2}{(k-1)(2k+1)}<\frac{c_{11}}{c_{44}}<1+\frac{k+1}{k^2(k-1)}. \tag{1.105}$$

The fourth region includes such crystals as Si, MgO, Ge, and LiF (see Figure 1.24). As the anisotropy parameter k–1 further increases, we cross curve 4 and reveal the fifth region. Here, the constant-energy surface for the mode t_2 at $\theta \ll 1$ becomes

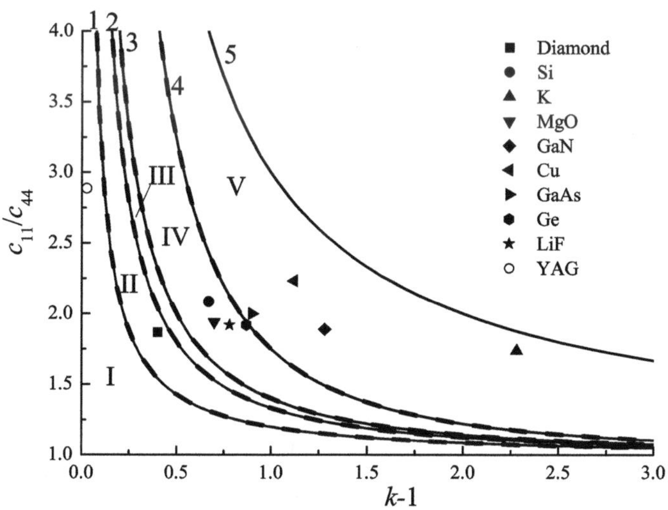

Figure 1.24: Dependencies of c_{11}/c_{44} on the parameter k – 1, defining five regions with different curvatures of the isoenergetic surfaces for slow transverse phonons for type-I cubic crystals. Regions I, II, III, IV, and V are bounded by curves 1, 2, 3, 4, and 5, for which equations are given by the inequalities (1.101), (1.103), (1.104), (1.105), and (1.106), respectively. Solid curves are the results of the above analysis, dashed curves are the results of calculations by Every [82]. Symbols denote the parameters for type-I crystals.

concave for an arbitrary angle φ: both curvatures K_1 and K_2 become negative (see Table 1.12). From the expressions (1.99) and the condition $c_{11} > c_{12}$, we can deduce the relations responsible for curves 4 and 5:

$$1 + \frac{k+1}{k^2(k-1)} < \frac{c_{11}}{c_{44}} < 1 + \frac{2}{k-1}. \tag{1.106}$$

The fifth region incorporate Fe, Cu, K, Li, Na, and GaN crystals with maximum elastic energy anisotropy (see Figure 1.2).

In [82], Every A.G. run a numerical analysis of the isoenergetic surface curvature for cubic crystals using formulas (1.89)–(1.91) in the coordinates c_{11}/c_{44} and c_{12}/c_{44} without introducing the anisotropy parameter k–1. He derived relations for elastic moduli to divide crystals with different isoenergetic surface types. As seen in Figure 1.24, the results secured in the coordinates c_{11}/c_{44} and $k - 1$ (solid curves) are consistent with those of the analysis performed in [82] (dashed curves).

Against the findings of [82], the present Section (see also [84]) chiefly deals with the examination of how the discontinuity magnitudes of the enhancement factors of transverse modes in the vicinity of the [100] direction affect varying the anisotropy parameter and isoenergetic surface curvatures. It is seen from Table 1.12 that the $A^{t_2}(0,\varphi)$ factors are different in the vicinity of the [100] directions at $\varphi = 0$ and $\varphi = \pi/4$ for all crystals, i.e. they undergo finite discontinuities. As the above analysis showed, these discontinuities are caused by different isoenergetic surface curvatures for wave vectors in the cube face and diagonal planes. Therewith, for the first-type crystals, when passing from the cube face plane to the diagonal plane, not only the principal curvatures for slow transverse modes change but also their sign (curvature type) (see Table 1.12). However, for the second-type crystals, isoenergetic surfaces for fast transverse modes remain convex, only curvature values vary at $\varphi = 0$ and $\varphi = \pi/4$ (see Table 1.12). The $A^{t_2}(0,\varphi)$ coefficients have maximum discontinuities for the first-type crystals entering the fourth region. So, for example, for Ge as $\theta \to 0$, the factors $A^{t_2}(0,\varphi) \cong 38$ and $A^{t_2}(0,\pi/4) \cong 0.4$ differ by two orders of magnitude. It should be pointed out that, for the type-I crystals, there is no correlation between the anisotropy parameter k–1 and the discontinuity in the enhancement factors. However, such correlation can be clearly traced for the type-II crystals: the higher the absolute magnitude of the anisotropy parameter $|k–1|$, the higher the relative discontinuities in the enhancement factors are in the [100]-type directions (see Table 1.12). Since the enhancement factors for them are less than unity, the absolute discontinuity magnitudes of $A^{t_1}(0,\varphi)$ are small.

To eliminate the ambiguousness of the $A^{t_2}(0,\varphi)$ coefficient in the vicinity of the [100] direction at $\varphi = 0$ and $\varphi = \pi/4$, Maris H.J. averaged the phonon flux over a circular shape detector and came up with the average values of the enhancement factor [17]. In [73], Lax M., etc. averaged the $A^{t_2}(0,\varphi)$ coefficient for transverse phonons over a rectangular shape detector for all symmetric direction in crystals. A GaAs crystal served as an example to demonstrate the change in the enhancement factors of transverse phonons by several times depending on the size and shape of a detector. Moreover, for the [100] and [111] degenerate directions, the resulting data depends on the detector shape even of its infinitely small dimensions (see [73], Tables 4,6). Since averaging was carried out over a small neighborhood of symmetric directions, these methods [17, 73] could not solve the problem of divergence of the enhancement factor at zero curvature points in the slowness surface. This problem was considered in [75]. The authors of [81] averaged the phonon flux density over the ΔS area of the detector for an arbitrary direction in the crystal. As a result, they succeeded in avoiding the enhancement factor divergence. Also, they computed the phonon flux density in the {100} plane for a GaAs crystal using square detectors ($(4 \times 4) \times 10^{-6}$ and $(15.6 \times 15.6) \times 10^{-6}$ mm^2). The resulting findings turned out to be in qualitative agreement with the data of [79].

Thus, a closed analytical expression is obtained for the enhancement factor of the phonon flux in cubic crystals with various types of elastic energy anisotropy. The features of the angular dependencies of the enhancement factor, as well as the types of curvature of the isoenergetic surfaces of acoustic modes on the magnitude and sign of the anisotropy parameters, are analyzed.

Summarizing the analysis of the influence of focusing on the propagation of phonon modes in elastically anisotropic crystals, it can be argued that the enhancement-factor-based approach [73–75, 79, 80–82, 84] allows making no quantitative estimates for the densities of states of quasi-transverse modes. However, it is widely used in analyzing ballistic phonon transport and constructing phonon images [73–75, 79, 80–82, 84]. On the other hand, in [69, 71], a simple and intuitive method for estimating the density of phonon states (DPS) for phonon focusing and defocusing regions is proposed, with it also utilizing an isotropic medium as a reference system. The above method gives a rather rough estimate of the influence of focusing on the DPS, but it brings to naught the afore-specified enhancement factor disadvantages. A strong argument in favor of the method for estimating the DPS, which is developed in [69, 71], is a good agreement between the DPS angular dependencies and mean free paths: the sectors of maxima and minima for both quantities correspond to the phonon focusing and defocusing regions, respectively. Therefore, in the future, it would be reasonable to apply this method when analyzing the influence of focusing of transverse phonons on phonon transport.

1.7 Conclusion

The main results of the first chapter can be formulated as follows:

1. An analysis of the Christoffel equations for the dynamic characteristics of elastic waves has shown that all cubic crystals, in accordance with the sign of the anisotropy parameter k-1, can be divided into two types: crystals with positive (type-I) and negative (type II) anisotropy of the second-order elastic moduli.

For crystals of the same type, the directions of focusing and defocusing of the vibrational modes coincide, whereas in crystals of various types they are opposite: the focusing directions in type-I crystals turn into the defocusing directions in type-II crystals.

2. A method of approximating the phonon spectrum of cubic crystals over the entire Brillouin zone has been developed, with the spectrum obtaining from data on inelastic neutron scattering for symmetric directions.
3. An analysis of the influence of focusing on the angular distribution of the density of phonon states has shown that the maximum and minimum DPS in elastically anisotropic crystals are achieved in phonon focusing and defocusing regions, respectively. Therefore, the directions of the DPS maxima in the first type crystals become the directions of the minima in the crystals of the second type.
4. The features of the propagation of phonon pulses in cubic crystals with various types of anisotropy of elastic energy have been examined. A closed analytical expression has been derived for the enhancement factor of the phonon flux in cubic crystals with various types of elastic energy anisotropy. Peculiarities of the angular dependencies of the enhancement factor, as well as the types of curvature of the isoenergetic surfaces of acoustic modes on the magnitude and sign of the anisotropy parameters have been analyzed.

Chapter 2
Phonon Relaxation Times during Diffuse Scattering at the Boundaries of monocrystalline Finite-Length Samples

The present chapter covers the influence of elastic energy anisotropy on the relaxation characteristics of a phonon gas during diffuse scattering at the boundaries of monocrystalline samples. At sufficiently low temperatures, when the phonon mean free path in at least one of the directions is greater than or comparable with the characteristic size of the nanostructure, the thermal conductivity is determined by the nature of the surface–phonon interaction. Such a situation, when diffuse boundary scattering is a dominant mechanism of phonon relaxation, is customary to call the boundary scattering regime or the Knudsen flow of the phonon gas. Later, we analyze the influence of phonon focusing on the Knudsen phonon gas flow in monocrystalline silicon samples in the framework of the Casimir–McCurdy theory [13, 19]. Casimir [13] calculated the thermal conductivity of a dielectric infinite-length rod within the model of an isotropic medium. He suggested that all phonons, when collided with the surface, are absorbed and then re-emitted isotropically into half-space towards inside the sample with an intensity that depends on the surface temperature in accordance with the theory of black-body radiation. Casimir's theory [13] has been generalized in [19] to the case of elastically anisotropic crystals. To the assumption of Casimir, the authors have added two more:

1. Heat flux and temperature distribution are uniform along the length of the sample.
2. A plane of mirror symmetry perpendicular to the axis of the sample is assumed to exist.

The last assumption about the existence of a plane of mirror symmetry is not critical. The fact is that the plane perpendicular to the [111] direction is not a plane of mirror symmetry. However, the calculated values of the phonon mean free paths for this direction are consistent with experimental data with the same error as for other directions [19]. The authors of [19] have not succeeded in deriving analytical expressions for the phonon relaxation times during diffuse scattering at the sample boundaries and in analyzing the temperature dependencies of the thermal conductivity for monocrystalline silicon samples with different directions of the heat flux relative to the crystallographic axes. Therefore, the present chapter explores the Knudsen phonon gas flow in finite-length samples and calculates the phonon relaxation times, focusing within the Casimir–McCurdy theory [13, 19].

Section 2.1 deals with expressions deduced for the relaxation times and mean free paths during diffuse phonon scattering at the boundaries of round, square-shaped,

https://doi.org/10.1515/9783110670509-003

and rectangular cross-section infinite-length samples. Section 2.2 calculates the phonon relaxation times at the boundaries and the mean free path of phonons in finite-length samples. Section 2.3 analyzes the anisotropy of the phonon mean free path in bulk silicon samples with circular and square-shaped cross-sections. It is shown that the estimated data are in good agreement with the findings of [19]. Section 2.4 discusses the anisotropy of phonon mean free paths in silicon samples with a rectangular cross-section at low temperatures and compares them with experimental results of [19].

2.1 Phonon Relaxation at the Boundaries of Infinite-Length Samples with Circular, Square-Shaped, and Rectangular Cross-Section

To begin with, we look at phonon transport in infinite-length samples with circular, square-shaped, and rectangular cross-sections in the framework of the Casimir–McCurdy theory [13, 19] and determine the phonon relaxation times during diffuse boundary scattering. We choose the direction of the temperature gradient along the axis of the sample and denote this direction by X_3. A cross-section of the sample should be drawn with a plane perpendicular to the axis X_3 and passing through point $X_3 = 0$ (see Figures 2.1 and 2.2). Consider a surface element dS in the neighborhood of a point with the coordinates(X_1, X_3).

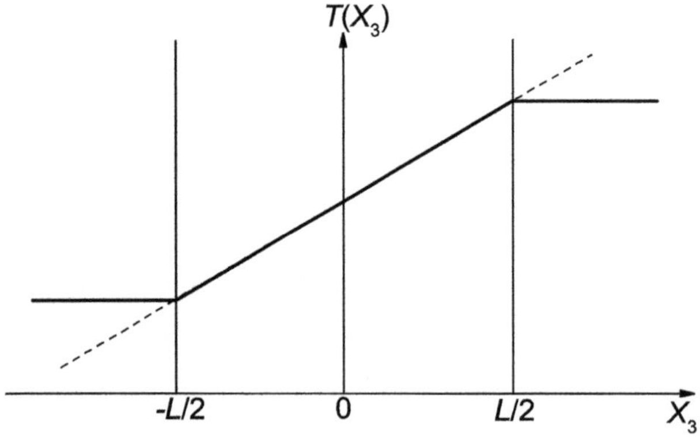

Figure 2.1: Temperature distribution in a sample of length L (solid line) and in an infinite sample (dashed line). The figure is taken from [19].

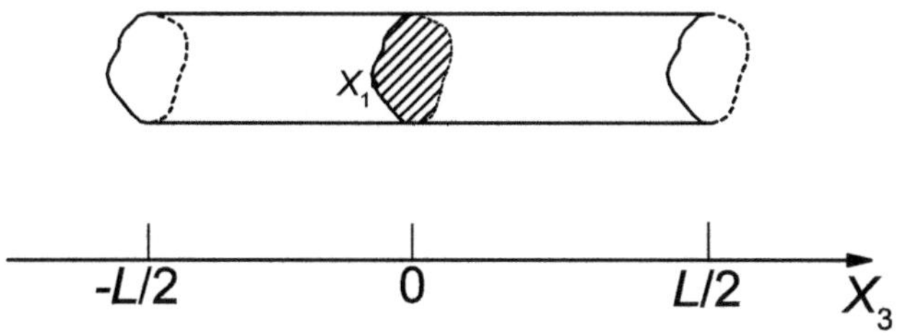

Figure 2.2: Schematic representation of a sample of length L with an arbitrary cross-section. The figure is taken from [19].

We designate a unit vector perpendicular to the surface of the sample and pointing towards the inside of the sample by $\mathbf{m}(X_1)$. Phonons with a wave vector $\mathbf{q}$ and polarization λ leave this surface element. Then, the phonon flux is equal to $(\mathbf{m}(X_1)\mathbf{V}_g^\lambda)N_{q\lambda}^{(0)}(X_3,\mathbf{q})dS$, where V_g^λ is the group velocity of phonons and $N_{q\lambda}^{(0)}(X_3,\boldsymbol{q}) = (\exp(\hbar\omega_q^\lambda/k_B T(X_3))-1)^{-1}$ is the Planck distribution function corresponding to the $T(X_3)$ temperature of the element at hand. Phonons that leave the surface element dS travel along the rod until they collide with the surface.

Let the vector $\boldsymbol{\Lambda}^\lambda(X_1,\mathbf{q})$ control the direction and mean free path of a phonon with the wave vector $\mathbf{q}$ and polarization λ. Assuming that the temperature and heat flux distributions are uniform throughout the sample, the authors of [19] calculated the heat flux passing through the plane $X_3 = 0$ from right to left and vice versa. As a result, they came up with the following expression for the thermal conductivity:

$$\kappa_\infty(T) = \frac{1}{2S_c}\sum_{\lambda,q}\hbar\omega_{q\lambda}\frac{dN_{q\lambda}^{(0)}}{dT}\int_{X_1} dX_1\left(\mathbf{m}(X_1)\mathbf{V}_g^\lambda\right)\left(\Lambda_3^\lambda(X_1,\mathbf{q})\right)^2 = \sum_{\lambda,q}\hbar\omega_{q\lambda}\frac{dN_{q\lambda}^{(0)}}{dT}I_\infty^\lambda, \quad (2.1)$$

$$I_\infty^\lambda = \frac{1}{2S_c}\int_{X_1}\left(\mathbf{m}(X_1)\mathbf{V}_g^\lambda\right)\left(\Lambda_3^\lambda(X_1,q)\right)^2 dX_1, \quad (2.2)$$

where S_c is the area of the transverse cross-section of the sample and $\Lambda_3^\lambda(X_1,q)$ is the projection of the mean free path of a phonon with the moment $\mathbf{q}$ and polarization λ on the temperature gradient direction.

Figure 2.3 illustrates a scheme for evaluating the integral over the contour X_1 for cylinder-shaped infinite-length samples, where R is the radius of the cylinder.

The phonon leaves the surface element at the point X_1 towards the group velocity vector V_g^λ. The group velocity components V_{g3}^λ and $V_{g\perp}^\lambda = \sqrt{(V_g^\lambda)^2-(V_{g3}^\lambda)^2}$ are the projections of the group velocity vector on the cylinder axis and cross-section plane, respectively. We denote the projection of the phonon mean free path on the cross-section plane as $\Lambda_\perp^\lambda(X_1,\mathbf{q})$. As can be seen from Figure 2.3, it is equal to the length

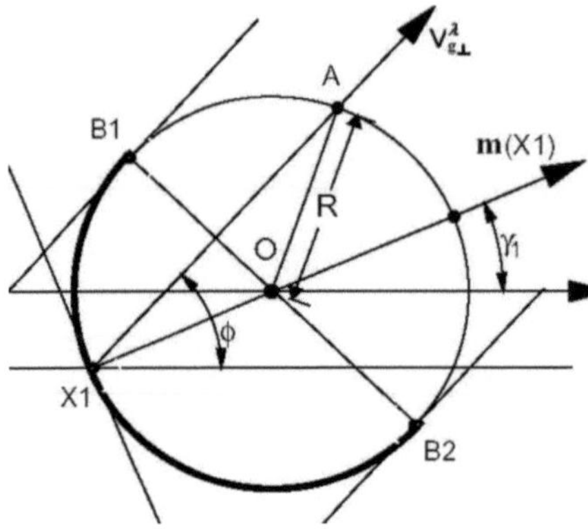

Figure 2.3: Scheme that illustrates the calculation of the contour integral for cylindrical samples. (The cross-section of the cylinder by a plane perpendicular to the cylinder axis is shown.) The bold line marks the part of the contour X_1 over which the integral is calculated.

of the X_1A segment. The $\Lambda_3^\lambda(X_1, \mathbf{q})$ projection of the mean free path on the cylinder axis is related to $\Lambda_\perp^\lambda(X_1, \mathbf{q})$ by the relation

$$\Lambda_3^\lambda(X_1, \mathbf{q}) = \frac{V_{g3}^\lambda}{V_{g\perp}^\lambda} \Lambda_\perp^\lambda(X_1, \mathbf{q}), \quad V_{g\perp}^\lambda = \sqrt{\left(V_{g1}^\lambda\right)^2 + \left(V_{g2}^\lambda\right)^2}. \tag{2.3}$$

Next, we enter the γ_1 angle between the Y-axis and the $\mathbf{m}(X_1)$ normal to the circumference at the point X_1 and as well the ϕ angle between the $V_{g\perp}^\lambda$ component and the Y-axis. The geometry of the picture brings to $\Lambda_\perp^\lambda(X_1, \mathbf{q}) = 2R\cos(\phi - \gamma_1)$. Then, we have

$$\Lambda_3^\lambda(X_1, \mathbf{q}) = \frac{V_{g3}^\lambda}{V_{g\perp}^\lambda} 2R\cos(\phi - \gamma_1), \quad \left(\mathbf{m}(X_1)\mathbf{V}_g^\lambda\right) = V_{g\perp}^\lambda \cos(\phi - \gamma_1). \tag{2.4}$$

For a circular cross-section sample, we set that $dX_1 = Rd\gamma_1$. The contribution to the contour integral ($A1$) is given only by phonons reflected by the surface element inside the sample. Therefore, for a given angle φ, we integrate over the γ_1 angles only within the sector between the $B1$ and $B2$ points. (The sector is indicated by a bold line in the figure.) So, from the physical analysis of the problem, it follows that the contour integrals over dX_1 are not calculated over the entire contour as is done in [19] but must cover only half of the contour. Therefore, for an infinite-length sample, the integration area X_1 and appropriate values of the γ_1 angles are limited by the range $-\pi/2 < \gamma_1 - \phi < \pi/2$ (see Figure 2.3). For an infinite-length sample with a circular cross- section, we easily find the relaxation function

$$I_\lambda^\infty = \frac{4R^3}{2S_c}\frac{\left(V_{g3}^\lambda\right)^2}{V_{g\perp}^\lambda}\int_{X1}\cos^3(\phi-\gamma_1)d\gamma_1 = \frac{8R}{3\pi}\frac{\left(V_{g3}^\lambda\right)^2}{V_{g\perp}^\lambda}. \tag{2.5}$$

Equation (2.5) coincides with that obtained in [19].

Now consider a rod of infinite length with a rectangular cross-section $D\times\mu D$ (see Figure 2.4). Let V_{g3}^λ be the projection of the group velocity on the direction of the temperature gradient, with this direction and the axis of the rod coinciding, $V_{g\perp}^\lambda$ be the projection of the group velocity on the plane of the cross-section of the rod, and V_{g1}^λ and V_{g2}^λ be the projections of the group velocity on the side faces of the sample. Figure 2.4 presents the case when $tg\phi = |V_{g2}^\lambda|/|V_{g1}^\lambda| < \mu$. In evaluating the integral over the contour X_1, we select three regions A, B, and C (see Figure 2.4). The A region is the ODF triangle, the B region is outlined by the OFHG parallelogram, and the C region involves the GH(μD) triangle. As in the case of circular cross-section samples, the contour integrals over dX_1 are not calculated over the entire rectangular contour but only for half.

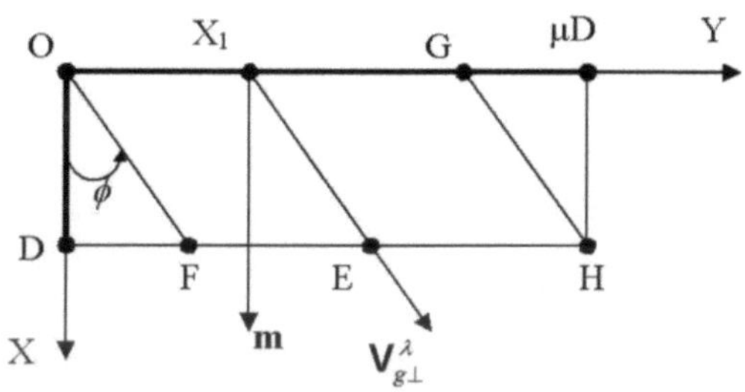

Figure 2.4: Schematic diagram that illustrates the calculation of the contour integral for a rod with a rectangular cross-section. The cross-section of the rod by a plane perpendicular to the long axis for $tg\phi = |V_{g2}^\lambda|/|V_{g1}^\lambda| > 1/\mu$ is shown. Bold lines indicate the part of the contour X_1 over which the integral is calculated.

The fact is that in the case of diffuse phonon-boundary scattering for a fixed angle ϕ, we should perform the integration only over those sides of the sample cross-section where the speed of the scattered phonon is directed inside the sample (bold lines in Figure 2.4). We denote the projection of the phonon mean free path on the cross-section plane as $\Lambda_\perp^\lambda(X_1,\mathbf{q})$. The latter is equal to the segment length X_1E in the B region. The $\Lambda_3^\lambda(X_1,\mathbf{q})$ projection of the mean free path on the axis rod relates through the relation (2.3). In the case of $tg\varphi = \left|V_{g2}/V_{g1}\right| < \mu$ in the A region, we have

$$(\mathbf{m}(X_1)\mathbf{V}_g^\lambda) = V_{g\perp}^\lambda\cdot\sin\phi,\quad \Lambda_1^\lambda(X_1,\mathbf{q}) = \tfrac{D-x}{\cos\phi},\quad \Lambda_3^\lambda(X_1,\mathbf{q}) = \Lambda_1^\lambda(X_1,\mathbf{q})\frac{V_{g3}^\lambda}{V_{g\perp}^\lambda}. \tag{2.6}$$

Given the expression (2.6), for the integral over the A region, we get the following:

$$I_{\infty}^{\lambda}(A) = \frac{1}{2S_c}\int_{0}^{D} V_{g\perp}^{\lambda} \cdot \sin\phi \left\{\frac{D-x}{\cos\phi} \cdot \frac{V_{g3}^{\lambda}}{V_{g\perp}^{\lambda}}\right\}^2 dx = \frac{D^3}{2S_c}\left(\frac{V_{g3}^{\lambda}}{V_{g1}^{\lambda}}\right)^2 \frac{\left|V_{g2}^{\lambda}\right|}{3}. \tag{2.7}$$

Analogously, for the B and C regions, we find the following:

$$I_{\infty}^{\lambda}(B) = \frac{D^3}{2S_c}\left(\frac{V_{g3}^{\lambda}}{V_{g1}^{\lambda}}\right)^2 \left(\mu\left|V_{g1}^{\lambda}\right| - \left|V_{g2}^{\lambda}\right|\right), \quad I_{\infty}^{\lambda}(C) = I_{\infty}^{\lambda}(A). \tag{2.8}$$

Having summed up over all the regions when $\left|V_{g2}^{\lambda}\right| < \mu\left|V_{g1}^{\lambda}\right|$ for the relaxation function $I_{\infty}^{\lambda}(\theta,\varphi)$, we arrive at the following:

$$I_{\infty}^{\lambda}(\theta,\varphi) = \frac{D}{6\mu}\left[\left(\frac{V_{g3}^{\lambda}}{V_{g1}^{\lambda}}\right)^2 \left(3\mu\left|V_{g1}^{\lambda}\right| - \left|V_{g2}^{\lambda}\right|\right)\right]. \tag{2.9}$$

The case of $tg\phi = \left|V_{g2}^{\lambda}\right| / \left|V_{g1}^{\lambda}\right| > \mu$ is considered in the same way:

$$I_{\infty}^{\lambda}(\theta,\varphi) = \mu\frac{D}{6}\left[\left(\frac{V_{g3}^{\lambda}}{V_{g2}^{\lambda}}\right)^2 \left(3\left|V_{g2}^{\lambda}\right| - \mu\left|V_{g1}^{\lambda}\right|\right)\right]. \tag{2.10}$$

Formulas (2.9) and (2.10) coincide with those deducted in [19]. To go over to square-shaped cross-section samples, it is sufficient to set $\mu = 1$ in formulas (2.9) and (2.10). So, the method proposed here to calculate the relaxation functions $I_{\infty}^{\lambda}(\theta,\varphi)$ for circular, square-shaped, and rectangular cross-section infinite-length samples results in the same data as in [19]. Next, we demonstrate that this method allows one to solve the problem of the Knudsen phonon gas flow in finite-length samples.

After comparing formula (2.1) with the standard expression for thermal conductivity [9, 85] in the relaxation time approximation

$$\kappa(T) = k_B \sum_{q,\lambda} \tau^{\lambda}(q)(V_{g3}^{\lambda})^2 \left(\frac{\hbar\omega_q^{\lambda}}{k_B T}\right)^2 N_{q\lambda}^0(N_{q\lambda}^0 + 1), \tag{2.11}$$

we can compute the phonon relaxation time during diffuse scattering at the boundaries of infinite-length samples in the following way:

$$\tau_{B\infty}^{\lambda}(\theta,\varphi) = \frac{I_{\infty}^{\lambda}(\theta,\varphi)}{\left(V_{g3}^{\lambda}(\theta,\varphi)\right)^2}. \tag{2.12}$$

From formulas (2.5), (2.9), and (2.10), we can derive the findings for circular and rectangular cross-section samples:

$$\tau^{\lambda}_{B\infty}(\theta,\varphi) = \frac{8R}{3\pi}\cdot\frac{1}{\sqrt{\left(V^{\lambda}_{g}\right)^{2}-\left(V^{\lambda}_{g3}\right)^{2}}}, \tag{2.13}$$

$$\tau_{B\infty}{}^{\lambda}(\theta,\varphi) = \begin{cases} \frac{D}{6\mu}\frac{\left(3\mu\left|V^{\lambda}_{g1}\right|-\left|V^{\lambda}_{g2}\right|\right)}{\left(V^{\lambda}_{g1}\right)^{2}}, & \text{if } \left|V_{g2}\right|<\mu\left|V_{g1}\right|; \\ \mu\frac{D}{6}\frac{\left(3\left|V^{\lambda}_{g2}\right|-\mu\left|V^{\lambda}_{g1}\right|\right)}{\left(V^{\lambda}_{g2}\right)^{2}}, & \text{if } \left|V_{g2}\right|>\mu\left|V_{g1}\right|. \end{cases} \tag{2.14}$$

The expression (2.13) for the phonon-boundary relaxation time for the infinite-length samples with a circular cross-section coincides with that inferred earlier in [86].

Now, we determine the average phonon mean free path Λ_{∞} for the infinite-length samples in the regime of the Knudsen phonon gas flow. For this, the range of temperatures far below the Debye temperature $(T << T_D)$ should be looked into using the anisotropic continuum model. We apply the known expression for the kinetic theory of gases for thermal conductivity:

$$\kappa = \frac{1}{3}C_V\,\bar{S}\,\Lambda_{\infty}. \tag{2.15}$$

We reduce formula (2.11) for thermal conductivity to the form of (2.15). In doing so, we distinguish the heat capacity C_V per unit volume and the average phonon velocity $\bar{S}$:

$$C_V = \frac{2\pi^2 k_{\mathbf{B}}^4}{5\hbar^3}T^3\frac{1}{3}\sum_{\lambda}\left\langle\left(S^{\lambda}\right)^{-3}\right\rangle,\ \bar{S} = \sum_{\lambda}\left\langle\left(S^{\lambda}\right)^{-2}\right\rangle\left\{\sum_{\lambda}\left\langle\left(S^{\lambda}\right)^{-3}\right\rangle\right\}^{-1}, \tag{2.16}$$

where $\langle(S^{\lambda})^{-3}\rangle = \int d\Omega_q (S^{\lambda})^{-3}/4\pi$ and $\langle(S^{\lambda})^{-2}\rangle = \int d\Omega_q (S^{\lambda})^{-2}/4\pi$. Then, the expression for the mean free path $\Lambda_{B\infty}$ can be represented in the form as follows:

$$\Lambda_{\infty} = \frac{3}{4\pi}\cdot\frac{1}{\sum_{\lambda 1}\left\langle\left(S^{\lambda 1}\right)^{-2}\right\rangle}\sum_{\lambda}\int_{-1}^{1}dx\int_{0}^{2\pi}d\varphi\frac{l^{\lambda}_{\infty}(\theta,\varphi)}{\left(S^{\lambda}(\theta,\varphi)\right)^{3}},\quad x=\cos\theta. \tag{2.17}$$

The values of the average phonon velocities and heat capacity of Si crystals are listed in Table 2.1.

The expressions (2.15)–(2.17) imply that, in the Debye approximation, the thermal conductivity is proportional to heat capacity and obeys the T^3 dependence at low temperature, which conforms to the Debye law. This outcome attracted the attention of researchers to the Casimir theory [13]. It is worth stressing, however, that the formula for the thermal conductivity in [13] contains a known error, but namely,

Table 2.1: The values of the parameters that govern the mean free path of phonons for silicon crystals at low temperatures.

Mode	$\langle (S^\lambda(\theta,\varphi))^{-2} \rangle$, 10^{-12} (s^2 cm^{-2})	$\bar{S}^\lambda$, 10^5 (cm s^{-1})	C_V^λ/T^3, erg cm^{-3} K^{-4}	$\bar{S}$, 10^5 (cm s^{-1})	C_V/T^3, erg cm^{-3} K^{-4}
L	1.213	9.069	0.5478		
T1	3.081	5.688	2.2156	5.668	5.899
T2	3.873	5.058	3.1360		

the extra multiplier $\pi/2$. The phonon mean free paths for each vibrational mode can be determined analogously. For this purpose, we represent the thermal conductivity as the additive sum of all vibrational modes:

$$\kappa_\infty(T) = \sum_\lambda \kappa_\infty^\lambda(T) = \sum_\lambda \frac{1}{3} C_V^\lambda \, \bar{S}^\lambda \Lambda_\infty^\lambda. \tag{2.18}$$

Consequently, the mean free path $\Lambda_{B\infty}$ for phonons of the λ branch appears as follows:

$$\Lambda_\infty^\lambda = \frac{3}{4\pi} \cdot \frac{1}{\left\langle (S^\lambda)^{-2} \right\rangle} \int_{-1}^{1} dx \int_0^{2\pi} d\varphi \, \frac{I_\infty^\lambda(\theta,\varphi)}{(S^\lambda(\theta,\varphi))^3}. \tag{2.19}$$

The expressions (2.5), (2.9), and (2.10) should be taken as I_∞^λ for circular and rectangular cross-section samples, respectively. So, the phonon mean free paths at low temperatures during boundary scattering within the anisotropic continuum model are calculated through a double angular integral.

2.2 Phonon Relaxation Times during Diffuse Scattering at the Boundaries of Finite-Length Samples with Circular, Square-Shaped, and Rectangular Cross-Sections

In this section, we discuss the Knudsen flow of a phonon gas in finite-length samples with circular, square-shaped, and rectangular cross-sections and determine the relaxation times for diffuse phonon-boundary scattering in the framework of the theory of McCurdy et al. [19]. The theory discussed earlier suggests that the contribution to the thermal conductivity of finite-length samples is made only by those phonons that collide with the surface of the sample limited by its length. Phonons that will collide with the surface of the sample beyond its length are said to make no contribution to the thermal conductivity. For them, the projection of the mean free path on the X_3-axis is determined by the inequality

$$\left|\Lambda_3^\lambda(X_1, q)\right| \geq \frac{L}{2}. \tag{2.20}$$

We designated the contribution of such phonons to the thermal conductivity by $\Delta\kappa(T)$. Then, the thermal conductivity of a rod with a length L is equal to that of an infinite-length rod after subtracting $\Delta\kappa(T)$:

$$\kappa\,(T) = \kappa_\infty(T) - \Delta\kappa(T). \tag{2.21}$$

In accordance with [19], the $\Delta\kappa(T)$ correction for the sample with a length L has the following form:

$$\Delta\kappa(T) = \sum_{\lambda,q}{}^{*}\hbar\omega_q^\lambda \frac{dN_{q\lambda}^{(0)}}{dT} \cdot \frac{1}{2S_C}\int\limits_{X_1} dX_1\left(\mathbf{m}(X_1)\mathbf{V}_g^\lambda\right)\left\{\left|\Lambda_3^\lambda(X_1,\mathbf{q})\right| - \frac{L}{2}\right\}^2 = \sum_{\lambda,q}{}^{*}\hbar\omega_q^\lambda \frac{dN_{q\lambda}^{(0)}}{dT}\Delta I^\lambda, \tag{2.22}$$

$$\Delta I^\lambda(\theta,\varphi) = \frac{1}{2S_C}\int\limits_{X_1} dX_1\left(\mathbf{m}(X_1)\mathbf{V}_g^\lambda\right)\left\{\left|\Lambda_3^\lambda(X_1,\mathbf{q})\right| - \frac{L}{2}\right\}^2. \tag{2.23}$$

The asterisk next to the summation sign over the wave vectors in formula (2.22) stands for the contribution of only those phonons to the $\Delta\kappa(T)$ correction, for which the inequality (2.20) holds. When calculating the integral over the contour X_1, we should also take the inequality (2.20) into account. From the expressions (2.21)–(2.23), it follows that the thermal conductivity of the finite-length samples can be given by

$$\kappa(T) = \sum_{\lambda,q}{}^{*}\hbar\omega_q^\lambda \frac{dN_{q\lambda}^{(0)}}{dT} I^\lambda(\theta,\varphi), \quad I^\lambda(\theta,\varphi) = I_\infty^\lambda(\theta,\varphi) - \Delta I^\lambda(\theta,\varphi). \tag{2.24}$$

The inequality (2.20) imposes restrictions on the region of integration over wave vectors in the expression for the correction to the thermal conductivity. Therefore, the authors of [19] did not succeed in directly deriving analytical expressions for the $\Delta I^\lambda(\theta,\varphi)$ quantity and determining the relaxation times during diffuse phonon-boundary scattering. The calculation of thermal conductivity for cubic crystals in the boundary scattering regime was performed by the numerical method only for symmetric directions at a temperature of 3 K.

To calculate the temperature dependencies of thermal conductivity using the relaxation method, it is necessary first and foremost to determine the phonon relaxation times for all relevant scattering processes, including phonon scattering at the sample boundaries and find the total relaxation time according to the Matthiessen rule. This allows exploring the change in the contributions of various vibrational modes to the thermal conductivity with increasing temperature, as well as analyzing the dependencies of the relaxation characteristics on the orientations of the temperature gradient relative to the crystal axes without limiting oneself by the anisotropic continuum model. Since boundary phonon scattering dominates in the entire temperature range below the maximum of thermal conductivity, neglecting the focusing effect may lead to a significant error in interpreting experimental data for all elastically anisotropic crystals. Because of the absence of analytical calculations of relaxation times, over the past 40 years since the publication of [19], none of the research works have been conducted on the temperature dependencies of the thermal conductivity of dielectric crystals accounting for the phonon focusing. In this regard, we disclose some details of the calculation of the $\Delta I^{\lambda}(\theta,\varphi)$relaxation functions during diffuse phonon scattering at the boundaries of circular and rectangular cross-section finite-length samples (see [20, 21]).

For a cylindrical sample of a length L when calculating the contour integral

$$\Delta I^{\lambda} = \frac{1}{2S_c}\int\limits_{X_1} dX_1 \left(\mathbf{m}(X_1)\mathbf{V}_g^{\lambda}\right)\left\{\left|\Lambda_3^{\lambda}(X_1,\mathbf{q})\right| - \frac{L}{2}\right\}^2 \tag{2.25}$$

we look at Figure 2.3 and determine the quantities, which the figure includes (see formulas (2.4)):

$$\Lambda_3^{\lambda}(X_1,\mathbf{q}) = \frac{V_{g3}^{\lambda}}{V_{g\perp}^{\lambda}}2R\cos(\phi-\gamma_1), \quad \left(\mathbf{m}(X_1)\mathbf{V}_g^{\lambda}\right) = V_{g\perp}^{\lambda}\cos(\phi-\gamma_1) \tag{2.26}$$

Substitution of the expression (2.4) into the inequality $\left|\Lambda_3^{\lambda}(X_1,\mathbf{q})\right| \geq L/2$ leads to the result:

$$\Delta^{\lambda} = \frac{L}{4R}\frac{V_{g\perp}^{\lambda}}{\left|V_{g3}^{\lambda}\right|} < 1. \tag{2.27}$$

With condition (2.27), the integral (2.25) is evaluated similarly to the previously considered case of infinite-length samples (see Figure 2.3) and we can obtain

$$\Delta I^{\lambda}(\theta,\varphi) = \frac{3}{2}I_{\infty}^{\lambda}(\theta,\varphi)\left[\left(1+(\Delta^{\lambda})^2\right)\sqrt{1-(\Delta^{\lambda})^2} - \frac{1}{3}\left(\sqrt{1-(\Delta^{\lambda})^2}\right)^3 \right.$$
$$\left. - \Delta^{\lambda}\left(\arccos\Delta^{\lambda} + \Delta^{\lambda}\sqrt{1-(\Delta^{\lambda})^2}\right)\right]. \tag{2.28}$$

If the opposite inequality, $\Delta^{\lambda}(\theta,\varphi) > 1$, is fulfilled, we have $\Delta I^{\lambda}(\theta,\varphi) = 0$. Thus, for the $\Delta I^{\lambda}(\theta,\varphi)$ function, we arrive at [20]:

$$I^{\lambda}(\theta,\varphi)=\begin{cases}\dfrac{4R\left(V_{g3}^{\lambda}\right)^2}{\pi V_{g\perp}^{\lambda}}\left[1-\left(1+(\Delta^{\lambda})^2\right)\sqrt{1-(\Delta^{\lambda})^2}-\dfrac{1}{3}\left(\sqrt{1-(\Delta^{\lambda})^2}\right)^3\right.\\ \left.+\Delta^{\lambda}\left(\arccos\Delta^{\lambda}+\Delta^{\lambda}\sqrt{1-(\Delta^{\lambda})^2}\right)\right],\ \text{if}\ \Delta^{\lambda}(\theta,\varphi)\le 1;\\ I_{\infty}^{\lambda}(\theta,\varphi),\ \text{if}\ \Delta^{\lambda}(\theta,\varphi)>1.\end{cases}\tag{2.29}$$

The earlier analysis argues that, in calculating analytically the $\Delta I^{\lambda}(\theta,\varphi)$ and $I^{\lambda}(\theta,\varphi)$ quantities for circular cross-section finite-length samples, the restrictions imposed by the condition (2.20) in (2.21) and (2.22) boil down to a system of inequalities between the geometric parameter $\kappa_0 = L/2D$ and the ratios of phonon group velocity components. Since these inequalities can be involved in the definition of the $\Delta I^{\lambda}(\theta,\varphi)$ and $I^{\lambda}(\theta,\varphi)$ relaxation functions, we get the possibility of computing the phonon relaxation times during diffuse boundary scattering. This can be done by the expression:

$$\tau_B^{\lambda}(\theta,\varphi)=\frac{I^{\lambda}(\theta,\varphi)}{\left(V_{g3}^{\lambda}(\theta,\varphi)\right)^2}.\tag{2.30}$$

The phonon relaxation time at the boundaries of cylindrical samples, as follows from expression (2.29), can be represented as piecewise-smooth functions as follows:

$$\tau_B^{\lambda}(\theta,\varphi)=\begin{cases}\dfrac{4R}{\pi V_{g\perp}^{\lambda}}\left[1-\left(1+(\Delta^{\lambda})^2\right)\sqrt{1-(\Delta^{\lambda})^2}-\dfrac{1}{3}\left(\sqrt{1-(\Delta^{\lambda})^2}\right)^3\right.\\ \left.+\Delta^{\lambda}\left(\arccos\Delta^{\lambda}+\Delta^{\lambda}\sqrt{1-(\Delta^{\lambda})^2}\right)\right],\ \text{if}\ \Delta^{\lambda}(\theta,\varphi)\le 1;\\ \tau_{B\infty}^{\lambda}(\theta,\varphi)=(8R/3\pi)/\sqrt{\left(V_g^{\lambda}\right)^2-\left(V_{g3}^{\lambda}\right)^2},\ \text{if}\ \Delta^{\lambda}(\theta,\varphi)>1.\end{cases}\tag{2.31}$$

For cylinder-shaped infinite-length samples, $\tau_B^{\lambda}(\theta,\varphi)$ comes from the expression (2.13) if the inequality $\Delta^{\lambda}(\theta,\varphi)>1$ is met and $\Delta I^{\lambda}(\theta,\varphi)=0$.

Next, we address the Knudsen flow of a phonon gas in samples with a length L and a rectangular cross-section $D\times\mu D$. In this case, the calculation of the contour integral

$$\Delta I^{\lambda}=\frac{1}{2S_c}\int\limits_{X_1} dX_1\left(\mathbf{m}(X_1)\mathbf{V}_g^{\lambda}\right)\left\{\left|\Lambda_3^{\lambda}(X_1,\mathbf{q})\right|-\frac{L}{2}\right\}^2\quad \text{for}\quad \left|\Lambda_3^{\lambda}(X_1,\mathbf{q})\right|\ge\frac{L}{2}.\tag{2.32}$$

is carried out, as in the case of infinite-length samples (see Figure 2.4). To integrate over the X_1 contour, we choose three regions A, B, and C. It should be pointed out that the contour integrals over dX_1 in formulas (2.25) and (2.32) are not evaluated for the entire rectangular contour as done in [19] but only for half of rectangular contour of infinite-length samples. For the A region (see Figure 2.4), the $|\Lambda_3^\lambda(X_1, \boldsymbol{q})| \geq L/2$ inequalities yield the following restriction on the integration domain:

$$0 \leq x \leq x_{\max}, \quad x_{\max} = D - \frac{L}{2} \cdot \frac{V_{g\perp}^\lambda \cos\phi}{V_{g3}^\lambda} = D\left(1 - k_0 \cdot \frac{V_{g1}^\lambda}{V_{g3}^\lambda}\right), \quad k_0 = \frac{L}{2D}. \tag{2.33}$$

Given (2.32) and (2.33), we get the following for the integral over the A region:

$$\Delta I_\lambda(A) = \frac{1}{2S_c} \int_0^{x_{\max}} V_{g\perp}^\lambda \cdot \sin\phi \left\{ \left| \frac{D - x}{\cos\phi} \cdot \frac{V_{g3}^\lambda}{V_{g\perp}^\lambda} \right| - L/2 \right\}^2 dx. \tag{2.34}$$

The condition $x_{\max} > 0$ implies that, in the case at hand $(tg\phi = |V_{g2}/V_{g1}| < \mu)$, the $|\Lambda_3^\lambda(X_1, \mathbf{q})| \geq L/2$ inequality reduces to the following ratio between the group velocity components and the geometric parameter κ_0:

$$\left| V_{g3}^\lambda / V_{g1}^\lambda \right| \geq k_0. \tag{2.35}$$

Directly calculating the $\Delta I_\lambda(A)$ integral leads to the following result:

$$\Delta I_\lambda(A) = \frac{D}{6\mu} \left\{ \frac{\left(V_{g3}^\lambda\right)^2}{|V_\perp^\lambda|} \cdot \frac{1}{\cos^2\phi} - k_0 \cdot \left|V_{g3}^\lambda\right| \cdot \frac{1}{3\cos\phi} + \frac{(k_0)^2}{3} |V_\perp^\lambda| - (k_0)^3 \cdot \frac{(V_\perp^\lambda)^2}{\left|V_{g3}^\lambda\right|} \cdot \cos\phi \right\}. \tag{2.36}$$

It is not hard to make sure that in integrating over the B and C regions, the $|\Lambda_3^\lambda(X_1, \mathbf{q})| \geq L/2$ inequality also boils down to the expression (2.35). Directly taking the integrals over these domains gives the following:

$$\Delta I_\lambda(B) = \frac{D}{2\mu} \left\{ \frac{\left(V_{g3}^\lambda\right)^2}{\left|V_{g\perp}^\lambda\right|} \cdot \frac{1}{\cos\phi} - 2k_0 \cdot \left|V_{g3}^\lambda\right| + k_0{}^2 \cdot \left|V_{g\perp}^\lambda\right| \cdot \cos\phi \right\} (\mu - tg\phi), \quad \Delta I_\lambda(C) = \Delta I_\lambda(A). \tag{2.37}$$

Next, we sum up over all three domains and account for that $tg\phi = |V_{g2}^\lambda| / |V_{g1}^\lambda|$. Then, if the inequalities $|V_{g2}^\lambda| < \mu |V_{g1}^\lambda|$ and $|V_{g3}^\lambda / V_{g1}^\lambda| \geq \kappa_0$ are fulfilled, we find the relaxation function

$$I^{\lambda}(\theta,\varphi)=I^{\lambda}_{\infty}(\theta,\varphi)-\Delta I^{\lambda}(\theta,\varphi)=Dk_0\left|V^{\lambda}_{g3}\right|\left\{1-\frac{k_0}{2\mu}\frac{\left(\left|V^{\lambda}_{g2}\right|+\mu\left|V^{\lambda}_{g1}\right|\right)}{\left(V^{\lambda}_{g3}\right)}+\frac{(k_0)^2}{3\mu}\frac{\left|V^{\lambda}_{g1}\right|\left|V^{\lambda}_{g2}\right|}{\left(V^{\lambda}_{g3}\right)^2}\right\}. \tag{2.38}$$

If the inequalities $|V^{\lambda}_{g3}/V^{\lambda}_{g2}|<\kappa_0/\mu$ and $|V^{\lambda}_{g2}|<\mu|V^{\lambda}_{g1}|$ are fulfilled, the expression for $I^{\lambda}(\theta,\varphi)$ exactly coincides with (2.38). If $|V^{\lambda}_{g3}|/|V^{\lambda}_{g1}|<\kappa_0$ or $|V^{\lambda}_{g3}/V^{\lambda}_{g2}|<\kappa_0/\mu$ and $|V^{\lambda}_{g2}|>\mu|V^{\lambda}_{g1}|$, then $\Delta I^{\lambda}(\theta,\varphi)=0$ and the functions $I^{\lambda}(\theta,\varphi)$ are controlled by eqs. (2.9) and (2.10) derived earlier for infinite-length samples [19]. If we set that $\mu=1$, it will be true for square-shaped samples.

From the analysis performed earlier (see formula (2.38)), it can be inferred that in analytical calculating the restrictions imposed by the condition (2.20) can be also included in the definition of the functions $\Delta I^{\lambda}(\theta,\varphi)$ and $I^{\lambda}(\theta,\varphi)$ for finite-length samples with rectangular cross-sections. Consequently, we get the possibility of determining the relaxation times of phonons of different polarizations during diffuse boundary scattering in terms of piecewise-smooth functions for various ranges of the θ and φ angles. Expressions for them have the following form:

$$\tau^{\lambda}_{B}(\theta,\varphi)=\begin{cases}\dfrac{D}{\left|V^{\lambda}_{g3}\right|}k_0\left\{1-\dfrac{k_0}{2}\dfrac{\left(\left|V^{\lambda}_{g2}\right|+\mu\left|V^{\lambda}_{g1}\right|\right)}{\mu\left|V^{\lambda}_{g3}\right|}+\dfrac{(k_0)^2}{3}\dfrac{\left|V^{\lambda}_{g1}\right|\left|V^{\lambda}_{g2}\right|}{\mu\left(V^{\lambda}_{g3}\right)^2}\right\}, \\ \quad \text{if } \left|\dfrac{V^{\lambda}_{g3}}{V^{\lambda}_{g1}}\right|\geq k_0 \text{ and } \mu\left|V^{\lambda}_{g1}\right|>\left|V^{\lambda}_{g2}\right| \text{ or } \left|\dfrac{V^{\lambda}_{g3}}{V^{\lambda}_{g2}}\right|\geq\dfrac{k_0}{\mu} \text{ and } \mu\left|V^{\lambda}_{g1}\right|<\left|V^{\lambda}_{g2}\right|; \\ \tau^{\lambda}_{B\infty}, \text{ if } \left|\dfrac{V^{\lambda}_{g3}}{V^{\lambda}_{g1}}\right|<k_0 \text{ and } \mu\left|V^{\lambda}_{g1}\right|>\left|V^{\lambda}_{g2}\right| \text{ or } \left|\dfrac{V^{\lambda}_{g3}}{V^{\lambda}_{g2}}\right|<\dfrac{k_0}{\mu} \text{ and } \mu\left|V^{\lambda}_{g1}\right|<\left|V^{\lambda}_{g2}\right|.\end{cases} \tag{2.39}$$

For square-shaped cross-section samples, it is sufficient to set $\mu=1$ in formula (2.39). Thus, during diffuse scattering at the boundaries of circular and rectangular cross-section finite-length samples, the phonon relaxation times are determined by piecewise-smooth functions for various intervals of the angles. The latter depend on the ratios between the group velocity components and the geometric parameters of the samples.

Expressions for the mean free paths $\Lambda$and corresponding magnitudes of Λ^{λ} for each branch of the phonon spectrum for finite-length samples can be deduced analogously to the case of infinite-length samples (see section 2.1). Finally, we arrive at the following:

$$\Lambda=\frac{3}{4\pi}\cdot\frac{1}{\sum\limits_{\lambda 1}\left\langle\left(S^{\lambda 1}\right)^{-2}\right\rangle}\sum_{\lambda}\int\limits_{-1}^{1}dx\int\limits_{0}^{2\pi}d\varphi\,\frac{I^{\lambda}(\theta,\varphi)}{\left(S^{\lambda}(\theta,\varphi)\right)^{3}}, \tag{2.40}$$

$$\Lambda^{\lambda} = \frac{3}{4\pi} \cdot \frac{1}{\left\langle (S^{\lambda})^{-2} \right\rangle} \int_{-1}^{1} dx \int_{0}^{2\pi} d\varphi \, \frac{I^{\lambda}(\theta, \varphi)}{(S^{\lambda}(\theta, \varphi))^{3}}, \quad x = \cos\theta, \tag{2.41}$$

where the $I^{\lambda}(\theta, \varphi)$ relaxation functions for circular, square-shaped, and rectangular cross-section samples with a length L come from formulas (2.5), (2.9), (2.10), and (2.38). As in the case of infinite-length samples, these functions are given by a double angular integral within the anisotropic continuum model for boundary scattering.

From formulas (2.39) and (2.40), it follows that the thermal conductivity coefficients $\tilde{\kappa}(T) = \kappa(T)/D$ and, accordingly, the mean free paths $\tilde{\Lambda} = \Lambda(L, W, D)/D$ normalized to the film thickness D depend only on two ratios, the parameters $\mu = W/D$ and $\kappa_0 = L/2D$

$$\kappa(T, L, W, D) = D\tilde{\kappa}(T, k_0, \mu), \quad \Lambda(L, W, D) = D \cdot \tilde{\Lambda}(k_0, \mu) \tag{2.42}$$

rather than on three sample geometric parameters such as D, W, and L.

It is obvious that the phonon mean free paths normalized to the cross-section side D depend only on the κ_0 parameter

$$\Lambda(L, D, D) = D \cdot \tilde{\Lambda}(k_0). \tag{2.43}$$

Formulas (2.42) and (2.43) are of interest for experimental validity of applicability of the theory described earlier to explore phonon transport in elastically anisotropic crystals.

2.3 Anisotropy of Mean Free Paths of Phonons in Silicon Samples with Circular and Square-Shaped Cross-Sections at Low Temperatures

Let us consider phonon transport in bulk silicon samples at temperatures much lower than the Debye temperature ($T \ll T_D$), when phonon scattering at the boundaries dominates. We compare the calculation results obtained within the anisotropic continuum model with the experimental data of [19] for symmetric directions. The gas-kinetic formula for thermal conductivity, $\kappa = 1/3 C_V \bar{S} \Lambda$, reads that the anisotropy of thermal conductivity is determined by the phonon mean free path Λ, since the specific heat C_V and the average phonon velocity S are independent of the direction of heat flux. We first analyze the phonon mean free paths in the model of an isotropic medium and make sure that formulas (2.5), (2.9), and (2.10) give the well-known results for the Casimir lengths $\Lambda_C = \Lambda_\infty$ in the limiting cases of infinite-length samples.

In isotropic media, the phonon phase velocities S^{λ} do not depend on the angles θ and φ, and the directions of the phase and group velocities coincide, $V_{gi}^{\lambda} = S^{\lambda} n_i$, where $\mathbf{n} = \mathbf{q}/q = \{\sin\theta\cos\varphi, \sin\theta\sin\varphi, \cos\theta\}$ is the unit wave vector. Therefore, formulas (2.40)–(2.41) imply that the mean free paths of phonons of different polarizations are equal to each other and average mean free path:

$$\Lambda^L = \Lambda^t = \Lambda = D\frac{3}{4\pi}\cdot\int d\Omega_q \tilde{I}(\theta,\varphi), \quad \tilde{I}(\theta,\varphi) = \frac{I^\lambda(\theta,\varphi)}{DS^\lambda}, \quad \tilde{\Lambda} = \Lambda/D. \tag{2.44}$$

For circular cross-section samples with a diameter $D = 2R$ and a length L, the $\tilde{I}(\theta,\varphi)$ relaxation functions are converted to the form:

$$\tilde{I}(\theta) = \frac{4\cos^2\theta}{3\pi\sin\theta}, \quad \text{for}\,\Delta \geq 1, \quad \Delta = \tfrac{L}{4R}tg\theta, \tag{2.45}$$

$$\tilde{I}(\theta) = \frac{4\cos^2\theta}{3\pi\sin\theta}\left\{1 - \frac{3}{2}\left[(1+\Delta^2)\sqrt{1-\Delta^2} - \frac{1}{3}\left(\sqrt{1-\Delta^2}\right)^3 - \Delta\left(\arccos\Delta + \Delta\sqrt{1-\Delta^2}\right)\right]\right\},$$
$$\text{for}\,\Delta < 1. \tag{2.46}$$

For infinite-length rods with a circular cross-section, the $\tilde{I}(\theta,\varphi)$ relaxation function is defined by the expression (2.45). Substitution of the formula (2.45) in the expression (2.44) yields the following:

$$\Lambda_C = \Lambda_\infty = 2R\cdot\frac{3}{4\pi}\cdot 2\pi\int_{-1}^{1} d(\cos\theta)\frac{4\cos^2\theta}{3\pi\sin\theta} = 2R. \tag{2.47}$$

Thus, for a dielectric infinite-length rod with a circular cross-section, we have arrived at the result of Casimir [13]: the mean free path of phonons is equal to the diameter of the rod.

For samples with a square-shaped cross-section with side D and length L, the relaxation functions (2.9), (2.10), and (2.38) are transformed to the form:

$$\tilde{I}(\theta,\varphi) = \begin{cases} (n_3/n_1)^2(3|n_1| - |n_2|)/6, & \text{if}\,|n_1| > |n_2| \text{ and } |n_3/n_1| < k_0, \\ (n_3/n_2)^2(3|n_2| - |n_1|)/6, & \text{if}\,|n_1| < |n_2| \text{ and } |n_3/n_2| < k_0; \end{cases} \tag{2.48}$$

$$\tilde{I}(\theta,\varphi) = k_0|n_3|\left\{1 - \frac{k_0}{2}\frac{|n_2|+|n_1|}{|n_3|} + \frac{(k_0)^2}{3}\frac{|n_1||n_2|}{(n_3)^2}\right\}, \; if \begin{cases} (|n_1| > |n_2| \text{ and } |n_3/n_1| \geq k_0)\text{ or} \\ (|n_1| < |n_2| \text{ and } |n_3/n_2| \geq k_0). \end{cases} \tag{2.49}$$

Plugging the formula (2.48) into the expression (2.44) for the Casimir length in samples with a square-shaped cross-section provides the known result (see, for example, [9]):

$$\Lambda_C = \frac{D}{\pi}\int_0^\pi \cos^2\theta d\theta\int_0^{\pi/4} d\varphi\left(\frac{3}{\cos\phi} - \frac{\sin\varphi}{\cos^2\varphi}\right) = \frac{D}{2}\left[3\ln\left(\sqrt{2}+1\right) - \sqrt{2}+1\right] \cong 1.115D. \tag{2.50}$$

From the expressions (2.44)–(2.50) it is known that the phonon mean free paths in isotropic media do not depend on elastic moduli but are determined completely by geometric dimensions of samples. However, in elastically anisotropic crystals, the mean free paths are different for phonons of different polarizations and depend not only on geometric parameters of samples but also on the directions of heat flux in crystals. Therefore, the findings obtained for isotropic media can be utilized as a convenient comparison system to predict the angular dependencies of mean free paths of phonons of various vibrational modes in elastically anisotropic crystals.

Let us examine the angular dependencies of mean free paths of phonons during scattering at the boundaries of Si samples with circular and square-shaped cross-sections at low temperatures. The spectrum and group velocities of the phonons should be determined in the edge-cube coordinate system. We consider the rotation of heat flux (an axis of the sample) in two planes:

- in the plane of the YZ-cube face;
- in the diagonal plane.

Let the angle ψ assign the deviation of heat flux from the Z-axis pointing along the cube edge. Next, we set a coordinate system with the 3-axis along the heat flux direction.

$$\nabla_r T = \{\nabla_x T, 0, \nabla_z T\} = |\nabla_r T|\{0, -\sin\psi, \cos\psi\}, V^\lambda_{g3} = -V^\lambda_{gy}\sin\psi + V^\lambda_{gz}\cos\psi, \quad (2.51)$$

$$\nabla_r T = \left\{\nabla_x T, \nabla_y T, \nabla_z T\right\} = |\nabla_r T|\left\{-\sin\psi/\sqrt{2}, \sin\psi/\sqrt{2}, \cos\psi\right\},$$
$$V^\lambda_{g3} = \left(-V^\lambda_{gx} + V^\lambda_{gy}\right)\sin\psi/\sqrt{2} + V^\lambda_{gz}\cos\psi. \quad (2.52)$$

The $\Lambda_B(\psi)$ dependencies on the ψ angle are governed by the dependencies of the group velocity components V^λ_{g1}, V^λ_{g2}, and V^λ_{g3}, which enter the relaxation functions $I^\lambda(\theta,\varphi)$. For cylindrical samples, it is sufficient to prescribe the phonon group velocity component parallel to the heat flux. For the cases of rotation of the temperature gradient in the cube face plane or the diagonal plane, we have the following:

$$(1)\ V^\lambda_{g3} = -V^\lambda_{gy}\sin\psi + V^\lambda_{gz}\cos\psi, \quad V^\lambda_{g\perp} = \sqrt{\left(V^\lambda_{g}\right)^2 - \left(V^\lambda_{g3}\right)^2},$$
$$(2)\ V^\lambda_{g3} = \left(-V^\lambda_{gx} + V^\lambda_{gy}\right)\sin\psi/\sqrt{2} + V^\lambda_{gz}\cos\psi. \quad (2.53)$$

The phonon group velocity components in a Cartesian coordinate system take the form

$$
\begin{aligned}
V_{gx}^{\lambda}(\theta,\varphi) &= S^{\lambda}(\theta,\varphi)\left\{\sin\theta\cos\varphi + S_{\theta}^{\lambda}\cos\theta\cos\varphi - S_{\varphi}^{\lambda}\sin\varphi\right\},\\
V_{gy}^{\lambda}(\theta,\varphi) &= S^{\lambda}(\theta,\varphi)\left\{\sin\theta\sin\varphi + S_{\theta}^{\lambda}\cos\theta\sin\varphi + S_{\varphi}^{\lambda}\cos\varphi\right\},\\
V_{gz}^{\lambda}(\theta,\varphi) &= S^{\lambda}(\theta,\varphi)\left\{\cos\theta - S_{\theta}^{\lambda}\sin\theta\right\}.
\end{aligned} \tag{2.54}
$$

For samples with a square-shaped cross-section, the value of thermal conductivity also depends on the orientation of the side faces. To analyze the angular dependencies, we choose the orientation of the faces as follows. In the case of rotation of the temperature gradient in the YZ cube face plane, the X-axis remains stationary. Therefore, we put one of the side faces to be perpendicular to the X-axis: $V_{g1}^{\lambda} = V_{gx}^{\lambda}$. Further, we take the direction of V_{g2}^{λ} as the axis lying in the YZ-plane and perpendicular to the temperature gradient

$$V_{g2}^{\lambda} = V_{gy}^{\lambda}\cos\psi + V_{gz}^{\lambda}\sin\psi. \tag{2.55}$$

In the case of rotation of the temperature gradient in the diagonal plane, the [110] direction perpendicular to this plane is a stationary axis. Therefore, we choose it as the direction of V_{g1}^{λ}

$$V_{g1}^{\lambda} = \left(V_{gx}^{\lambda} + V_{gy}^{\lambda}\right)/\sqrt{2}. \tag{2.56}$$

As the direction of V_{g2}^{λ}, we mark the axis lying in the diagonal plane and perpendicular to the temperature gradient

$$V_{g2}^{\lambda} = \left(-V_{gx}^{\lambda} + V_{gy}^{\lambda}\right)\cos\psi/\sqrt{2} - V_{gz}^{\lambda}\sin\psi. \tag{2.57}$$

Let us look at the angular dependencies of the Casimir lengths for circular and square-shaped cross-section samples, calculated using formulas (2.17) and (2.19). With the equality of the areas of the circular and square-shaped cross-sections ($D^2 = \pi R^2$), we normalize the Λ_C Casimir lengths to the side D for the square-shaped cross-section sample $\left(\tilde{\Lambda}_C(\psi) = \Lambda_C(\psi)/D\right)$ and to $\sqrt{\pi}R$ for the sample of a circular cross-section with a radius R $\left(\tilde{\Lambda}_C(\psi) = \Lambda_C(\psi)/\sqrt{\pi}R\right)$. In this case, the $\tilde{\Lambda}_C(\psi)$ angular dependencies of the mean free paths for the samples specified above differ by 1%. The maximum difference of the phonon mean free path values does not exceed 1.6% for a fast transverse mode. On the scale of Figure 2.5, these curves are indiscernible. Therefore, Figure 2.5 refers virtually to infinite-length samples with both circular and square-shaped cross-sections.

As can be seen from Figure 2.5a, in the vicinity of the [100] directions at the angles $\psi = \pm 0.06 + n\pi/2$ (n is an integer) for the slow transverse mode, the Casimir lengths reach their maximum values. The latter are two times larger than those of $\Lambda_C^{t_1}$ for the fast transverse mode and three times larger than Λ_C^{L} for longitudinal

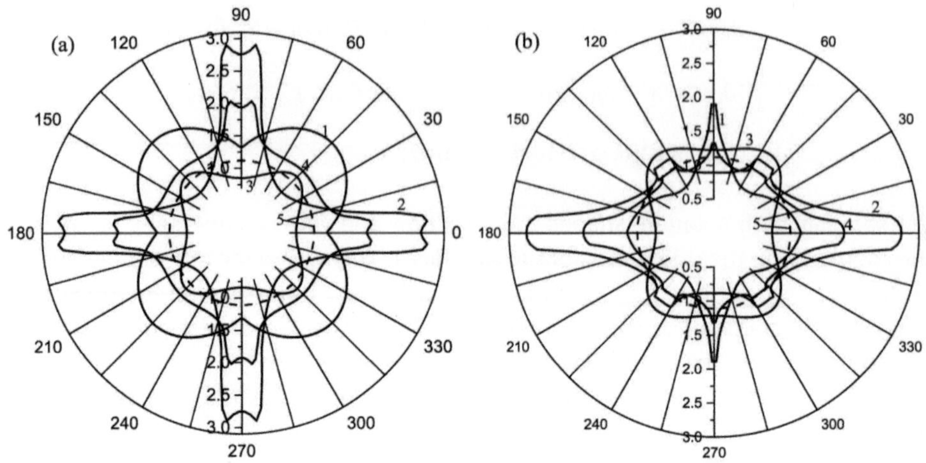

Figure 2.5: The angular dependencies of the reduced Casimir lengths $\tilde{\Lambda}_C^{\lambda}(\psi) = \Lambda_C^{\lambda}(\psi)/D$ (curves 1–3) for square-shaped cross-section samples in cases when the temperature gradient rotates in the plane of the cube face (a) and diagonal plane (b): 1 is for a fast transverse mode, 2 is for a slow transverse mode, 3 is for longitudinal modes, 4 is the average Casimir length, and 5 is the Casimir length in the model of an isotropic medium.

phonons. When going over to the [110] directions, the situation changes: the Casimir lengths of the slow transverse mode become three times less and take their minimum values. Along the [110] directions, the Casimir lengths for the fast transverse mode reach their maximum values. They are 2.8 and 1.6 times larger than slow transverse and longitudinal phonons, respectively. In the [111] directions, the Casimir lengths are observed maximal in the event of longitudinal phonons, but namely, they are 1.8 and 1.1 times larger than fast and slow transverse phonons, respectively (see Figure 2.5b). For symmetric directions, the ratio of the average Casimir lengths amounts to $\Lambda_C^{[100]} : \Lambda_C^{[110]} : \Lambda_C^{[111]} = 1.74 : 1.2 : 1$. From comparison of the results obtained for Si crystals and the isotropic medium model, it is seen that the Casimir lengths for all modes turn out to be larger in the focusing directions and less in the defocusing directions than those within the isotropic medium model (see Figure 2.5).

The expressions derived earlier for the phonon mean free paths in the boundary scattering regime make it possible to describe the experimental data on the thermal conductivity anisotropy of square-shaped cross-section samples at low temperatures [19]. Figure 2.6 outlines the angular dependencies of the phonon mean free paths in Si crystals with a length of $L = 2.9$ cm and square-shaped cross-section of $D = 0.293$ cm, calculated by formulas (2.40) and (2.41). For all symmetric directions, the calculation results agree well with the experimental data of [19]. Figures 2.6 show that the phonon mean free paths reach their maximum values for each vibrational mode along the focusing directions, exceeding both the mean

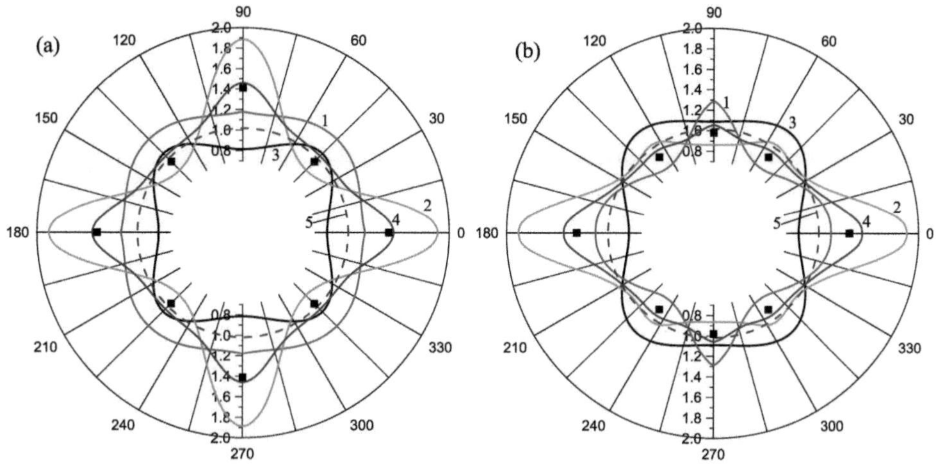

Figure 2.6: Angular dependencies of reduced mean free paths $\tilde{\Lambda}^{\lambda}(\psi) = \Lambda^{\lambda}(\psi)/D$ (curves 1–3) and average mean free paths $\tilde{\Lambda}(\psi) = \Lambda(\psi)/D$(curve 4) for samples with a length of L = 2.9 cm with a square-shaped cross-section of D = 0.293 cm in cases when the temperature gradient lies in the plane of the cube face (a) and in the diagonal plane (b) for: 1 – a fast transverse mode, 2 – a slow transverse mode, 3 – a longitudinal mode, 4 – an average mean free path, and 5 – a mean free path in the model of an isotropic medium. Symbols are experimental data of [19].

free paths of other vibrational modes and the mean free paths within the isotropic medium model.

In the defocusing directions, they become minimal and turn out to be less than those within the isotropic medium model. So, for example, phonons of a slow transverse mode are focused in the [100] direction. Their mean free path exceeds 1.6 and 2.4 times the mean free path of fast and longitudinal modes, respectively, and 1.3 times the average phonon mean free path. Phonons of a fast transverse mode are focused in the [110] directions. Their mean free path has a maximum value, which is 1.5 and 1.2 times greater than that of the slow transverse and longitudinal modes, respectively, and is 1.2 times greater than the average mean free path. In this direction, the phonon mean free paths of the slow and fast transverse modes diminish 1.8 and 1.5 times as much regarding the [100] direction, respectively, whereas Λ^{L} for the longitudinal mode rises by 1.5 times. Thus, the mean free path of the longitudinal phonons appears to be 1.6 and 1.2 times larger than that of the fast and slow transverse modes. In this direction, it exceeds 1.3 times the average phonon mean free path. For symmetric directions, the ratio of the average mean free paths amounts to $\Lambda^{[001]} : \Lambda^{[101]} : \Lambda^{[111]} = 1.50 : 1.08 : 1$. Thus, the thermal conductivity of Si samples has maximum and minimum values for the [100] and [111] directions, respectively, with the maximum being provided by a slow transverse mode. When transitioning from the infinite-length to finite-length samples, the anisotropy of the phonon mean free paths and, accordingly, the thermal conductivity decreases considerably.

It should be stressed that the calculated values of thermal conductivity of Si $(\kappa_{calc}(T_0))$ and the average phonon mean free paths exceed the experimental data $(\kappa_{exp}(T_0))$ at $T_0 = 3$K by 4% for the [100] direction and by 8% for the [110] and [111] directions (see Figure 2.6). To explain such divergence, the concept of a defective surface layer, proposed in [32, 87, 88], was used in the works [20, 63]. The point is that the surface of the sample was processed with sandpaper before measuring to provide phonon diffuse scattering. Such processing forms a heat-isolative layer directly beneath the sample surface. It contains microcracks with appropriate deformation fields and a network of dislocations [88]. Accounting for this layer and, accordingly, the effective thickness of the sample achieves the predicted and experimental findings at a temperature of 3 K. However, this approach left aside the phonon scattering by isotopic disorder. As will be shown in Section 3.3, the mechanism described above aligns the theoretical outcomes concerning the thermal conductivity with the experiment in the temperature range from 3 K to 15 K within the experimental error. Therefore, in the case at hand, the role of the defective surface layer turned out to be small and its influence can be ignored.

2.4 The Influence of Focusing on Phonon Transport in Silicon Crystals with a Rectangular Cross-Section at Low Temperatures

Let us discuss phonon transport in finite-length silicon samples with a rectangular cross-section at temperatures much lower than the Debye temperature. To begin with, we calculate the angular dependencies of the phonon mean free path in the Knudsen flow regime when phonon scattering at the boundaries is dominant. Next, we compare the obtained findings with the experimental data of [19] for silicon samples with a rectangular cross-section of $W \times D$ ($D = 0.185$ cm, $W = \mu D = 0.638$ cm, $\mu = 3.45$) and a length of $L = 3.5$ cm. We normalize the phonon mean free paths to the thickness of the sample:

$$\hat{\Lambda}(\psi) = \Lambda(\psi)/D. \tag{2.58}$$

We define a coordinate system associated with the sample as follows: the axis “3” should be directed along the axis of the sample and the direction of heat flux; the axis “1” marks the axis of rotation (see Figure 2.7). The axes “1” and “2” are perpendicular to the sample side faces. Consider the rotation of the temperature gradient in the plane of the cube face for two cases:

(A) The axis of rotation is perpendicular to the narrow face of the sample.

$$V_{g3}^{\lambda} = V_{gx}^{\lambda} \sin\psi + V_{gz}^{\lambda} \cos\psi, \quad V_{g1}^{\lambda} = V_{gy}^{\lambda}, \quad V_{g2}^{\lambda} = V_{gx}^{\lambda} \cos\psi - V_{gz}^{\lambda} \sin\psi. \tag{2.59}$$

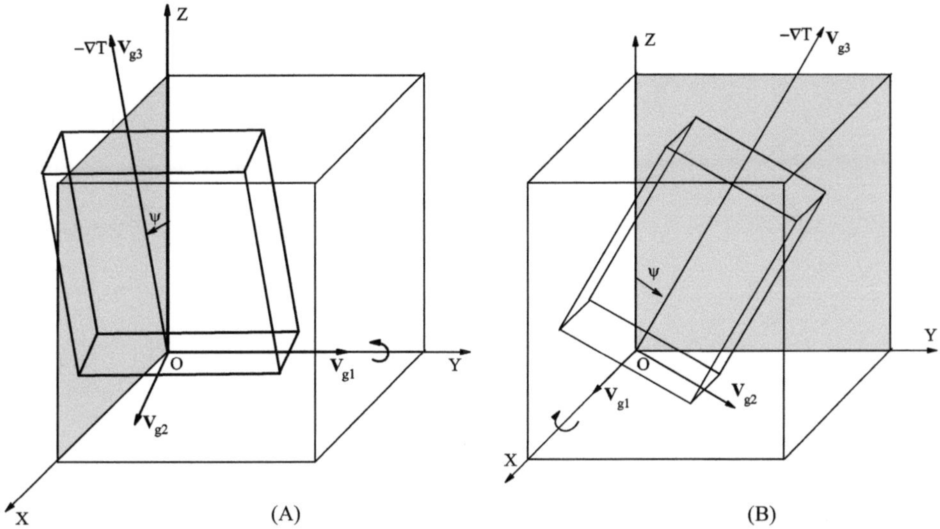

Figure 2.7: Scheme that illustrates the rotation of a temperature gradient in the plane of a cube face for the cases: (A) Axis of rotation is perpendicular to the narrow face of the sample. (B) Axis of rotation is perpendicular to the wide face of the sample.

(B) The axis of rotation is perpendicular to the wide face of the sample.

$$V_{g3}^{\lambda} = -V_{gy}^{\lambda}\sin\psi + V_{gz}^{\lambda}\cos\psi\,,\quad V_{g2}^{\lambda} = V_{gy}^{\lambda}\cos\psi + V_{gz}^{\lambda}\sin\psi\,,\quad V_{g1}^{\lambda} = V_{gx}^{\lambda}\,. \tag{2.60}$$

Here V_{g3}^{λ}, V_{g2}^{λ}, and V_{g1}^{λ} are the projections of the group velocity on the temperature gradient direction and the side faces of the sample, respectively. The angle ψ defines the deviation of the heat flux from the Z-axis.

The maximum values of the average phonon mean free paths for the cases (A) and (B) coincide and are achieved in the [001] directions. Their minimum values are observed in the [101] directions and they are different for the cases (A) and (B) (see Figure 2.8a). Therefore, the angular dependencies of the average mean free paths of phonons and, accordingly, the thermal conductivities significantly distinguish between each other due to the phonon focusing effect. It holds true for the cases of rotation of the narrow and wide sample faces. Besides, the angular dependencies mentioned above differ from those predicted for the square-shaped cross-section samples (see Figure 2.8, curves 1b, 1a, and 1). With rotation of the heat flux in the wide face of the {100} rectangle, the ratio of the mean free paths amounts to 1.21 $\left(\Lambda_{[100]}^{\{100\}}/\Lambda_{[110]}^{\{100\}} = 1.21\right)$ and is less than that for the square-shaped cross-section samples$\left(\Lambda_{[100]}^{\{100\}}/\Lambda_{[110]}^{\{100\}} \cong 1.4\right)$. However, when the heat flux rotates in the narrow face of the rectangle, the anisotropy of thermal conductivity increases. In this case, it turns out to be larger for the square-shaped cross-section samples$\left(\Lambda_{[100]}^{\{100\}}/\Lambda_{[110]}^{\{100\}} \cong 1.57\right)$

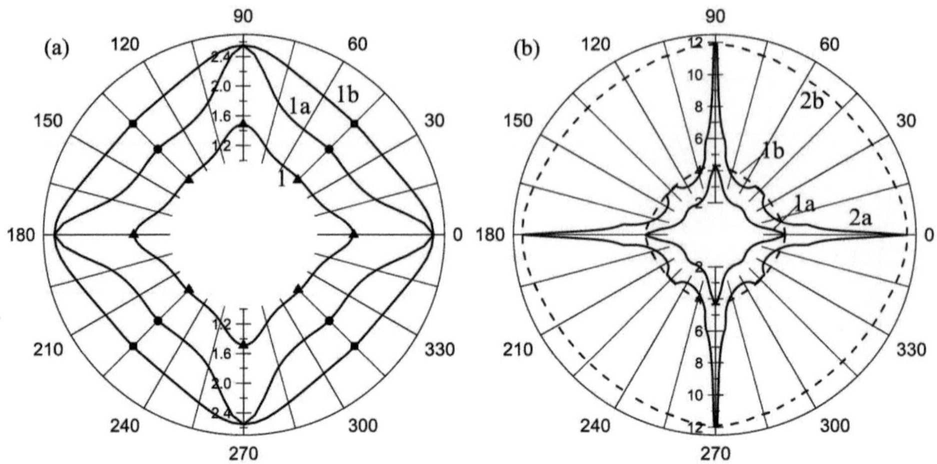

Figure 2.8: Angular dependencies of the average mean free paths $\tilde{\Lambda}(\psi)$ in Si crystals in the case when heat flux rotates in the {100} cube face plane. (a) curve 1 is for samples of [19] with a square cross-section ($D = 0.293$ cm and $L = 2.9$ cm); curves 1a and 1b refer to samples of [19] with a rectangular cross-section ($L = 3.5$ cm and $D = 0.185$ cm, $\mu D = 0.638$ cm); curve 1a is rotation in the narrow face of the sample; curve 1b is rotation in the wide face of the sample. The experimental data of [19] are denoted by symbols; (b) for samples with a thickness of $D = 0.185$ cm and length of $L = 5.5\ \mu D$: $\mu = 10$ (curves 1a and 1b), $\mu = 100$ (curves 2a and 2b). The rotation of the temperature gradient in the narrow (curves 1a and 2a) and wide (curves 1b and 2b) faces of the sample {100}.

(Figure 2.8a, curve 1a). Figure 2.8a shows that the average phonon mean free paths calculated in [21] for rectangular cross-section samples in the [101] direction match well the experimental data of [19]. Here, it should be emphasized that we have come up with an interesting experimental result: for rectangular cross-section samples, the thermal conductivity (and mean free paths) depend not only on the direction of the temperature gradient but also on the orientation of the wide faces of the sample [19]. From comparison of the thermal conductivities of two identical samples in the [110] temperature gradient direction, it becomes understandable that, at low temperatures, the thermal conductivity of the sample with the {001} wide and {110} narrow faces appears to be 33% higher than that of the sample with the {110} wide and {001} narrow faces [19]. For this, the calculation in [21] using formulas (2.38) and (2.40) gives the ratio: $\Lambda_{[110]}^{\{100\}} : \Lambda_{[110]}^{\{110\}} = 1.30$, which is in good agreement with the data of [19].

Let us explore how this ratio changes as the width of the sample increases. In doing so, we fix the sample thickness of $D = 0.185$ cm and the ratio of length and width as $L = 5.49$ and $W = 5.49\ \mu D$. Then, we examine the thermal conductivity and phonon mean free paths for $\mu = 10$ and $\mu = 100$. As can be seen from Figure 2.8b, the angular dependencies of the average mean free paths for the two cases under consideration change qualitatively (see Figure 2.8b). When the heat flux rotates in a wide face of the rectangle, an increase in the μ parameter leads to isotropization of the $\Lambda^{\{100\}}(\psi)$ dependence, and it becomes almost isotropic for $\mu = 100$.

In contrast, the rotation of the heat flux in a narrow face of the rectangle leads to an increase in the anisotropy of thermal conductivity (see Figure 2.8b). The ratio $\Lambda^{\{100\}}_{[100]}(\psi):\Lambda^{\{100\}}_{[110]}(\psi)$ rises to 1.9 and 3.0 for $\mu = 10$ and $\mu = 100$, respectively. In the [100] directions, the mean free paths coincide for both cases of rotation of the heat flux. However, in the [110] direction, the ratio of the thermal conductivities of two identical samples with the orientations of the {001} and {110} wide faces for $\mu = 10$ and $\mu = 100$ amounts to 1.7 and 2.8, respectively. Since the $\Lambda^{\{100\}}(\psi)$ dependence becomes almost isotropic for large values of μ during rotation in a wide face of the sample, these ratios differ little from $\Lambda^{\{100\}}_{[100]}(\psi):\Lambda^{\{100\}}_{[110]}(\psi)$.

The angular dependencies of the thermal conductivity and, accordingly, the phonon mean free paths look like interesting enough (see Figure 2.9), compared to square-shaped cross-section samples, in the event of coinciding the rotation axis and temperature gradient with the Z-axis when the rotation occurs in the XY-plane. In this case, the phonon group velocity components parallel and perpendicular to the heat flux are determined by the expressions:

$$V^{\lambda}_{g3} = V^{\lambda}_{gz}, \quad V^{\lambda}_{g1} = V^{\lambda}_{gx}\cos\psi + V^{\lambda}_{gy}\sin\psi, \quad V^{\lambda}_{g2} = -V^{\lambda}_{gx}\sin\psi + V^{\lambda}_{gy}\cos\psi, \tag{2.61}$$

where ψ is the angle between the X-axis and the side face. In this situation, the phonon mean free paths for square-shaped cross-section samples are isotropic to an accuracy of 1.3% (see Figure 2.9a, curve 1). However, an increase in the plate width causes an increase in the mean free paths and their anisotropy. Figure 2.9a illustrates that the

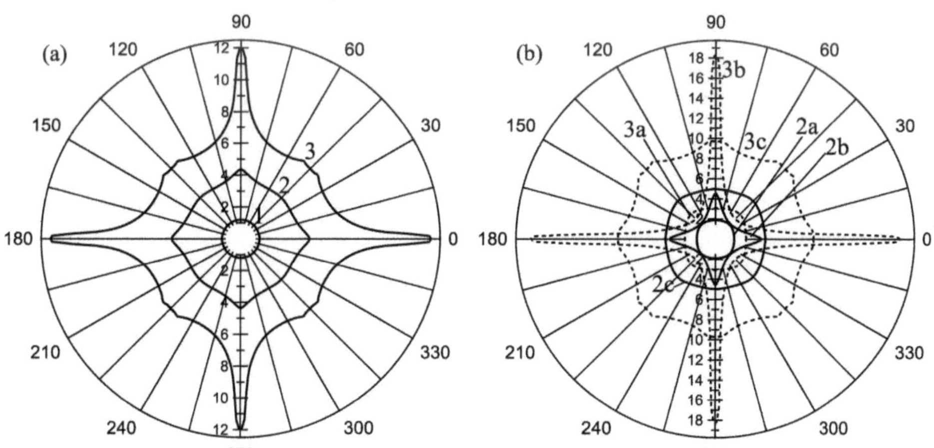

Figure 2.9: Angular dependencies of the mean free paths $\tilde{\Lambda}(\psi)$ and $\tilde{\Lambda}^{\lambda}(\psi)$ for samples having a rotation axis and a temperature gradient in the [100] direction. The length of the samples is $L = 5.49\ \mu D$, the width is $W = \mu D$, and the thickness is $D = 0.185$ cm. (a) $\tilde{\Lambda}(\psi)$ for $\mu = 1$ (curve 1), $\mu = 10$ (curve 2), and $\mu = 100$ (curve 3); (b) $\tilde{\Lambda}^{\lambda}(\psi)$ for a longitudinal mode (curves 2a and 3a), for fast (curves 2b and 3b) and slow (curves 2c and 3c) transverse modes. Curves 2a, 2b, and 2c are for $\mu = 10$ and curves 3a, 3b, and 3c are for $\mu = 100$.

mean free paths reach their maximum values in the [100] directions. With an increase in the μ parameter from 1 to 100, the maximum values of thermal conductivity and average mean free path of phonons become greater by an order of magnitude (see Figure 2.9a, curves 1 and 3). In the [101] directions, a local maximum takes place, and the minimum value is reached at the angles $\psi = (\pi/2)\, n \pm \pi/6$, where n is an integer. In this case, the $\Lambda_{[100]}^{\{100\}}(\psi)/\Lambda_{[110]}^{\{100\}}(\psi)$ ratio of the phonon mean free paths in the symmetric directions rises from 1.2 for $\mu = 10$ to 1.9 for $\mu = 100$. As shown in Figure 2.9b, in the [100] directions, the mean free path of a fast transverse mode reaches its maximum values and ensures maximum thermal conductivity. An increase in the μ parameter from 10 to 100 leads to an increase in $\Lambda_{[100]}^{t_1\{100\}}(\mu)$ by four times. The mean free path of a slow transverse mode reaches maximum values in the [101] directions and provides a local maximum of thermal conductivity in this direction.

Thus, the conducted analysis showed that the anisotropy of thermal conductivity (average mean free path of phonons) for rectangular cross-section samples can dramatically differ from that of square-shaped cross-section samples. In contrast to square-shaped cross-section samples, the values of thermal conductivity and phonon mean free paths at low temperatures for fairly wide samples with a rectangular cross-section are largely determined by the orientation of the wide face of the sample relative to the crystal axes. The phonon mean free paths of the fast and slow transverse modes change most significantly.

2.5 Conclusion

The main results of Chapter 2 can be formulated as follows:

1. An analytical solution has been given to the Knudsen phonon gas flow problem for finite-length samples with circular, square-shaped, and rectangular cross-sections. Relaxation times in the form of piecewise-smooth functions for various intervals of angles controlled by the relations between the group velocity components and the geometric parameters of the samples have been determined for diffuse phonon-boundary scattering regime.
2. As to circular and square-shaped cross-section samples, the mean free paths of phonons have been shown to reach their maximum values for each vibrational mode at low temperature in focusing directions, exceeding those of other vibrational modes in these directions. The phonon mean free paths calculated for symmetric directions align experimental data.
3. As to square-shaped cross-section samples, the values of thermal conductivity and phonon mean free paths at low temperatures are determined mainly by the direction of heat flux, and their dependence on the orientation of the side faces is weak. However, for rectangular cross-section samples, they largely depend on the orientation of the wide face of the sample relative to the axis of the crystal.

Chapter 3
Anisotropy and Temperature Dependencies of Thermal Conductivity of Bulk Silicon Samples

The experimental studies of [19] showed that, at low temperatures, when the mean free path of phonons is greater than the transverse size of a sample, phonon focusing leads to two effects in the thermal conductivity of Si crystals. The former is the dependence of the thermal conductivity of square-shaped cross-section samples on the direction of the temperature gradient relative to the crystallographic axes. The latter involves the dependence of the thermal conductivity of Si rectangular cross-section samples on the orientation of wide faces. In [19], two rectangular cross-section samples having identical geometric parameters were explored in the [110] gradient temperature direction. It turned out that, at low temperatures, the thermal conductivity of the sample with the {001} wide and {110} narrow faces 33% higher than that of the sample with {110} and narrow {001} faces.

In [19], as in a row of other monographs, thermal conductivity was measured by the method of stationary longitudinal heat flux [35, 89]. Before measuring, Si bulk samples were preliminarily processed with sandpaper to provide diffuse phonon scattering at the boundaries [19]. A typical experiment (e.g. [87]) suggests gluing a heater directly to the butt of the sample, with the opposite end secured at the cold sink. Thermometers are mounted at a distance L from each other, and the reading difference ΔT is recorded (see Figure 3.1). Once a steady state of temperature distribution is established, the thermal conductivity coefficient κ is determined as the division of ΔQ heat power liberated inside the heater by the area $S_c$of the transverse cross-section of the sample and by the temperature gradient $(\Delta T/L = (T_2 - T_1)/L)$:

$$\kappa = \Delta QL/\{(T_2 - T_1)S_c\}.$$

If the ΔT difference is not too large, the value of thermal conductivity corresponds to the average temperature between the thermometers. The method of stationary longitudinal heat flux can be used if the heat radiation through the side faces is small and can be neglected. Then all the heat received from the heater will reach the cold sink.

It was shown in [22] that correctly accounting for the phonon focusing effects to calculate the phonon relaxation rates at the boundaries of monocrystalline samples requires introduction of two orientation parameters. These parameters take into account the dependencies of relaxation characteristics on the direction of the heat flux [*I*] and the orientation of the sample wide-face {*J*} relative to the crystallographic axes. Therefore, unlike an isotropic medium, in computing the kinetic characteristics of monocrystalline samples, we should bear in mind their dependence on the orientation parameters

https://doi.org/10.1515/9783110670509-004

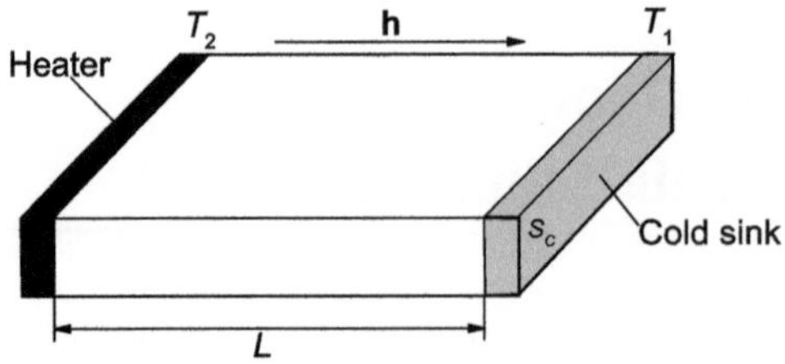

Figure 3.1: Scheme for measuring thermal conductivity by the method of stationary longitudinal heat flux.

$$\kappa(T) \Rightarrow \kappa^{\{J\}}_{[I(\psi)]}(T) \text{ and } \Lambda \Rightarrow \Lambda^{\{J\}}_{[I(\psi)]}.$$

Further, it will be shown that orientation parameters such as the direction of the heat flux $[I(\psi)]$ and the plane $\{J\}$, in which the angle ψ varies, can be defined through the group velocity components parallel and perpendicular to the heat flux direction. The proposed method of accounting for phonon focusing is relevant because a significant number of publications devoted to the study of phonon transport, both in bulk single crystals and nanostructures [5–7, 25, 26, 37, 90], completely do not taken into account for these effects.

Section 3.1 deals with the influence of normal phonon–phonon scattering processes on the lattice thermal conductivity of cubic crystals within the three-mode Callaway model taking into account the anisotropy of elastic energy. Section 3.2 develops a method of accounting for phonon focusing in calculating the temperature dependencies of the thermal conductivity of monocrystalline samples. Section 3.3 demonstrates that the phonon relaxation times at the boundaries [20, 21], calculated on the pages of the monograph, help adequately describe experimental data on the thermal conductivity of silicon crystals for various directions of the temperature gradient and orientations of their side faces. Section 3.4 analyses the change in thermal conductivity anisotropy with temperature. Section 3.5 presents a physical interpretation of the McCurdy effects in the thermal conductivity of cubic crystals.

3.1 Normal Phonon–Phonon Scattering Processes and Lattice Thermal Conductivity of Cubic Crystals

Consider the influence of normal phonon–phonon scattering processes on the lattice thermal conductivity of cubic crystals, taking into account the focusing of phonons. A set of kinetic equations for non-equilibrium phonon distribution functions N^{λ}_{q} in three-mode Callaway model has the form [91–94]:

$$\mathbf{V}^{\lambda}_{g}(q) \cdot \nabla_r N^{\lambda}_{q} = -(N^{\lambda}_{q} - N^{(0)}_{q\lambda})\nu^{\lambda}_{R} - \left(N^{\lambda}_{q} - N(\mathbf{q}, \mathbf{u}_{\lambda})\right)\nu^{\lambda}_{N}. \tag{3.1}$$

Here $\nu_N^\lambda(q)$ and $\nu_R^\lambda(q)$ are the phonon relaxation rates in normal (N-processes) and resistive scattering processes. Resistive scattering processes cause the momentum of the phonon system to relax. Among them are phonon–phonon scattering in the Umklapp processes $(\nu_U^\lambda(q))$, scattering by defects $(\nu_{iso}^\lambda(q))$, and sample-boundary scattering $\left(\nu_{B[I(\psi)]}^{\lambda\{J\}}\right)$. Therefore, we have $\nu_{R[I(\psi)]}^{\lambda\{J\}} = \nu_{B[I(\psi)]}^{\lambda\{J\}} + \nu_{iso}^\lambda(q) + \nu_U^\lambda(q)$. The role of N-processes in phonon–phonon scattering has been studied quite well within the theory of lattice thermal conductivity [9, 36, 91–96]. In these processes, the momentum of phonons participating in the collisions is conserved. Therefore, N-processes do not directly contribute to the phonon momentum relaxation and, accordingly, to the thermal resistance. However, they re-distribute the energy and momentum between various phonon modes and tend to establish a drift locally non-equilibrium distribution. The latter is described by the shifted Planck function [91–94]:

$$N(\mathbf{q}, \mathbf{u}_\lambda) = \left(\exp\left(\frac{\hbar\omega_{q\lambda} - \hbar\mathbf{q}\mathbf{u}_\lambda}{k_B T}\right) - 1\right)^{-1} \cong N_{q\lambda}^0 + \frac{\hbar\mathbf{q}\mathbf{u}_\lambda}{k_B T} N_{q\lambda}^0 (N_{q\lambda}^0 + 1), \tag{3.2}$$

where $N_{q\lambda}^0$ is the Planck function. In this case, the non-equilibrium phonon system is described by nine parameters such as relaxation rates of normal and resistive scattering processes, as well as average drift velocities u_λ for each branch of the phonon spectrum. Accounting for a special role of N-processes is necessary under conditions when the $\nu_N^\lambda(q)$ phonon relaxation frequency is greater or comparable to the relaxation frequency in resistive scattering processes [91–94]. In the opposite case, they can be taken into account additively with resistive scattering processes, as is done in [97].

Let us represent the phonon distribution function as a sum of the Planck function and a non-equilibrium addition $g_\lambda(\mathrm{q})$. Then, eq. (3.1) implies that

$$g_\lambda(\mathbf{q}) = -\frac{N_{q\lambda}^0(N_{q\lambda}^0 + 1)}{\nu_{ph}^\lambda(q)} \frac{\hbar\omega_q^\lambda}{k_B T^2} (\mathbf{V}_g^\lambda \nabla_r T) + \frac{\hbar\mathbf{q}\mathbf{u}_\lambda}{k_B T} N_{q\lambda}^0 (N_{q\lambda}^0 + 1) \frac{\nu_N^\lambda(q)}{\nu^\lambda(q)} = g_\lambda^{dif}(\mathbf{q}) + g_\lambda^{dr}(\mathbf{q}). \tag{3.3}$$

The first term in the expression (3.3) is governed by the diffuse motion of phonons exposed to the temperature gradient. The second term is responsible for the drift motion of phonons and associated with normal phonon–phonon scattering processes. To find the drift velocities $\mathbf{u}_l$, the set of kinetic eq. (3.1) should be added by a phonon momentum balance equation. The latter comes from the multiplication of eq. (3.1) by the phonon momentum vector $\hbar\mathbf{q}$ and summation over all $\mathbf{q}$ vectors.

$$\frac{1}{V}\sum_{q,\lambda} \hbar\mathbf{q}\nu_N^\lambda(q) \frac{N_{q\lambda}^0(N_{q\lambda}^0 + 1)}{k_B T \nu^\lambda(q)} \left[\frac{\hbar\omega_q^\lambda}{T}(\mathbf{V}_g^\lambda \nabla_r T) + \hbar(\mathbf{q}\mathbf{u}_\lambda)\nu_R^\lambda(q)\right] = 0. \tag{3.4}$$

It was shown in [94] that the solution of eq. (3.4) can be sought for two limiting cases. The first one is that if relaxation processes involving momentum redistribution between phonons of different polarization dominate in N-processes, they tend to establish the same drift velocity for all phonons. In this case, the drift velocity $\mathbf{u}^{(\lambda)} = \mathbf{u}^{(1)}$ should be equal for all polarizations. The second one includes independent relaxation of phonons of each mode conditional upon the phonon re-distribution that occurs only inside each vibrational branch. We, thus, virtually return to the original single-mode Callaway model [91]. This relaxation event was considered in [95] and was called the generalized Callaway model. In this case, the drift velocities are different for phonons of different polarizations: $\mathbf{u}_L^{(2)} \neq \mathbf{u}_t^{(2)}$. The first type of N-processes comprises Herring [38] and Landau-Rumer [98] relaxation mechanisms, as well as a number of other processes. The Herring mechanism [38] merges a longitudinal phonon and a slow transverse phonon (*ST*) followed by creation of a fast phonon (*FT*),$(L_1 + ST_2 \rightarrow FT_3)$:

$$\nu_N^L \cong B_{LTT} T^3 \omega_L^2 \tag{3.5}$$

According to [9, 36, 94–97], the main relaxation mechanism for transverse phonons in N-scattering processes is the Landau-Rumer mechanism [98], in which the merging of transverse and longitudinal phonons generates a longitudinal phonon $(T_1 + L_2 \rightarrow L_3)$

$$\nu_N^t \cong B_{TLL} T^4 \omega_t. \tag{3.6}$$

This phonon relaxation mechanism in N-scattering processes and its role in thermal conductivity were analysed in [94]. For the second type of N-processes, Simons [99] showed that N-processes of the second type ($L_1 + L_2 \rightarrow L_3$, $T_1 + T_2 \rightarrow T_3$) can happen in isotropic media only when allowance for phonon state attenuation is made, with only collinear phonons taking part in them. However, it was shown in [100] that accounting for cubic anisotropy leads to the fulfillment of the energy conservation law in relaxation mechanisms involving transverse phonons (*TTT*). In the long-wave approximation, these mechanisms contribute chiefly to the relaxation of transverse phonons as compared to the Landau-Rumer mechanism. For the *TTT* mechanism, the expression (3.6) that carries the coefficient B_{TTT}^N instead of B_{TLL}^N yields the ν_N^t relaxation rate. So, the first version of N-processes contains an independent drift velocity; the second one holds the momentum conservation law for each branch of the phonon spectrum.

It is worth emphasizing that the directions of group and phase velocities coincide within the model of an isotropic medium. Therefore, the diffuse and drift summands of the $g_\lambda(\mathbf{q})$ function can be combined, and the effective frequency of phonon relaxation can be introduced [94]:

$$g_\lambda(\mathbf{q}) = -\frac{N_{q\lambda}^0(N_{q\lambda}^0+1)}{\tilde{\nu}_{ph}^{\lambda}(q)}\frac{\hbar\omega_{q\lambda}}{k_B T^2}(\mathbf{v}_q^\lambda \nabla T), \quad \tilde{\nu}_{ph}^{\lambda(1,2)}(q) = \nu_{ph}^{\lambda}(q)\left(1 + \frac{\nu_N^\lambda(q)}{(S^\lambda(\theta,\varphi))^2}B^{(1,2)}(T)\right)^{-1}. \tag{3.7}$$

When taking the anisotropy of the phonon spectrum into account, these directions do not coincide. Bearing in mind the phonon focusing [23], directly calculating the $\mathbf{u}^{(1,2)}$ drift velocities and $g_\lambda(\mathbf{q})$ for both versions results in:

$$\begin{aligned} g_\lambda^{(1,2)}(\mathbf{q}) &= -k_B \frac{N^0_{q\lambda}(N^0_{q\lambda}+1)}{k_B T \nu^\lambda(q)} \left(\frac{\hbar\omega_q^\lambda}{k_B T}\right) \left\{ (\mathbf{V}_g^\lambda \nabla_r T) + \left(\left(\frac{\mathbf{q}}{q}\right)\nabla_r T\right) \frac{\nu_N^\lambda(q)}{(S^\lambda(\theta,\phi))} B^{(1,2)}(T) \right\} \\ &= g_\lambda^{dif}(\mathbf{q}) + g_\lambda^{dr}(\mathbf{q}), \\ \mathbf{u}^{(1,2)} &= -k_B \nabla_r T B^{(1,2)}(T)/k_B T. \end{aligned} \tag{3.8}$$

When taking the dispersion of thermal phonons into account, the $B^{(1,2)}(T)$ coefficients appear as [23]:

$$\begin{aligned} &\mathrm{B}^{(1)}(T) = \sum_\lambda \Psi_N^\lambda \Big/ \sum_\lambda \Psi_{NR}^\lambda, \ \mathrm{B}^{(2)}(T) = \Psi_N^\lambda \Big/ \Psi_{NR}^\lambda, \\ &\Psi_N^\lambda = 3\int_{-1}^{1} \cos\theta d(\cos\theta) \int_0^{2\pi} d\varphi (y)^4 \int_0^1 \frac{V_{gz}^\lambda z^\lambda x^3}{(sh(z^\lambda/2))^2} \frac{\nu_N^\lambda}{\nu^\lambda} dx, \\ &\Psi_{NR}^\lambda = \int_{-1}^{1} d(\cos\theta) \int_0^{2\pi} d\varphi (y)^5 \int_0^1 \frac{x^4}{(sh(z^\lambda/2))^2} \frac{\nu_N^\lambda \nu_R^\lambda}{\nu^\lambda} dx, \quad x = \frac{q}{q_{\max}(\theta,\varphi)}, \\ &y(T,\theta,\varphi) = \frac{q_{\max}(\theta,\varphi)}{q_T}, \quad q_T = \frac{k_B T}{\hbar}, \quad Z^\lambda(x,\theta,\varphi) = \frac{\hbar\omega_q^\lambda(x,\theta,\varphi)}{k_B T}. \end{aligned} \tag{3.9}$$

The maximum wave vector $q_{\max}(\theta,\varphi)$ is defined by the expression (1.44). Thus, the non-equilibrium phonon distribution functions can be represented within the three-mode Callaway model as an additive sum of the diffuse and drift motion of phonons.

3.2 Relaxation Mechanisms and Thermal Conductivity of Silicon Crystals

In contrast to studies conducted earlier on thermal conductivity of bulk materials within the model of an isotropic medium [9, 36, 91–96] when analysing phonon transport in monocrystalline Si samples, the present section takes into account the phonon focusing and resulting dependence of the thermal conductivity on the direction of heat flux [I (ψ)] and orientation {J} of a wide face of the sample relative to the axis of the crystal. The three-mode Callaway model represents it as an additive sum of the diffuse and drift contributions [23]:

$$\kappa_{dif[I(\psi)]}^{\{J\}}(T) = \frac{k_B q_T^3}{4(2\pi)^3} \sum_\lambda \int_{-1}^{1} d(\cos\theta) \int_0^{2\pi} d\varphi y^3 \int_0^1 \frac{\left(V_{g3}^\lambda\right)^2 z_\lambda^2 x^2}{\nu_{[I(\psi)]}^{\lambda\{J\}} (sh(z_\lambda/2))^2} dx, \tag{3.10}$$

$$\kappa_{dr[I(\psi)]}^{\{J\}}(T) = \frac{k_B q_T^3}{12(2\pi)^3}\sum_\lambda B_{[I(\psi)]}^{\{J\}}(T)\Psi_{N[I(\psi)]}^{\lambda\{J\}}(T), \quad B_{[I(\psi)]}^{\{J\}}(T) = \sum_\lambda \Psi_{N[I(\psi)]}^{\lambda\{J\}} \Big/ \sum_\lambda \Psi_{NR[I(\psi)]}^{\lambda\{J\}},$$

$$\Psi_{N[I(\psi)]}^{\lambda\{J\}} = 3\int_{-1}^{1}\cos\theta d(\cos\theta)\int_0^{2\pi} d\varphi y^4 \int_0^1 \frac{V_{gz}^\lambda x^3 z_\lambda}{(sh(z_\lambda/2))^2}\frac{\nu_N^\lambda}{\nu_{[I(\psi)]}^{\lambda\{J\}}}dx,$$

$$\Psi_{NR[I(\psi)]}^{\lambda\{J\}} = \int_{-1}^{1} d(\cos\theta)\int_0^{2\pi} d\varphi y^5 \int_0^1 \frac{x^4}{(sh(z_\lambda/2))^2}\frac{\nu_N^\lambda \nu_{R[I(\psi)]}^{\lambda\{J\}}}{\nu_{[I(\psi)]}^{\lambda\{J\}}}dx. \tag{3.11}$$

Here, $z_\lambda = \hbar\omega_q^\lambda/k_B T, x = \frac{q}{q_{\max}(\theta,\varphi)}, y(T,\theta,\varphi) = \frac{q_{\max}(\theta,\varphi)}{q_T}$, $q_{\max}(\theta,\varphi)$ is a maximum wave vector (see formula (1.44)), the (θ,φ) angles define the direction of the wave vector of a phonon, V_{g3}^λ and V_{gz}^λ are the projections of the group velocity on the temperature gradient direction and Z-axis, respectively, ω_q^λ is the frequency of a phonon with polarization accounting for the dispersion of thermal phonons, $\nu_{[J(\psi)]}^{\lambda\{I\}} = \nu_{[J(\psi)]R}^{\lambda\{I\}}(q) + \nu_N^\lambda(q)$ is the total phonon relaxation rate. Phonon relaxation rates in resistive $\left(\nu_{[I(\psi)]R}^{\lambda\{J\}}\right)$ and normal $(\nu_N^\lambda(q))$ scattering processes are computed in Section 3.1. Normal phonon–phonon and resistive scattering processes are responsible for drift and diffuse contributions to thermal conductivity, respectively. For calculating the temperature dependencies of the thermal conductivity of bulk Si samples, the approximation spectrum of phonons is used [63] (see Section 1.2). In this case, the phonon spectrum for symmetric directions, extracted from data on inelastic neutron scattering [61, 62], is approximated by a seventh-degree polynomial in the reduced wave phonon vector x. When extrapolated over the entire Brillouin zone, the spectrum is expanded in cubic harmonics (see Section 1.2). The approximation above saves the cubic symmetry and allows analysing the change in the phonon focusing when transitioning from the long-wave $(x << 1)$ to short-wave $(x \sim 1)$ limits. In the long-wave limit, it is examined within the anisotropic continuum model that specifies the anisotropy through the $S_0^\lambda(\theta,\varphi)$ phase velocity (see (1.28)).

Let L, D, and $W = \mu D$ be the length, thickness, and width of rectangular cross-section samples, respectively. For square-shaped cross-section samples, we have $\mu = 1$. To make allowance for the influence of the phonon focusing on the thermal conductivity anisotropy of monocrystalline samples, it is sufficient to express the orientation parameters through group velocity components. These parameters are the heat flux direction $[I(\psi)]$ and the orientation $\{J\}$ of the plane in which the ψ angle is measured from the Z-axis.

For this purpose, we explore the rotation of heat flux in two symmetric planes:

(1) A wide face of the sample and the YZ-plane of the cube face coincide, $\{J\} = \{100\}$;
(2) A wide face of the sample and the diagonal plane coincide, $\{J\} = \{110\}$ (see Figure 3.2).

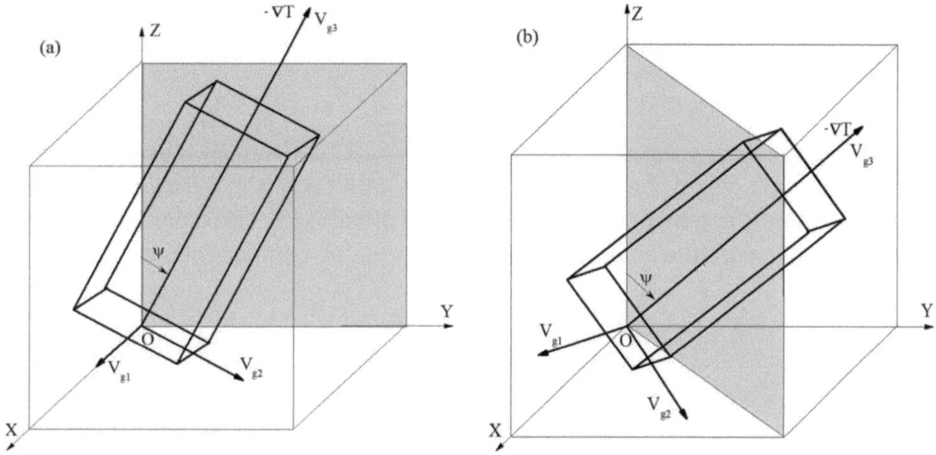

Figure 3.2: Diagram that illustrates the change in the angle ψ during rotation of the temperature gradient: (a) in the ZY {J} = {100} plane of the face of the cube and (b) in the {J} = {110} diagonal plane. V^{λ}_{g1}, V^{λ}_{g2}, and V^{λ}_{g3} are the group velocity components.

Next, we define a coordinate system related to the sample as follows. Let the axis 3 be directed along the sample axis being in line with the heat flux direction. Axis 1 (the axis of rotation) should be turned perpendicularly to the wide face or one of the faces of the square-shaped cross-section sample. Axis 1 assigns the plane orientation {J}. Axis 2 is perpendicular to the narrow face of the sample. Further, we take into consideration that the phonon spectrum is found in the cube-edge coordinate system. Then, the phonon group-velocity components in the accepted coordinate system can be represented in the form:

$$(1)\ V^{\lambda}_{g3} = -V^{\lambda}_{gy}\sin\psi + V^{\lambda}_{gz}\cos\psi,\quad V^{\lambda}_{g2} = V^{\lambda}_{gy}\cos\psi + V^{\lambda}_{gz}\sin\psi,\quad V^{\lambda}_{g1} = V^{\lambda}_{gx}, \tag{3.12a}$$

$$(2)\ V^{\lambda}_{g3} = \left(V^{\lambda}_{gx} + V^{\lambda}_{gy}\right)\sin\psi/\sqrt{2} + V^{\lambda}_{gz}\cos\psi,\quad V^{\lambda}_{g2} = \left(V^{\lambda}_{gx} + V^{\lambda}_{gy}\right)\cos\psi/\sqrt{2} - V^{\lambda}_{gz}\sin\psi$$

$$V^{\lambda}_{g1} = \left(V^{\lambda}_{gx} - V^{\lambda}_{gy}\right)/\sqrt{2}, \tag{3.12b}$$

From Figure 3.2 it is seen that, for the cross-section {J} of cubic crystals, the dependence of the heat flux direction on the ψ angle is controlled by the V^{λ}_{g3} group velocity component. Because of the projection of the V^{λ}_{g1} component is the rotation axis, it does not depend on the ψ angle. It identifies the {J} plane or the wide face. The V^{λ}_{g2} component is ψ-angle-dependent and orientates two other sample faces. For rectangular cross-section samples investigated in [19], the first case, when the wide face coincides with {100}, the narrow face coincides with {110}, and the temperature gradient is pointed along [110], is described by the expression (3.12a) with the angle $\psi = \pi/4$ (see Figure 3.2a). Then, for the V^{λ}_{gi} components, we have:

$$V^{\lambda}_{g3} = (-V^{\lambda}_{gy} + V^{\lambda}_{gz})/\sqrt{2}, \quad V^{\lambda}_{g2} = \left(V^{\lambda}_{gy} + V^{\lambda}_{gz}\right)/\sqrt{2}, \quad V^{\lambda}_{g1} = V^{\lambda}_{gx}. \tag{3.13a}$$

The second case, when the wide face coincides with {110}, the narrow one merges with {100}, and the temperature gradient is set along [110], corresponds to the expression (3.12b) with the angle $\psi = \pi/2$. Then, for the V^{λ}_{gi} components, we get:

$$V^{\lambda}_{g3} = \left(V^{\lambda}_{gx} + V^{\lambda}_{gy}\right)/\sqrt{2}, \quad V^{\lambda}_{g2} = -V^{\lambda}_{gz}, \quad V^{\lambda}_{g1} = \left(V^{\lambda}_{gx} - V^{\lambda}_{gy}\right)/\sqrt{2}. \tag{3.13b}$$

So, we have demonstrated that the $[I(\psi)]$ and $\{J\}$ orientation parameters can be evaluated through the group velocity components parallel and perpendicular to the heat flux. This holds for an arbitrary direction of the heat flux relative to the crystal axes. In the Cartesian coordinate system, the phonon group-velocity components required for calculating the temperature dependencies of the thermal conductivity have the form:

$$\begin{aligned}
V^{\lambda}_{gx}(x,\theta,\varphi) &= S^{\lambda}_{0}(\theta,\varphi)\left\{V^{\lambda}_{n}(x,\theta,\varphi)\sin\theta\cos\varphi + S^{\lambda}_{\theta}(x,\theta,\varphi)\cos\theta\cos\varphi - S^{\lambda}_{\varphi}(x,\theta,\varphi)\sin\varphi\right\},\\
V^{\lambda}_{gy}(x,\theta,\varphi) &= S^{\lambda}_{0}(\theta,\varphi)\left\{V^{\lambda}_{n}(x,\theta,\varphi)\sin\theta\sin\varphi + S^{\lambda}_{\theta}(x,\theta,\varphi)\cos\theta\sin\varphi + S^{\lambda}_{\varphi}(x,\theta,\varphi)\cos\varphi\right\},\\
V^{\lambda}_{gz}(x,\theta,\varphi) &= S^{\lambda}_{0}(\theta,\varphi)\left\{V^{\lambda}_{n}(x,\theta,\varphi)\cos\theta - S^{\lambda}_{\theta}(x,\theta,\varphi)\sin\theta\right\}.
\end{aligned} \tag{3.14}$$

The $V^{\lambda}_{n}(x,\theta,\varphi)$, $S^{\lambda}_{\theta}(x,\theta,\varphi)$, and $S^{\lambda}_{\varphi}(x,\theta,\varphi)$ functions normalized to the phase plane are the phonon group-velocity components in a spherical coordinate system [63]. Accounting for the dispersion of phonons leads to the conclusion that the group velocity, as well as the relaxation rates, depend neither on the angles ψ, θ, and φ nor the reduced wave vector x. In other words, $\mathbf{V}^{\lambda}_{g} = \mathbf{V}^{\lambda}_{g}(x,\psi,\theta,\varphi)$ and $\nu^{\lambda\{J\}}_{B[I(\psi)]} = \nu^{\lambda}_{B[I(\psi)]}(x,\theta,\varphi)$.

Focusing affects thermal conductivity via phonon-boundary scattering. In this mechanism, phonon relaxation rates can be regarded as piecewise-smooth functions for various angle ranges governed by the ratios between group velocity components and the geometric parameters $k_0 = L/2D$ and $\mu = W/D$. For the heat flux in the direction of $[I(\psi)]$, where the ψ angle is measured from the Z-axis in the plane $\{J\}$, the phonon relaxation rate can be expressed as:

$$\nu^{\lambda\{J\}}_{B[I(\psi)]}(x,\theta,\varphi) = \frac{\left|V^{\lambda}_{g3}\right|}{k_0 D}\left\{1 - \frac{k_0}{2}\frac{\left(\left|V^{\lambda}_{g2}\right| + \mu\left|V^{\lambda}_{g1}\right|\right)}{\mu\left|V^{\lambda}_{g3}\right|} + \frac{(k_0)^2}{3}\frac{\left|V^{\lambda}_{g1}\right|\left|V^{\lambda}_{g2}\right|}{\mu\left(V^{\lambda}_{g3}\right)^2}\right\}^{-1}, \quad k_0 = \frac{L}{2D} \tag{3.15}$$

if the inequalities $\mu|V_{g1}^{\lambda}|>|V_{g2}^{\lambda}|$ and $|V_{g3}^{\lambda}/V_{g1}^{\lambda}|\geq k_0$ or $\mu|V_{g1}^{\lambda}|<|V_{g2}^{\lambda}|$ and $|V_{g3}^{\lambda}/V_{g2}^{\lambda}|\geq k_0/\mu$ are realized. Conditional upon the opposite inequalities are fulfilled, we arrive at:

$$\nu_{B[I]}^{\lambda\{J\}}(x,\theta,\varphi)=\nu_{B\infty[I]}^{\lambda\{J\}}(x,\theta,\varphi)=\begin{cases}\frac{6\mu}{D}\frac{\left(V_{g1}^{\lambda}\right)^2}{\left(3\mu\left|V_{g1}^{\lambda}\right|-\left|V_{g2}^{\lambda}\right|\right)}, \text{if } \left|V_{g2}\right|<\mu\left|V_{g1}\right| \text{ and } \left|V_{g3}^{\lambda}\middle/V_{g1}^{\lambda}\right|<k_0;\\ \frac{6}{\mu D}\frac{\left(V_{g2}^{\lambda}\right)^2}{\left(3\left|V_{g2}^{\lambda}\right|-\mu\left|V_{g1}^{\lambda}\right|\right)}, \text{if } \left|V_{g2}\right|>\mu\left|V_{g1}\right| \text{ and } \left|V_{g3}^{\lambda}\middle/V_{g2}^{\lambda}\right|<\frac{k_0}{\mu}.\end{cases} \tag{3.16}$$

The dependencies of thermal conductivity on the temperature gradient direction $[I(y)]$ and orientation $\{J\}$ of side faces of the sample relative to the crystal axes are determined by the magnitudes of the phonon group-velocity components. They directly enter the thermal conductivity and relaxation rates $\nu_{B[I(\psi)]}^{\lambda\{J\}}$.

According to [101–104], the expression for the relaxation rate of phonons by isotopic disorder in cubic crystals has the form:

$$\nu_{iso}(q_1,\lambda_1)=\frac{\pi}{6}gV_0\omega_{q_1\lambda_1}^2D\left(\omega_{q_1\lambda_1}\right),\quad D\left(\omega_{q_1\lambda_1}\right)=\frac{1}{V}\sum_{\mathbf{q}_2\lambda_2}\delta(\omega_{q_1\lambda_1}-\omega_{q_2\lambda_2}). \tag{3.17}$$

Here, V_0 is the volume per atom, g is the factor of isotopic disorder, $g=\sum_i C_i(\Delta M_i/\bar{M})^2$, where $\Delta M=M_i-\bar{M}$, M_i is the mass of i-th isotope, $\bar{M}=\sum_i C_iM_i$ is the average mass of the isotopic composition, C_i is the concentration of i-th isotope, $D(\omega)$ is the density of phonon states, V is the normalizing volume. The anisotropic continuum model offers the expression (3.17) as [104]:

$$\nu_{iso}^{\lambda}\cong A_{iso}(T\cdot z_\lambda)^4,\quad A_{iso}^{\lambda}=\frac{gV_0}{12\pi}(k_B/\hbar)^4\left\langle\left(S_0^{\lambda}\right)^{-3}\right\rangle,\ \left\langle\left(S_0^{\lambda}\right)^{-3}\right\rangle=\frac{1}{4\pi}\int d\Omega\left(S_0^{\lambda}(\theta,\varphi)\right)^{-3},$$

$$A_{iso}=\begin{cases}A_{iso}^{L}+A_{iso}^{t1}+A_{iso}^{t2}=0.46, & \text{if } \omega\leq\omega_{\max}^{t2};\\ A_{iso}^{L}+A_{iso}^{t1}=0.21, & \text{if } \omega_{\max}^{t2}<\omega\leq\omega_{\max}^{t1};\\ A_{iso}^{L}=0.04, & \text{if } \omega_{\max}^{t1}<\omega\leq\omega_{\max}^{L};\\ 0, & \text{if } \omega>\omega_{\max}^{L}.\end{cases} \tag{3.18}$$

The magnitude of the isotopic disorder parameter for Si^{nat} with the natural isotope composition is $g=2.01\cdot10^{-4}$. By accounting for the dispersion of phonons, the density of phonon states, $D(\omega)$, can be obtained from the appearance of the approximation spectrum (see Section 1.2).

According to [9, 36, 38, 91, 94–97], the expression for the phonon relaxation rate in the Umklapp processes has the form:

$$\nu_U^\lambda = A_U^\lambda \cdot z_\lambda^2 \cdot T^3 \cdot \exp\left(-\frac{C_U^\lambda}{T}\right). \tag{3.19}$$

The parameters of anharmonic scattering, A_U^λ and C_U^λ, for the phonon relaxation rate in the Umklapp processes are fitting parameters. The latter can be found from matching calculation data for the temperature dependencies of thermal conductivity and experimental outcomes. In accordance with generally accepted beliefs [9, 36, 97], the Herring relaxation mechanism for longitudinal phonons [38] and the Landau-Rumer mechanism [98] for transverse phonons are the major ones of N-scattering processes in Si crystals (see formulas (3.5) and (3.6)). These processes re-distribute the momentum of phonons between various vibrational branches. Therefore, we should utilize the first version of N-scattering processes to calculate the drift rates (see formulas (3.8)). Comparison of the predicted data on the temperature dependencies of thermal conductivity with the experimental findings of [19] yields parameters that control the phonon relaxation rates in anharmonic phonon–phonon scattering processes (Table 3.1).

Table 3.1: Parameters that determine phonon relaxation rates in the anharmonic processes of phonon–phonon scattering for Si crystals, $A_{iso} = 0.46\text{K}^{-4}\text{sec}^{-1}$.

A_N^L, K^{-5} sec^{-1}	A_N^t, K^{-5}sec^{-1}	A_u^L, K^{-3}sec^{-1}	A_u^t, K^{-3}sec^{-1}	C_u^L, K	C_u^t, K
0.8	$2.0 \cdot 10^{-3}$	$2.0 \cdot 10^{3}$	$0.70 \cdot 10^{3}$	310	98

The fitting parameters from Table 3.1 testify that the relaxation frequency for transverse phonons in N-processes is two orders of magnitude less than that for longitudinal phonons. It is not difficult to verify that the inequality $\nu_N^t(q) < \nu_R^t(q)$ is met for transverse phonons over the entire temperature range. Their contribution to the thermal conductivity is due to the diffuse motion. For longitudinal phonons, the ratio between the relaxation rates is $\nu_N^L/\nu_R^L < 1$; in the temperature interval $1 < T < 12\text{K}$, it is less than unity. However, it turns out to be greater than unity at higher temperatures. Therefore, the contribution of the drift motion of phonons to the thermal conductivity is mandatory to estimate analogously to as done in [94].

3.3 Analysis of the Temperature Dependencies of Thermal Conductivity for Square-Shaped and Rectangular Cross-Section Si Samples

Further, we show the results calculated in [22] for thermal conductivity $k(T)$of finite-length Si samples with square-shaped and rectangular cross-sections using formulas (3.10)–(3.19). For boundary scattering, fitting parameters were not applied. Fitting of the calculation results was carried out by varying the parameters of the anharmonic scattering processes in order to achieve the best agreement between the predicted and the experimental data in the region of the κ (T) maximum. The calculated parameters derived using the data of [85, 91, 94–97] are listed in Table 3.1. First and foremost, we explore the influence of various relaxation mechanisms on the temperature dependencies of thermal conductivity of Si crystals for the fitting parameters from Table 3.1. For this, we take a square-shaped cross-section sample [19] $(D = 0.293\text{cm}, L = 2.9\text{cm})$; the temperature gradient is set along the [100] direction. Taking into consideration boundary scattering $(\nu^\lambda = \nu_B^\lambda)$, the thermal conductivity follows the $\kappa(T) \sim T^3$ dependence at the lowest temperature (see Figure 3.3, curve 2).

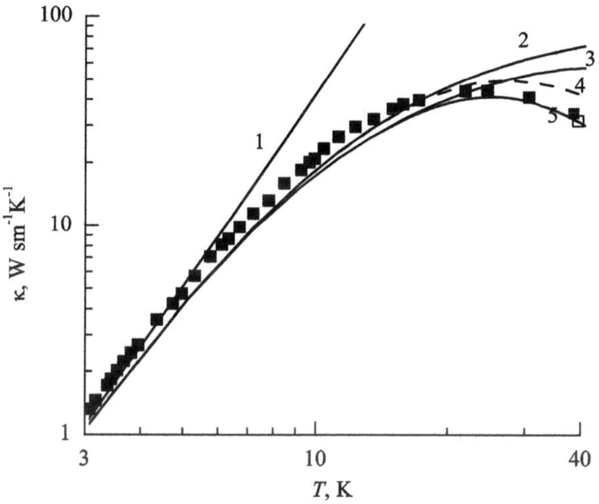

Figure 3.3: Temperature dependencies of thermal conductivity of square-shaped cross-section Si crystals (L = 2.9 cm, D = 0.293 cm) for the [100] direction, taking into account the following phonon relaxation mechanisms: curve 1 refers to boundary scattering, curve 2 is for boundary scattering and isotopic scattering disorder, curve 3 reflects boundary scattering, scattering by isotopic disorder and normal phonon–phonon scattering processes, curve 4 refers to boundary scattering, scattering by isotopic disorder, and the Umklapp phonon–phonon processes, curve 5 is for all mechanisms p relaxation of phonons. Symbols are the experimental data of [19].

In the low-temperature range $(T = 3\text{K})$, the calculated values of the thermal conductivity appear to be 4%–8% higher than those in the experiment. To explain this divergence, the papers [2, 23] exploited the concept of a defective surface layer, proposed in [32, 87, 88]. However, the papers [20, 63] disregarded isotopic scattering whose contribution amounts to 5% at T = 3 K. As can be understood from Figure 3.3 (curve 2), accounting for isotopic and boundary scattering $(\nu^\lambda = \nu_B^\lambda + \nu_{iso}^\lambda)$ aligns the predicted and experimental data on the temperature dependencies of thermal conductivity in the temperature range from 3 K to 15 K within the experimental error, without resorting to fitting parameters. At temperatures higher than 15 K, the normal phonon–phonon scattering and Umklapp processes become noticeable (see Figure 3.3, curves 3 and 4).

A total allowance for all scattering processes using the fitting parameters from Table 3.1, allows reconciling the findings of the calculation and experiment of [19] over the measuring temperature range up to 40 K (see Figure 3.3, curve 1). It should be stressed that the major role in the bending-down of the plotted temperature dependence of thermal conductivity is played by the Umklapp processes (compare curves 3 and 5 in Figure 3.3). Figure 3.4 sketches the temperature dependencies of thermal conductivity for three identical Si samples with a square-shaped cross-section in symmetric directions of heat flux. This graph is built based on the phonon relaxation times calculated for diffuse boundary scattering.

As can be observed in Figure 3.4, the theory developed above properly describes the anisotropy and temperature dependencies of thermal conductivity for square-shaped cross-section Si crystals, or otherwise speaking, the first of the discovered effects in [19]. Moreover, for samples located in different directions of heat flux relative to the crystallographic axes, the calculated results are in good agreement with the experiment in the temperature range from 3 K to 15 K, without applying fitting parameters (see Figure 3.4). The maximum thermal conductivity is attained for the samples when the heat flux direction coincides with [100]. At low temperatures, it exceeds the thermal conductivity of samples with the axis pointing along [110] and [111] by 36% and 50%, respectively.

These estimates match well the values of the experimental anisotropy of thermal conductivity [19]. Formulas (3.10), (3.11), as well as (3.13)–(3.16), for phonon-boundary scattering, give the possibility of satisfactorily describing the second McCurdy effect [19]. The latter consists in the dependence of the thermal conductivity magnitudes on the orientation of sample wide faces. Figure 3.5 shows that the temperature dependencies of thermal conductivity agree quite well with the experimental outcomes [19] (see Figure 3.5). This is true for two rectangular cross-section samples having the same area of the cross-section and the same temperature gradient direction. At $T = 3$ K, for the sample with {100} wide and {110} narrow faces, the thermal conductivity appears to be 31% higher than that of the {110} wide and {100} narrow faces of the sample.

Within the experiment error, this result finds good agreement with the experimental value of 33%. It should be underscored that, for both rectangular and square-shaped cross-section samples, the difference between the predicted and the experimental values [19] of $k(T)$ is not above the experiment error in the temperature range $3\text{K} < T < 15\text{K}$. Thus, the newly developed theory quantitatively describes, without using fitting parameters, both effects revealed by McCurdy in thermal conductivity of Si crystals in the temperature range where scattering at the boundaries and by isotopic disorder dominates. These findings testify that the analytic solutions

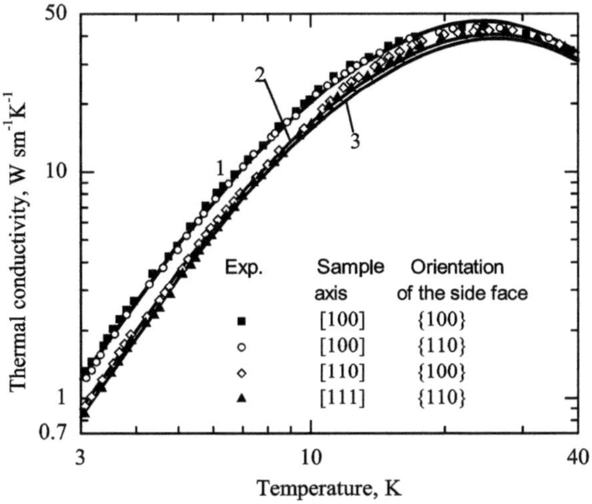

Figure 3.4: Temperature dependencies of thermal conductivity of square-shaped cross-section Si crystals ($L = 2.9$ cm, $D = 0.293$ cm) for symmetric directions of the temperature gradient: curve 1 is for the [100] direction, curve 2 is for the [110] direction and curve 3 is for the [111] direction. Symbols are the experimental values of [19].

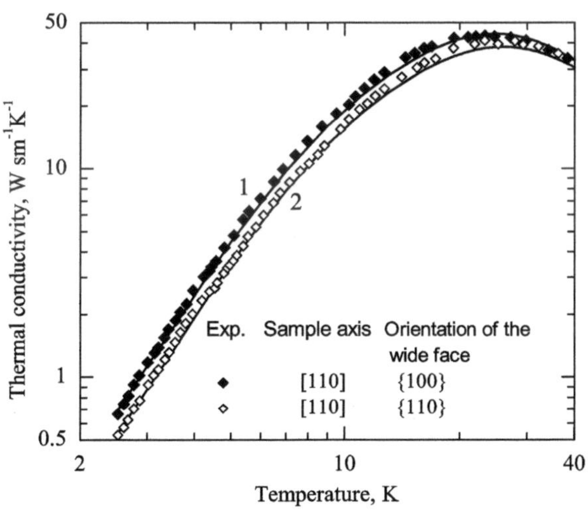

Figure 3.5: Temperature dependencies of thermal conductivity for rectangular cross-section samples ($L = 3.5$ cm, $D_1 = 0.185$ cm, $D_2 = \mu D_1 = 0.638$ cm, $\mu = 3.45$) and the orientation of the temperature gradient in the [110] direction. Curve 1 refers to samples with the {100} wide face, curve 2 refers to samples with the {110} wide face. Symbols are the experimental values of [19].

for the phonon relaxation rates during diffuse boundary scattering are quite adequate to be suitable for interpreting the features of phonon transport in different semiconductive and dielectric materials.

The experimental data of [19] show that the anisotropy of thermal conductivity quickly decays as the temperature rises when going over from the boundary scattering regime to the bulk relaxation mechanisms (above the maximum) (see Figures 3.4 and 3.5). The deviation of the experimental values of $k(T)$ for symmetric directions in the vicinity of the maximum does not exceed 6%. At $T = 40$ K, these values are equal to or within the experiment error, that is to say, the phonon focusing effect does not make itself felt in the thermal conductivity magnitudes. The maximum difference between the predicted and experimental data of [19] takes place near the maximum of the thermal conductivity. So, for example, at $T_{max} = 25$K, the thermal conductivity values are 5% higher than the experimental outcomes and 4% and 6% lower than the latter in the [101] and [111] directions, respectively. Although the deviation from the experimental results in the vicinity of the maximum is not great; however, the calculated data yield the maximum value of the anisotropy of 19%, which is noticeably larger than that of the experiment. At higher temperatures, the anisotropy of the thermal conductivity quickly diminishes. According to the calculation above, it becomes equal to the experiment error (4%) at $T = 60$K. As is mentioned earlier, the Matthiessen rule appears to be the possible reason for the divergence with the experiment in the region of the maximum when transitioning from boundary to bulk scattering. This approximation gives a too slow decrease in the anisotropy of the thermal conductivity and, apparently, is not a precise enough procedure in this case. This is explained by focusing.

Analysing the role of low-energetic phonons in thermal conductivity and thermopower of monocrystalline semiconductors, Herring brought to light the weakness of the Matthiessen rule at temperatures near the maximum of $k(T)$. This is because boundary scattering occurs near the surface of the samples, whereas bulk processes come about uniformly throughout the entire volume. Therefore, they cannot be additively summed up as it follows from the Matthiessen rule. Obviously, when phonon relaxation rates in bulk mechanisms and boundary scattering become equal, the maximum deviation from the Matthiessen rule can be observed. To correctly make allowance for combined action of boundary and bulk-mechanism scattering of phonons, the kinetic Boltzmann equation needs to be solved for the case of a non-uniform distribution of heat flux throughout the transverse cross-section of the sample. With accounting for phonon focusing, this problem is too involved and requires separate consideration.

Let us compare the role of boundary and bulk-mechanism scattering of phonons in the thermal conductivity of Si crystals (see Figure 3.6). In doing so, we calculate the temperature dependencies of thermal conductivity, $k(T)$, in the phonon-boundary scattering regime, $\nu^{\lambda}_{(2)}(q) = \nu^{\lambda}_{B}(q)$ (curve 2). In this case, we take into account only bulk phonon relaxation mechanisms, $\nu^{\lambda}_{(5)}(q) = \nu^{\lambda}_{iso}(q) + \nu^{\lambda}_{U}(q) + \nu^{\lambda}_{N}(q)$ (curve 5) (see Figure 3.6). At temperatures below the thermal conductivity maximum, scattering at the boundaries and by isotopic disorder dominantly contributes to thermal resistance. Moreover, at $T = 3$K, the boundary scattering provides 95% of the thermal resistance, but the isotopic scattering is responsible for only 5%. With rising

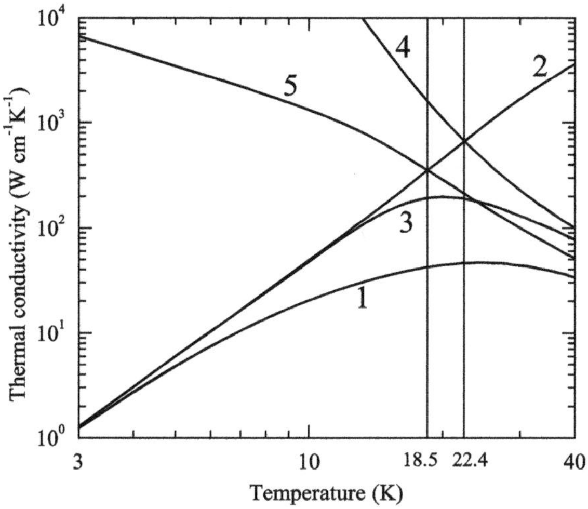

Figure 3.6: Temperature dependencies of full total thermal conductivity for the [100] direction (curve 1) and its dependencies for different phonon relaxation mechanisms: curve 2 is for thermal conductivity in the boundary scattering regime, curve 3 is for a monoisotopic Si^{28} sample, curve 4 corresponds to anharmonic scattering processes, curve 5 corresponds to for bulk scattering mechanisms.

temperature, the relative contribution of the boundary scattering diminishes and the contribution of the bulk relaxation mechanisms increases. At the temperature $T_{BV} = 18.5$K for Si^{nat} crystals with the natural isotopic composition, curves 2 and 5 intersect. This stands for equalizing the contributions of the bulk relaxation mechanisms and boundary scattering to thermal resistance. At $T = T_{BV}$, there takes place a transition from the boundary scattering regime to the dominant role of the bulk relaxation mechanisms (see Figure 3.6).

If one assumes that phonon relaxation rates in anharmonic scattering processes such as the Umklapp and normal scattering processes remain unchanged when transitioning to isotopically pure Si^{28} crystals, the isotopic effect in thermal conductivity of Si^{28} can be estimated, and the T_{BV} transition temperature can be predicted. In this case, curves 2 and 4 intersect at a higher temperature, $T_{BV} = 22.4$K. It should be underscored that curve 4 refers to anharmonic scattering processes, $\nu^{\lambda}_{(4)}(q) = \nu^{\lambda}_{U}(q) + \nu^{\lambda}_{N}(q)$. In this case, the maximum temperature of the thermal conductivity drops from 25 K for Si^{nat} to 20 K for isotopically pure Si^{28}. Maximum values of the thermal conductivity increase 3.5 times when transitioning from Si^{nat} crystals to isotopically pure Si^{28} crystals. This means that, at $T = T_{max}$, the contribution of isotopic scattering to the thermal resistance turns out to be 3.5 times more than the total contribution of the rest of the phonon relaxation mechanisms. The estimation of the isotopic effect in the thermal conductivity of Si crystals can be regarded as crude enough because it leaves aside a possible violation of the Matthiessen rule for boundary and bulk relaxation mechanisms, as well as interference of isotopic and anharmonic scattering processes.

Let us look into the influence of thermal phonons on the temperature-dependence of thermal conductivity, k(T), for Si crystals. Figure 3.7 displays that the role of dispersion is small for the [100] direction at temperatures below 20 K. This is traced from the fact that the difference in curves plotted accounting for dispersion (solid lines) and using the anisotropic continuum model (dashed lines) does not exceed the experiment error. As the temperatures become higher, the difference grows. At the temperatures of 25 K and 40 K, it amounts to 5% and 14%, respectively. Dispersion is the most powerful factor that affects the contribution of a slow transverse mode to the thermal conductivity: the latter lowers by 15% compared to that computed within the anisotropic continuum model. Noticeably less influence of the dispersion can be observed on longitudinal phonons and a fast transverse mode (see Figure 3.7). An analysis of the drift motion of longitudinal phonons showed that its contribution to the total thermal conductivity in the temperature range of $3K < T < 40K$ is small and can be neglected.

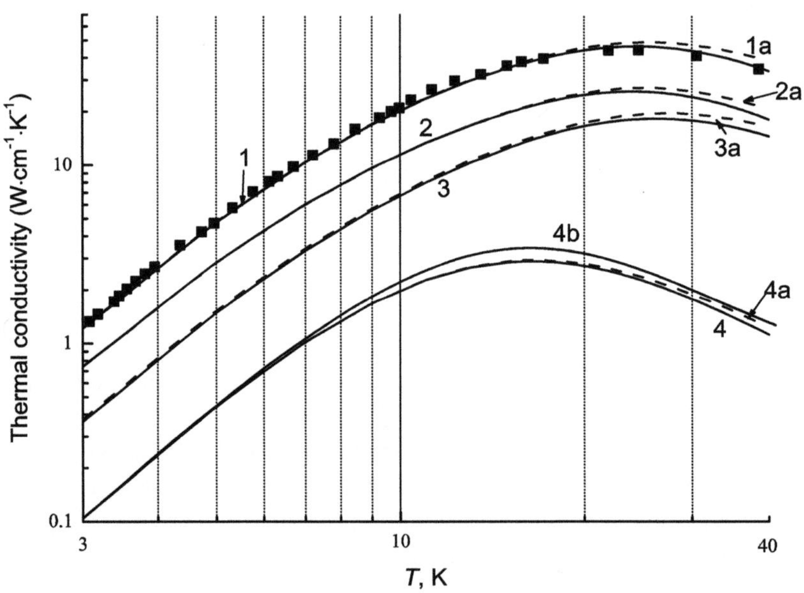

Figure 3.7: Dependencies of κ (T) (curves 1, 1a) and the contributions of slow (curves 2, 2a), fast (curves 3, 3a), transverse and longitudinal (curves 4, 4a, 4b) phonons for square-shaped cross-section samples placed in the [001] direction. Curves 1, 2, 3, 4 are calculated taking phonon dispersion into account, curves 1a, 2a, 3a, 4a are calculated within the anisotropic continuum model. Curve 4b is the contribution of longitudinal phonons with allowance for their drift motion. Symbols are the experimental values of [19].

This contribution reaches its upper limit of 18% in the region of the maximum of $\kappa^{L}_{[001]}(T)$ at $T^{L}_{max} = 17K$. As to the total thermal conductivity, the contribution of longitudinal phonons figures up to only 7.6%, whereas the contribution of their drift motion does not exceed 1.5%. Thus, for Si crystals with the natural isotopic composition, the drift motion of longitudinal phonons in the above temperature range can be disregarded.

3.4 Change in Anisotropy of Thermal Conductivity in Silicon Crystals with Temperature

Let us calculate the angular dependencies of thermal conductivity of square-shaped cross-section Si samples investigated in [19]; heat flux rotates in the {100} and {110} planes. We use formulas (3.10)–(3.19) to analyse the change in the thermal conductivity anisotropy with temperature. In Figure 3.7, the solid curves are meant for illustrating the calculated angular dependencies of thermal conductivity and the contributions of all branches of the phonon spectrum. All relevant phonon-relaxation mechanisms (3.10)–(3.19) occurring at the temperature of 3 K and 25 K are taken into consideration. The dashed lines in Figure 3.8a, b represent the angular dependencies of thermal conductivity at $T = 3K$ in the phonon boundary scattering regime when diffuse boundary scattering is the only phonon relaxation mechanism. In this case, the calculated findings turn out to be 4%–8% greater than the experiment values. Accounting for isotopic scattering whose contribution is 5%, at $T = 3K$, aligns the predicted and experimental data of [19] (see Figure 3.8a, b). As previously stated in Section 3.3, the deduced phonon relaxation times and making allowance for isotopic and boundary scattering ensure good agreement between the estimated and experimental data for all symmetric directions within the experiment error without fitting in the temperature range from 3 K to 15 K. The abovementioned conclusion is confirmed by evaluating the angular dependencies of the thermal conductivity of square-shaped cross-section Si samples (see Figure 3.8a, b).

Figure 3.8 claim that the thermal conductivity anisotropy is chiefly due to the contribution of slow transverse phonons that provide the maximum thermal conductivity when focused in the [001] directions.

The anisotropy of the thermal conductivity goes up to its maximum at $T = 3K$; the ratio of the thermal conductivity values in symmetric directions has the form: $\kappa_{[001]}:\kappa_{[110]}:\kappa_{[111]} \approx 1.5:1.1:1$. For the {110} plane in the [111] and [110] directions, the angular dependencies of thermal conductivity have local maxima caused by focusing of fast transverse and longitudinal modes, respectively (see Figure 3.8b). As the temperature rises, the thermal conductivity anisotropy diminishes. In the temperature range of 3 K–15 K, our calculations are in agreement with the data of [19] within the experiment error on both thermal conductivity magnitude and anisotropy. However, on approaching the maximum of the thermal conductivity, a

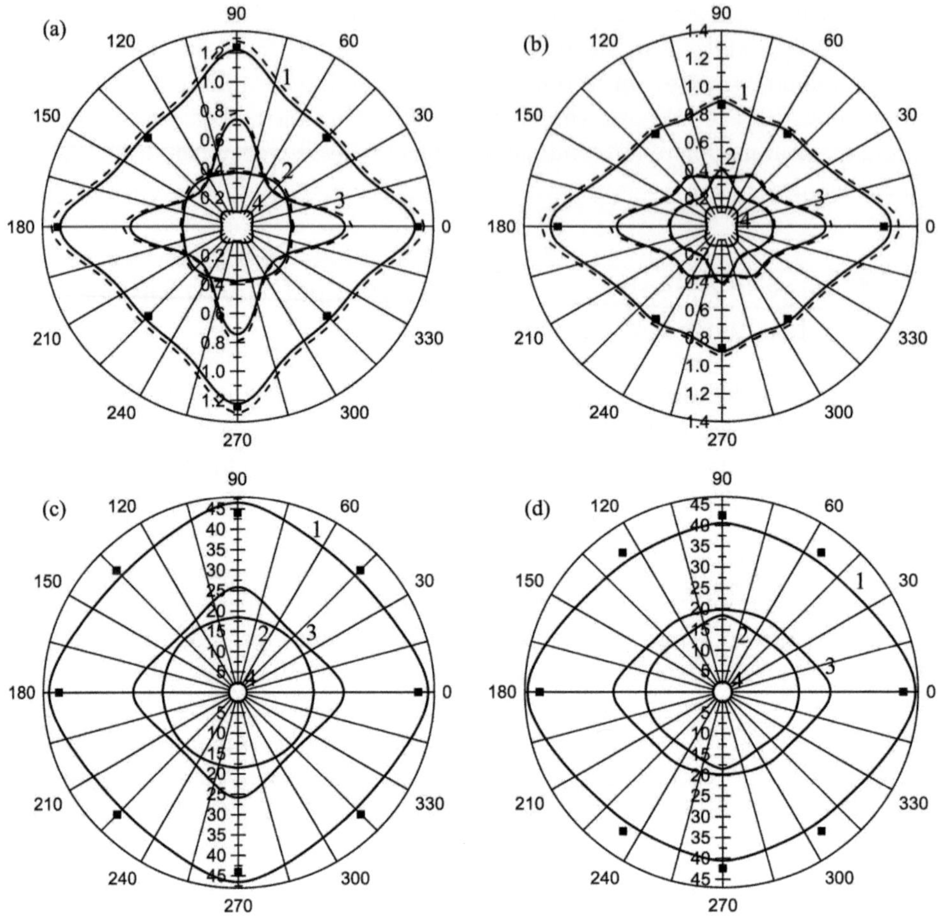

Figure 3.8: The angular dependencies of full thermal conductivity κ (ψ) (Wcm^{-1} K^{-1}) (curves 1) and the contributions to it from each of the branches of the phonon spectrum at temperatures of 3 K (a,b), 25 K (c,d) for square-shaped cross-section samples (L = 2.9 cm, D = 0.293 cm): curves 2 are for a fast transverse mode, curves 3 are for a slow transverse mode, curves 4 are for a longitudinal mode; for cases: (a,c) – the axis of the sample (and the temperature gradient) rotates in the plane of the face of the cube, (b,d) – the axis of the sample rotates in the diagonal plane. The dashed lines in Figures (a, b) show the predicted thermal conductivity values in the regime of boundary phonon scattering. Symbols are the experimental values of [19].

divergence of the predicted and experimental data increases (see Figure 3.8). At $T = 25$K, they become 6% larger and 5%–6% lesser than outcomes of [19] in the [001], [110], and [111] directions, respectively. Although the deviation of the theoretical dependencies from the experimental ones of [19] in the vicinity of the maximum is small (5%–6%), however, the calculation yields $\kappa_{[001]}:\kappa_{[111]} \approx 1.19$ instead of 1.06 as in the experiment. As already emphasized, the Matthiessen rule used for the phonon relaxation rates near the transition from boundary to bulk scattering is, probably, the very reason for this divergence.

For square-shaped cross-section samples placed in a fixed direction of the temperature gradient, the dependence of thermal conductivity on the orientation of the side faces is weak. For the [001] heat flux direction and {100} orientation of the side faces, the thermal conductivity magnitudes are only 1.3% greater than those for {110} orientation. For rectangular cross-section samples set in the [110] temperature gradient direction, the orientation of the side faces more essentially affects the thermal conductivity. At $T = 3$K, the thermal conductivity of the sample with the {100} wide face is 30% more in the [110] temperature gradient direction than that of the sample with the {110} wide face. As seen in Figure 3.9, this result meets the experiment [19]. An analysis showed that rectangular cross-section samples of the same geometric dimensions as in [19], placed in the [100] heat flux direction, produce thermal conductivity anisotropy significantly less: for the {100} wide face of the sample, it is 5% less than that for the {110} wide face.

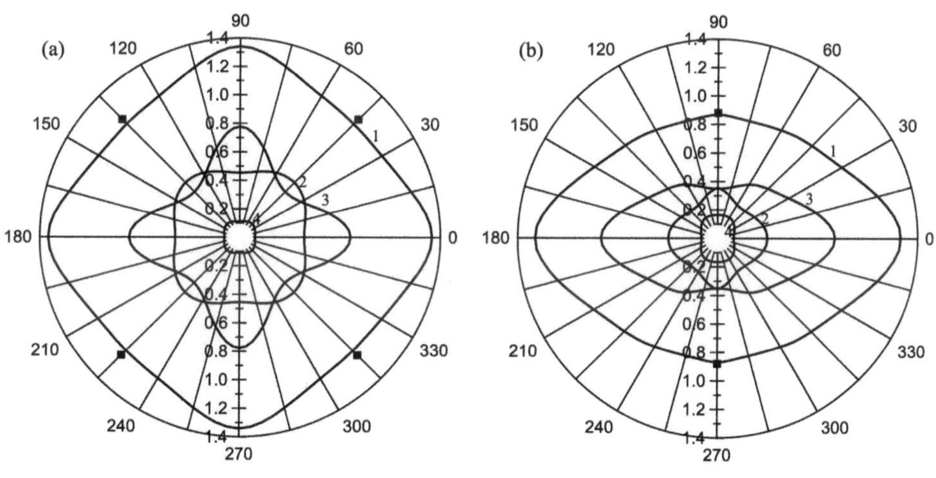

Figure 3.9: Angular dependencies of the full thermal conductivity κ (ψ) (Wcm^{-1} K^{-1}) (curves 1) and the contributions to it from each of the branches of the phonon spectrum at a temperature of 3 K for rectangular cross-section samples (L = 3.5 cm, D = 0.185 cm, and μ = 3.45). (a) The axis of the sample (and a temperature gradient) rotates in the plane of the cube face, (b) The axis of the sample rotates in the diagonal plane: curves 2 are for a fast transverse mode, curves 3 are for a slow transverse mode, curves 4 are for a longitudinal mode. Symbols are the experimental values of [19].

It is worth stressing that, in the temperature range for observations ($3\text{K} < T < 40\text{K}$), the dominant contribution to the thermal conductivity is governed by transverse phonons (see Table 3.2). Slow transverse phonons are focused in the [100] direction; their contribution is maximal. At $T = 3$K, it reaches 61% of the full thermal conductivity and the contribution of the fast transverse mode almost two times less and comes up to 31% (see Figure 3.7). With increasing the temperature to 40 K, the contribution of the slow transverse mode falls to 54% in the [001] direction, whereas the contribution of the fast transverse mode goes up to 43% (see Figure 3.7).

Table 3.2: Relative contributions of various branches of the phonon spectrum to the full thermal conductivity, $\tilde{\kappa}^{\lambda}_{[I]}(T) = \kappa^{\lambda}_{[I]}(T) / \kappa_{[I]}(T)$, for square-shaped cross-section samples for symmetric directions [I] at the temperatures of 3 K, 25 K, and 40 K.

$\tilde{\kappa}^{\lambda}_{[I]}(T)$	[001]			[101]			[111]		
	3 K	25 K	40 K	3 K	25 K	40 K	3 K	25 K	40 K
$\tilde{\kappa}^{t1}_{[I]}$	0.31	0.40	0.432	0.46	0.45	0.458	0.30	0.42	0.444
$\tilde{\kappa}^{t2}_{[I]}$	0.61	0.55	0.535	0.39	0.49	0.506	0.51	0.52	0.519
$\tilde{\kappa}^{L}_{[I]}$	0.08	0.05	0.033	0.15	0.06	0.036	0.19	0.06	0.037

For longitudinal phonons, the [001] direction corresponds to the defocusing direction, therefore, their contribution is minimal; at $T = 3\mathrm{K}$, it amounts to 8%. Longitudinal phonons are focused in the [111] directions and their contribution is maximal; it reaches 19% of the full thermal conductivity (see Table 3.2). With increasing the temperature from 3 K to 40 K, the total contribution of transverse phonons grows from 92% at $T = 3\mathrm{K}$ up to 97% at $T = 40\mathrm{K}$. At $T = 40\mathrm{K}$, the contribution of longitudinal phonons gradually drops from 8% to 3% in the [100] direction. In the [111] direction, it dramatically diminishes from 19% at $T = 3\mathrm{K}$ to 4% at $T = 40\mathrm{K}$ (see Table 3.2). Such a behaviour of the contributions of longitudinal and transverse phonons to the thermal conductivity of Si can be explained by different influence of anharmonic scattering mechanisms.

For longitudinal phonons in the vicinity of the thermal conductivity maximum, the main role is played by normal scattering processes, whereas by the Umklapp processes for transverse phonons. Therefore, the maximum temperature for transverse phonons is 8 K higher than that for longitudinal phonons. It should be pointed out that the transition from boundary to bulk relaxation mechanisms for longitudinal phonons occurs at $T_{\mathrm{B}} \approx 14\mathrm{K}$, whereas for transverse phonons it happens at $T_{\mathrm{B}} \approx 18\mathrm{K}$.

3.5 Physical Interpretation of McCurdy Effects in Thermal Conductivity of Cubic Crystals [116]

Let us analyse the influence of focusing on the propagation of phonons in monocrystalline square-shaped cross-section samples and pictorially explain the physical meaning of the anisotropy of thermal conductivity or the first McCurdy effect. The phonon focusing affects the thermal conductivity of elastically anisotropic crystals through a difference in wave vector and group velocity directions. It is worth recalling that the group velocity is responsible for the phonon propagation (see Figure 3.10). As an example, a t_2 mode can serve to illustrate the phonon focusing in Si crystals for wave vectors in the {100} plane (see Figure 3.10a). For this purpose, the isoenergetic surface needs to be constructed for this mode and the directions of the phonon group velocities should be identified. Characteristic angles $\theta_1, \theta_2, \theta_3,$ and θ_4 are determined in Section 1.5. In an isotropic medium, the directions of the propagation of a phonon and its wave vector coincide. Phonons with wave vectors in sector $-\theta_4 \le \theta \le \theta_4$ propagate in the same sector. However, as seen in Figure 3.10, the directions of wave vectors and group velocities differ for Si. Phonons of the slow transverse mode travel inside a significantly smaller sector, $-\theta_3 \le \theta \le \theta_3$. The latter is controlled by the directions of group velocities at zero curvature points in the isoenergetic surface (see details in [69, 71]). Let heat flux be pointing along the [100] direction in square-shaped cross-section Si crystals. Then, phonons with wave vectors in sector $-\theta_4 \le \theta \le \theta_4$ (for Si $\theta_4 = 28.6^o$) will deviate from the side faces towards the axis of the rod and move in sector $-\theta_3 \le \theta \le \theta_3$, where $\theta_3 = 6.8^o$ for Si (see Figure 3.10). As a result, for the focusing sector $-\theta_3 \le \theta \le \theta_3$, the average density of states of the t_2 mode in the ratio $\theta_4/\theta_3 \approx 4.2$ is larger than that in an isotropic medium. In contrast, for the defocusing sector $\pi/2 - \theta_3 \le \theta \le \theta_3$, it is 2.3 times less than that in an isotropic medium. In this case, the average density of states of the t_2 mode for the focusing sector is 9.9 times greater than that for the defocusing sector (see Section 1.5).

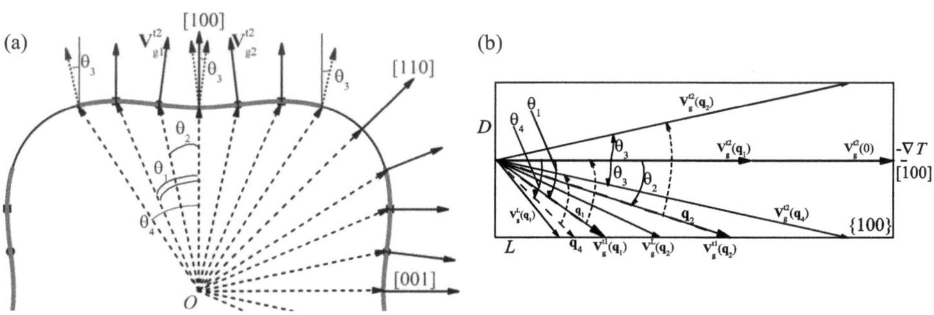

Figure 3.10: Schematic diagrams that illustrate (a) focusing of slow transverse modes in Si crystals for cutting the constant-energy surface by the {100} plane (arrows show wave vectors inside the surface and the corresponding phonon group velocities beyond it) and (b) the influence of focusing on the phonon propagation in Si samples with a length L and square cross-section D × D for wave vectors lying in the {100} plane with the angles $\theta_1^{t2} = 23.6^\circ$, $\theta_2^{t2} = 11.9^\circ$, $\theta_3^{t2} = 6.8^\circ$, $\theta_4^{t2} = 28.6^\circ$; the temperature gradient is set along the [100] axis.

Further, we explore in more detail the influence of focusing on the propagation of the phonon modes in square-shaped cross-section Si samples. This theme is expected to be useful for technical applications to design silicon electronic devices. Special attention should be drawn towards some peculiarities caused by affecting the focusing on traveling of phonons of the t_2 mode in the {100} plane. Firstly, phonons with wave vectors q_1(for Si $\theta_1 = 23.6^o$) will propagate in the [100] direction of heat flux (see Figure 3.10). In this case, the phonon mean free path limited by either bulk relaxation mechanisms or length of the sample. Secondly, since phonons of the t_2 mode with wave vectors in sector $-\theta_4 \leq \theta \leq \theta_4$ deviate from the side faces towards the axis of the rod and move at lesser angles (sector $-\theta_3 \leq \theta \leq \theta_3$), their mean free path can be substantially increased. *It is worth emphasizing that, in diffuse scattering regime, the mean free path of a phonon with the wave vector q is determined by a distance traveled by the phonon until it collides with the sample surface.* So, phonons with the wave vector $\pm q_4$ and at the angle $\theta_4^{t2} = 28.6^o$ propagate at a far much lesser angle ($\theta_3 = 6.8^o$). As a consequence, their mean free paths increase (see Figure 3.10). For the longitudinal mode, the [001] direction is a defocusing direction. Therefore, longitudinal phonons deviate from the sample axis towards the side face. Their mean free path is less than that offered by the model of an isotropic medium (see Figure 3.10). For wave vectors in the {100} plane, the fast transverse mode is isotropic; there is no focusing effect. Its mean free path coincides with the predicted one within the model of an isotropic medium (see Figure 3.10). Thus, the schemes depicted in Figure 3.10 show that the maximum thermal conductivity for type-I crystals is due to the contribution of slow transverse phonons.

Consider the propagation of phonon modes in the {100} axial cross-section of the sample when the temperature gradient lies along [110] direction. The latter matches the defocusing of the slow transverse mode t_2. Therefore, phonons of this mode deviate towards a sample face; their mean free path is less compared to that within the model of an isotropic medium (see Figure 3.11).

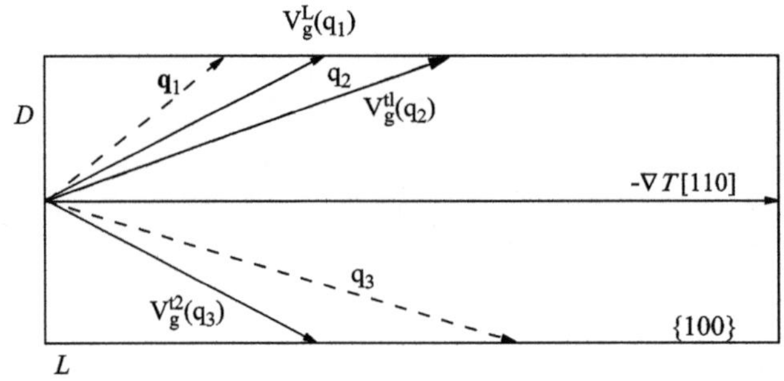

Figure 3.11: Scheme that illustrates the effect of phonon focusing in rectangular cross-section Si samples for wave vectors in the {100} plane and a temperature gradient along [110].

For wave vectors in the {100} plane, the fast transverse mode is isotropic; there exists no focusing effect. Its mean free path is equal to that calculated within the model of an isotropic medium (see Figure 3.11). Longitudinal phonons in this plane have a local focusing in the [110] direction. They deviate towards the sample axis and their mean free path is greater than within the model of an isotropic medium. It should be stressed that the role of longitudinal phonons in phonon transport for Si is small. This is because their thermal conductivity is an order of magnitude less than the total thermal conductivity of transverse phonons. Besides, the contribution of the longitudinal phonons to the thermal conductivity in the [001] and [110] directions amounts to 8% and 15%, respectively.

Let us examine the propagation of phonons for their wave vectors in the {110} plane when the temperature gradient is set along [110] direction. In the {110} plane, the fast transverse mode is focused in the [110] direction (see Figure 1.12b). Its phonons will move towards the sample axis, and their mean free path exceeds that computed within the model of an isotropic medium. However, the focusing effect, in this case, is not so evident as compared to the slow transverse mode in the [001] direction (see Figure 2.6). Therefore, the average, over modes, phonon mean free path, and, accordingly, the thermal conductivity in the [110] direction turns out to be 40% lesser than that in the [001] direction. Longitudinal phonons are focused in the [111] direction; their mean free path doubles compared with that in the [001] defocusing direction and takes a maximum value (see Figure 2.6). Here, the fast transverse mode defocuses, and its mean free path is shorter than within the model of an isotropic medium. Phonons of the slow transverse mode in this direction have a local maximum of the focusing; their mean free paths slightly exceed their value for the model of an isotropic medium (see Figure 2.6). Since the thermal conductivity of the longitudinal phonons have no significant magnitude, the full thermal conductivity and average phonon mean free path are minimal in the [111] direction.

Directly calculating the mean free paths confirms the qualitative speculations given above (see Figure 2.6). In the paper [20], the mean free paths of each vibrational mode are shown to reach their maximum values in focusing directions, exceeding the mean free paths of other vibrational modes. Their minimum values are recorded in defocusing directions and less than those predicted within the model of an isotropic medium. So, for example, in Si crystals, the mean free path of the slow transverse mode t_2 in the focusing direction exceeds by 1.6 and 2.4 times that of fast transverse and longitudinal modes, respectively.

Since the average density of states of the mode t_2 also increases due to the focusing effect, the contribution of slow transverse phonons, as a result of these two effects, to the thermal conductivity reaches 61% of the total thermal conductivity for the [100] direction at the temperature of $T = 3\text{K}$ and is twice the contribution of the fast transverse mode. Thus, the first McCurdy effect in Si samples is produced by the slow transverse mode, which focuses in the [001] direction and ensures maximum thermal conductivity in this direction.

Let us analyse the influence of focusing on the propagation of phonons in monocrystalline samples with a rectangular cross section and give a physical explanation of the second McCurdy effect. To start with, we consider samples with the {001} wide and {110} narrow faces. The main contribution to the boundary phonon scattering is made by scattering by wide faces of the sample (planes of films). Therefore, we dwell on the focusing and defocusing of phonons of different polarizations in the {110} plane perpendicular to the wide face of the sample and their deviation from the [110] temperature gradient direction (see Figure 3.12).

Phonons of the fast transverse mode are focused in the [110] direction; therefore, they will deviate from the wide face of the sample towards the temperature gradient direction (see Figure 3.12). Their mean free path is longer than that predicted within the model of an isotropic medium. Figure 3.12 outlines that, as heat flux moves along the [110] direction, the phonons of the t_1 mode with wave vectors in sector $-\theta_4 \le \theta \le \theta_4$ ($\theta_4 = 15.7^o$ for the t_1 mode in the {110} plane) deviate from the side faces towards the axis of the rod and travel in sector $-\theta_3 \le \theta \le \theta_3$, $\theta_3^{t1} = 0.86^o$ for Si. In the end, the density of states for the t_1 mode becomes much higher for the [110] direction and much lower for the defocusing direction than that for an isotropic medium (see Section 1.5). In this case, the phonons of the t_1 mode in the {110} plane and with the q_1 wave vector and θ_1 angle ($\theta_1^{t1} = 13.4^o$ for Si) propagate along the [110] heat flux direction (see Figure 3.12). At the same time, the phonons with the q_2 wave vector and θ_2 angle ($\theta_2^{t1} = 7.5^o$ for the t_1 mode) will propagate at the $-\theta_3$ angle ($\theta_3^{t1} = 0.86^o$) (see Figure 3.12). The [110] direction is a defocusing direction of the slow transverse mode, and its mean free path is shorter than in an isotropic medium (see Figure 3.12). Longitudinal phonons are focused in the [111] direction, and the [110] direction meets the defocusing direction for wave vectors in the {110} plane. Therefore, they deviate from the temperature gradient direction towards the wide face of the sample; their mean free path is less than in the model of an isotropic medium. As already noted earlier, the role of longitudinal phonons in phonon transport is small due to the low heat capacity.

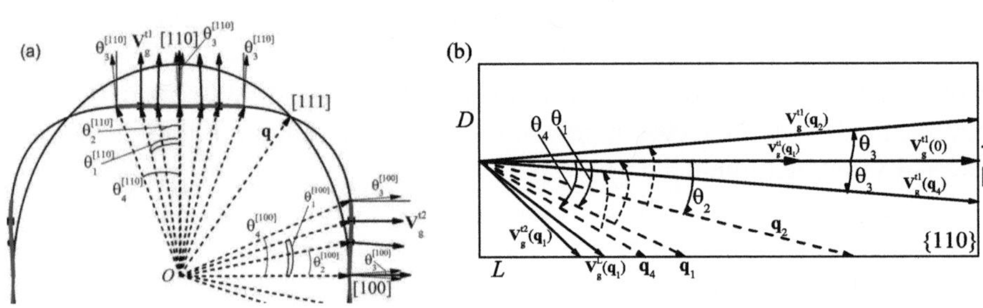

Figure 3.12: Schemes illustrating the effect of focusing on the phonon propagation in rectangular cross-section Si samples in the {110} plane and a temperature gradient along [110]. Phonons with wave vectors directed at the angles $\theta_1^{t1} = 13.4°$, $\theta_2^{t1} = 7.5°$, $\theta_3^{t1} = 0.86°$, and $\theta_4^{t1} = 15.7°$ are considered.

Another situation arises for samples with the {110} wide and {001} narrow faces. Similar to the previous case, we should investigate the phonon focusing in the plane perpendicular to a wide face, that is in the {001} plane. In it, the fast transverse mode is isotropic, there is no focusing effect, and its mean free path coincides with that obtained within the model of an isotropic medium. In this plane, as well as in the {110} plane, the slow transverse mode t_2 is defocused in the [110] direction. Therefore, the phonons of this mode deviate from the temperature gradient direction towards a wide face, and their mean free path, as in the previous case, is shorter than that in the model of an isotropic medium (see Figure 3.12). Longitudinal phonons in this plane have local focusing in the [110] direction and they deviate from the wide face of the sample to the temperature gradient direction. However, as noted above, their role in thermal conductivity is small. Obviously, in the case under consideration, the phonon mean free path and thermal conductivity for samples with the {001} wide face are greater than those for samples with the {110} wide face. The direct calculation yields an effect magnitude of 31%. The authors of [19] succeeded in choosing the [110] direction as a correct direction to set heat flux, since, in this direction, the slow transverse mode is defocused in both the {001} and {110} planes. In contrast, in the [110] direction, the fast transverse mode is focused in the {110} plane, and there is no focusing effect in the {001} plane. If one takes a rectangular cross-section Si sample of the same geometric parameters as in [19] but cuts it out with an axis coinciding with the [100] direction, the thermal conductivity of the sample with the wide {100} face is only 5% higher than that of the sample with the {110} wide face. So, from the above analysis, it follows that the second McCurdy effect [19] is generated by focusing of the fast transverse mode in the [110] direction for wave vectors in the diagonal plane.

3.6 Conclusion

The main results of the third chapter can be formulated in the following way:

1. A method of accounting for phonon focusing in calculating the thermal conductivity of monocrystalline samples has been proposed.
2. It has been shown that the phonon relaxation times calculated in this section for diffuse scattering at the boundaries of finite-length samples with square-shaped and rectangular cross-sections allow one to adequately describe the experimental data on the thermal conductivity of silicon crystals for different directions of the temperature gradient and orientations of the side faces of the samples over the entire temperature range studied. To date, when analysing the temperature dependencies of thermal conductivity for the relaxation rates of phonons at the boundaries, fitting parameters have been always introduced to take into account the influence of geometric parameters and orientations of the side faces of the samples, as well as the directions of heat flux on the thermal conductivity of monocrystalline samples. An analytical solution to the problem of the Knudsen

phonon gas flow in finite-length samples allows getting rid of fitting for boundary phonon scattering. It is shown that, in the temperature range from 3 K to 15 K, when scattering at the boundaries and by isotopic disorder dominates, the newly developed theory quantitatively describes, within the experimental error and without fitting parameters, the following aspects:
   - The temperature dependencies of the thermal conductivity of silicon crystals for all directions of heat flux in square-shaped cross-section samples;
   - The dependence on the orientation of the side faces for rectangular cross-section samples.
3. Anisotropy of the thermal conductivity of square-shaped cross-section silicon samples is shown to be mainly due to the contribution of slow transverse phonons, which are focused in the [001] directions and provide maximum thermal conductivity in this direction.
4. A physical interpretation of McCurdy effects in the thermal conductivity of cubic crystals has been given. An analysis of the degree to which focusing affects the phonon propagation in silicon single crystals shows that the first McCurdy effect in square-shaped cross-section samples is caused by focusing of a slow transverse mode, and the second McCurdy effect in rectangular cross-section samples is due to focusing of a fast transverse mode.

Chapter 4
Thermal Conductivity of monocrystalline Nanostructures with Various Types of Elastic Anisotropy Energy at Low Temperatures

As we have already seen in the previous chapter, the anisotropy of elastic properties in bulk silicon samples leads to the dependence of the thermal conductivity and phonon mean free paths on the direction of heat flux and the orientation of the side faces of the samples [20, 21]. Therefore, it is of great interest to study the influences of focusing on phonon transport in films and nanowires with various types of anisotropy of elastic energy. The present section deals with the case of low temperatures, when the thermal resistance of nanostructures is due to diffuse phonon scattering at the boundaries. An analysis performed in [4–7, 23, 24] shows that the thermal conductivity follows the $\kappa(T) \sim T^3$ dependence for nanowires with diameters greater than 50 nm and films with thicknesses greater than 20 nm in the temperature range from 20 K to 50 K. An analogous situation can be observed for heat capacity of bulk samples in the Debye theory. Therefore, under the indicated restrictions, the influence of spatial confinement on the spectrum of acoustic modes can be neglected and the anisotropic continuum model can be utilized to analyse phonon transport in such nanostructures.

This chapter discusses phonon transport in monocrystalline films and nanowires based on cubic crystals with various types of anisotropy of elastic energy. Let us look into nanostructures based on cubic crystals with positive ($k-1>0$) (GaN, GaAs, LiF Ge, MgO, Si, diamond, and YAG) and negative ($k-1<0$) (NaCl, CaF2, SrF2, and YIG) anisotropy of the second-order elastic moduli (see Table 1.2). At the beginning, the change in thermal conductivity magnitudes with decreasing parameter $|k-1|$ should be investigated for nanostructures of type-I crystals: from GaN ($k-1=1.275$) to YAG ($k-1=0.03$), as well as for type-II nanostructures: from NaCl ($k-1=-0.48$) to YIG ($k-1=-0.04$) (see Table 1.2). Further, we compare these data with those calculated within the model of an isotropic medium for which $k-1=0$ and demonstrate that the phonon mean free paths tend to the values offered by the model of an isotropic medium as the $|k-1|$ parameter decreases [27].

There are many important problems to be solved for the technical use of nanostructures in microelectronics. Among these are investigation of the dependencies of thermal conductivity and phonon mean free paths on the geometric parameters of nanostructures, directions of heat flux, and orientation of the film planes relative to the crystal axes. An analysis of the influence of the elastic energy anisotropy on the propagation of phonons in films and nanowires makes it possible to determine optimal orientations of the film planes and directions of heat flux, ensuring maximum and minimum thermal conductivity of the nanostructures.

https://doi.org/10.1515/9783110670509-005

The main objective of this section is to examine the anisotropy of thermal conductivity in the regime of the Knudsen phonon gas flow and its dependencies on the geometric parameters and directions of heat flux in elastically anisotropic nanostructures. Note that, in exploring the thermal conductivity in nanostructures, the papers [1–3] do not cover important issues such as the influence of elastic anisotropy of cubic crystals on the dependencies of phonon mean free paths on the geometric parameters of films, as well as on the heat flux directions and the orientation of the film planes relative to the axes. The solution of these problems is expected to identify the orientations of the film planes and the heat flux directions providing the maximum or minimum heat removal from the elements of semiconductor microcircuits. These challenges are relevant for semiconductor films widely used in microelectronics [1–3, 5–7] and are the subject of study in this section.

Section 4.1 presents equations for mean free paths of phonons in monocrystalline nanostructures. Section 4.2 analyses the dependencies of the phonon mean free paths on the geometric parameters in nanostructures with different types of anisotropy of elastic energy. Section 4.3 investigates the influence of focusing on the density of states and the phonon mean free paths in nanowires with various types of the elastic energy anisotropy. Section 4.4 discusses the change in the thermal conductivity anisotropy upon transition from nanowires to square-shaped films with different plane orientations. The content of this chapter is based on the papers [27, 71, 70,109, 117].

4.1 Phonon Mean Free Paths in monocrystalline Nanostructures

In the pages of this section, we will analyse phonon transport in rectangular cross-section films. Let L, D, and $W = \mu D$ be the length, thickness, and width, respectively. For square-shaped cross-section nanowires, the μ parameter is equal to unity ($\mu = 1$). We study the thermal conductivity of nanostructures in the regime of boundary scattering when the thermal resistance is due to diffuse scattering of phonons at the boundaries. As a preliminary, we take a temperature range much lower than Debye temperature ($T \ll T_{\mathrm{D}}$) when the anisotropic continuum model is applicable. The expression $\kappa^{\{J\}}_{[I(\psi)]}(T) = (1/3)C_V(T)\,\bar{S}\,\Lambda^{\{J\}}_{[I(\psi)]}$ for thermal conductivity implies that the anisotropy of thermal conductivity is governed by the mean free path of phonons. This is because the specific heat capacity C_V and the average phonon velocity $\bar{S}$ are independent on the direction of heat flux. With diffuse phonon boundary scattering, the average phonon mean free paths Λ, as well as the mean free paths Λ^{λ} of phonons with polarization λ in films, are found in [27]. Bearing in mind the phonon focusing for the heat flux direction $[I(\psi)]$ and orientation $\{J\}$ of a wide face of the film relative to the crystal axes, we can represent them in the form:

$$\Lambda_{[I(\psi)]}^{\{J\}} = \frac{3}{4\pi} \cdot \frac{1}{\sum\limits_{\lambda 1} \left\langle (S^{\lambda 1})^{-2} \right\rangle} \sum_{\lambda} \int d\Omega_q \frac{I_{[I(\psi)]}^{\lambda\{J\}}(\theta,\varphi)}{(S^{\lambda}(\theta,\varphi))^3}, \quad \Lambda_{[I(\psi)]}^{\lambda\{J\}} = \frac{3}{4\pi} \cdot \frac{1}{\left\langle (S^{\lambda})^{-2} \right\rangle} \int d\Omega_q \frac{I_{[I(\psi)]}^{\lambda\{J\}}(\theta,\varphi)}{(S^{\lambda}(\theta,\varphi))^3}, \tag{4.1}$$

When the inequalities $\mu|V_{g1}^{\lambda}| > |V_{g2}^{\lambda}|$ and $|V_{g3}^{\lambda}/V_{g1}^{\lambda}| \geq k_0 = L/2D$ or $\mu|V_{g1}^{\lambda}| < |V_{g2}^{\lambda}|$ and $|V_{g3}^{\lambda}/V_{g2}^{\lambda}| \geq k_0/\mu$ hold, the relaxation functions $I_{[I(\psi)]}^{\lambda\{J\}}(\theta,\varphi)$for rectangular cross-section films are defined by expressions [27]:

$$I_{[I(\psi)]}^{\lambda\{J\}}(\theta,\varphi) = \frac{Dk_0 \left|V_{g3}^{\lambda}\right|}{\mu} \left\{ \mu - \frac{k_0}{2} \frac{\left(\left|V_{g2}^{\lambda}\right| + \mu\left|V_{g1}^{\lambda}\right|\right)}{\left|V_{g3}^{\lambda}\right|} + \frac{(k_0)^2}{3} \frac{\left|V_{g1}^{\lambda}\right|\left|V_{g2}^{\lambda}\right|}{\left(V_{g3}^{\lambda}\right)^2} \right\}. \tag{4.2}$$

If $\mu|V_{g1}^{\lambda}| > |V_{g2}^{\lambda}|$ and $|V_{g3}^{\lambda}/V_{g1}^{\lambda}| < k_0$ or $\mu|V_{g1}^{\lambda}| < |V_{g2}^{\lambda}|$ and $|V_{g3}^{\lambda}/V_{g2}^{\lambda}| < k_0/\mu$ are satisfied, the relaxation functions $I_{[I(\psi)]}^{\lambda\{J\}}(\theta,\varphi)$ can be computed through the expressions for infinite-length samples:

$$\begin{aligned} I_{[I(\psi)]}^{\lambda\{J\}}(\theta,\varphi) &= \frac{D}{6\mu} \left[\left(\frac{V_{g3}^{\lambda}}{V_{g1}^{\lambda}} \right)^2 \left(3\mu\left|V_{g1}^{\lambda}\right| - \left|V_{g2}^{\lambda}\right| \right) \right], \quad \mu\left|V_{g1}^{\lambda}\right| > \left|V_{g2}^{\lambda}\right|, \\ I_{[I(\psi)]}^{\lambda\{J\}}(\theta,\varphi) &= \mu\frac{D}{6} \left[\left(\frac{V_{g3}^{\lambda}}{V_{g2}^{\lambda}} \right)^2 \left(3\left|V_{g2}^{\lambda}\right| - \mu\left|V_{g1}^{\lambda}\right| \right) \right], \quad \mu\left|V_{g1}^{\lambda}\right| < \left|V_{g2}^{\lambda}\right|. \end{aligned} \tag{4.3}$$

Here, $V_{g3}^{\lambda}(\theta,\varphi)$, $V_{g1}^{\lambda}(\theta,\varphi)$, and $V_{g2}^{\lambda}(\theta,\varphi)$ are the projections of the group velocity on the temperature gradient direction and directions perpendicular to it in a coordinate system related to the film. In this coordinate system, we set axis 3 along the heat flux direction $[I(\psi)]$. Next, we put axis 1 (the axis of rotation) to be perpendicular to film plane; it is responsible for the $\{J\}$ orientation of the plane. Axis 2 should be pointing perpendicular to two narrow side film faces. The ψ angle is measured from the Z-axis coinciding with the cube edge. To account for the influence of phonon focusing on the thermal conductivity anisotropy of monocrystalline films, the orientation parameters should be expressed through the group velocity components.

Unlike Chapter 3, the present chapter deals with the dependencies of phonon mean free paths on geometric and the orientation parameters when heat flux rotates in the film plane. The subject of the investigation is three cases listed below:

1. The film plane coincides with the $\{J\} = \{100\}$ plane of the YZ cube face;
2. The film plane coincides with the $\{J\} = \{110\}$ diagonal plane;
3. The film plane is perpendicular to the $\{J\} = \{111\}$ cube diagonal line.

The latter issue measures the ψ angle from the $[11\bar{2}]$ direction. For the cases in question, the phonon group velocity components $V_{g3}^{\lambda}(\theta,\varphi)$, $V_{g1}^{\lambda}(\theta,\varphi)$, and $V_{g2}^{\lambda}(\theta,\varphi)$ can be represented as follows [27]:

$$(1)\quad V_{g3}^{\lambda} = -V_{gy}^{\lambda}\sin\psi + V_{gz}^{\lambda}\cos\psi,\quad V_{g2}^{\lambda} = V_{gy}^{\lambda}\cos\psi + V_{gz}^{\lambda}\sin\psi,\quad V_{g1}^{\lambda} = V_{gx}^{\lambda},$$

$$(2)\quad V_{g3}^{\lambda} = \left(V_{gx}^{\lambda} + V_{gy}^{\lambda}\right)\sin\psi/\sqrt{2} + V_{gz}^{\lambda}\cos\psi,\quad V_{g2}^{\lambda} = \left(V_{gx}^{\lambda} + V_{gy}^{\lambda}\right)\cos\psi/\sqrt{2} - V_{gz}^{\lambda}\sin\psi,$$
$$V_{g1}^{\lambda} = \left(V_{gx}^{\lambda} - V_{gy}^{\lambda}\right)/\sqrt{2},$$

$$(3)\quad V_{g3}^{\lambda} = \left(V_{gx}^{\lambda} - V_{gy}^{\lambda}\right)\sin\psi/\sqrt{2} + \sqrt{\frac{2}{3}}\left(-\left(V_{gx}^{\lambda} + V_{gy}^{\lambda}\right)/2 + V_{gz}^{\lambda}\right)\cos\psi,$$
$$V_{g2}^{\lambda} = \left(-V_{gx}^{\lambda} + V_{gy}^{\lambda}\right)\cos\psi/\sqrt{2} + \sqrt{\frac{2}{3}}\left(-\left(V_{gx}^{\lambda} + V_{gy}^{\lambda}\right)/2 + V_{gz}^{\lambda}\right)\sin\psi, \tag{4.4}$$
$$V_{g1}^{\lambda} = \left(V_{gx}^{\lambda} + V_{gy}^{\lambda} + V_{gz}^{\lambda}\right)/\sqrt{3}.$$

The dependence of the heat flux direction on the ψ angle is governed by the $V_{g3}^{\lambda}(\theta,\varphi)$ group velocity component. The $V_{g1}^{\lambda}(\theta,\varphi)$ projection of the group velocity is independent on the ψ angle because it is the axis of rotation. This projection identifies the $\{J\}$ film plane. So, it can be inferred that formulas (4.1)–(4.4) elucidate the features of phonon transport in square-shaped cross-section nanofilms and nanowires at low temperatures. The applicability of the anisotropic continuum model is the necessary condition of legitimacy for such an approach to explore phonon transport in nanostructures. As already noted, for 50 nm diameter nanowires and 20 nm thick films, the influence of the spatial confinement on the spectrum of acoustic modes in nanostructures can be neglected. In this case, to study phonon transport, the anisotropic continuum model is utilized. Calculations of the temperature dependencies of thermal conductivity of Si films and nanowires confirm the conclusion above.

In isotropic media, the phonon phase velocities S^{λ} are independent on the angles θ and φ. Therefore, from formulas (4.1)–(4.4), it follows that the mean free paths of phonons of different polarization are equal to each other and average mean free path:

$$\Lambda_{iso}^{L} = \Lambda_{iso}^{t} = \Lambda_{iso} = D\frac{3}{4\pi}\cdot\int d\Omega_q\tilde{I}(\theta,\varphi),\quad \tilde{I}(\theta,\varphi) = \frac{I^{\lambda}(\theta,\varphi)}{DS^{\lambda}},\quad \tilde{\Lambda}_{iso} = \Lambda_{iso}/D,$$

$$\tilde{I}(\theta,\varphi) = k_0|n_3|\left\{1 - \frac{k_0}{2\mu}\frac{|n_2| + \mu|n_1|}{|n_3|} + \frac{(k_0)^2}{3\mu}\frac{|n_1||n_2|}{(n_3)^2}\right\},\ \text{if}\ \left[\begin{array}{l}(\mu|n_1| > |n_2|\text{and}|n_3/n_1| \ge k_0)\ or\\ (\mu|n_1| < |n_2|\text{and}|n_3/n_2| \ge k_0/\mu)\end{array}\right., \tag{4.5}$$

$$\tilde{I}(\theta,\varphi) = \begin{cases}(n_3/n_1)^2(3\mu|n_1| - |n_2|)/6\mu, & if\ \mu|n_1| > |n_2|\text{and}|n_3/n_1| < k_0,\\ (n_3/n_2)^2(3|n_2| - \mu|n_1|)\mu/6, & if\ \mu|n_1| < |n_2|\text{and}|n_3/n_2| < k_0/\mu.\end{cases}$$

Here, $n_1, n_2,$ and n_3 are the components of a unit wave vector $\mathbf{n}$. Expressions (4.5) imply that the phonon mean free paths in isotropic media do not depend on elastic moduli but are completely given by geometric parameters of samples. However, for elastically anisotropic crystals, they are different for phonons of different polarization and dependent on the direction of heat flux in the crystals. Expression (4.5) yields Smoluchowski's result [107] for the mean free path of molecules during the flow of a very rarefied gas through an infinite rectangular cross-section ($a \times b$) tube:

$$\Lambda = \frac{3}{4}(ab)^{1/2}\left\{\left(\frac{a}{b}\right)^{1/2}\ln\left[\frac{b}{a}+\sqrt{1+\left(\frac{b}{a}\right)^2}\right]+\left(\frac{b}{a}\right)^{1/2}\ln\left[\frac{a}{b}+\sqrt{1+\left(\frac{a}{b}\right)^2}\right]\right.$$
$$\left.+\frac{1}{3}\left[\left(\left(\frac{a}{b}\right)^{3/2}+\left(\frac{b}{a}\right)^{3/2}\right)-\left(\frac{a}{b}+\frac{b}{a}\right)^{3/2}\right]\right\}. \tag{4.6}$$

From formula (4.6) for the flow of the Knudsen phonon gas throughout an infinite rectangular cross-section rod, we can obtain the outcome of McCurdy et al. [19] for the Casimir length:

$$\Lambda_C = \frac{3D}{4}\left\{\ln\left[\mu+\sqrt{1+\mu^2}\right]+\mu\ln\left[\mu^{-1}+\sqrt{1+\mu^{-2}}\right]+\frac{1}{3}\left[\left(\mu^{-1}+\mu^2\right)-\mu^{1/2}\left(\mu^{-1}+\mu\right)^{3/2}\right]\right\}. \tag{4.7}$$

For the thermal conductivity of a cylinder-shaped infinite-length rod with a diameter d in the model of an isotropic medium, the result of Casimir follows [13]: $\Lambda_C = d$, which is consistent with that of Knudsen [12] for a rarefied molecule gas flowing throughout an infinite-length circular cross-section tube. The lack of divergence of the data is not accidental. In his monograph [8], J.M. Ziman has stressed: "This argument (and, indeed, the whole theory of boundary scattering) is clearly the same as the discussion of the flow of a very rarefied gas in a tube (Knudsen 1909, Smoluchowski, 1910)". Therefore, the boundary scattering regime is called the Knudsen phonon gas flow [8, 9, 35], and the phonon mean free path of infinite-length samples is the Casimir length.

4.2 The Influence of Geometric Parameters on Anisotropy of Phonon Mean Free Paths in Films and Nanowires

Let us consider the features of phonon transport in single-crystal films in the regime of the Knudsen phonon gas flow. Earlier, to estimate the thermal conductivity and mean free paths of phonons in monocrystalline films of cubic symmetry for diffuse boundary scattering, the model of an isotropic medium and Casimir theory were, as a rule, used [see, for example, 25, 26, 108]. Other papers (see, e.g. [5–7, 28, 29]) apply the model of an isotropic medium to account for boundary phonon scattering in sufficiently thin films ($D \ll L$, W) in the same way as it was done in the works of Fuchs [30] and

Sondheimer [31] in analysing the conductivity of thin metal films. It was assumed that the mean free paths and relaxation times of phonons in the films depend only on its thickness [5–7, 28, 29]. This conclusion based on Casimir's results and choice of dimensions, as well as the attempt to determine the phonon mean free paths in nanofilms and nanowires at low temperatures, as was done in [108] are incorrect. The authors of [108] suggested that the phonon mean free path in a film depends only on its thickness, and a cylindrical rod explored by Casimir is a one-dimensional system. Since the Casimir length for a cylindrical rod is $\Lambda_C = 1 \cdot D$ (where D is the rod diameter), and a film is a two-dimensional structure, the authors of [108] postulated that the Casimir length is twice the thickness, $\Lambda_C = 2 \cdot D$ (where D is the film thickness) for heat flux flowing along the film. In other words, it depends neither on the width nor length of the film. These results underlie the analysis of the temperature dependencies of thermal conductivity in silicon nanofilms and nanowires in [108], relying on the additivity of the contributions of boundary and bulk scattering mechanisms.

When examining the thermal conductivity of thin infinite-length films within the isotropic continuum model, H. J. Maris and S. Tamura have pointed out the fallacy of this result [25]. It was shown that the Casimir length values significantly depend not only on the film width but, moreover, diverge logarithmically as it tends to infinity. Earlier, this data divergence for films exploring in the model of an isotropic medium was mentioned in the article [32]. In fact, the finding of [25, 32] is elicited from formula (4.6):

$$\tilde{\Lambda}_C(\mu) = \Lambda_C(\mu)/D \cong A + B\ln\mu\,, \; \mu \gg 1, \quad A \approx 0.895\,, \quad B = 0.75\,, \tag{4.8}$$

where $\Lambda_C(\mu)$ is the Casimir length for films with a thickness D and width μD. As was underscored in [25, 32], the logarithmic divergence is due to phonons travelling almost parallel to the film plane. The papers [25, 32] leave aside the issue about the influence of a finite length of the film on the logarithmic divergence of the Casimir lengths, as well as other important challenges such as the influence of elastic anisotropy of cubic crystals on the phonon free path dependencies on geometric parameters of films and directions of heat flux. These problems are still relevant to manufacture semiconductive films for microelectronics [1–3].

4.2.1 Casimir Lengths in Elastically Anisotropic Nanostructures

A numerical analysis of thermal conductivity using expressions (4.1)–(4.3) showed that the Casimir lengths for monocrystalline films also diverge logarithmically when the film width tends to infinity [27]. However, their values for phonons of different polarizations, in contrast to an isotropic medium, differ significantly (see Figure 4.1). In this figure, curve 5 corresponds to the result of Maris and Tamura [25] for the model of an isotropic medium.

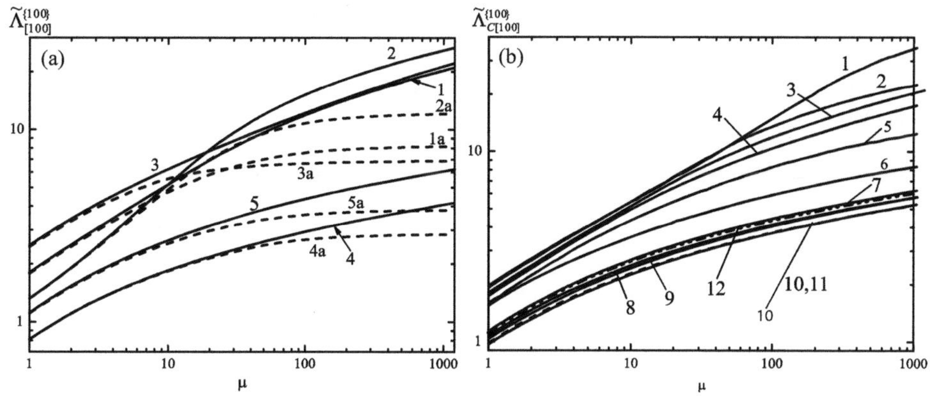

Figure 4.1: (a) The dependencies of the reduced phonon mean free paths $\tilde{\Lambda}^{\{100\}}_{[100]}(\mu) = \Lambda^{\{100\}}_{[100]}(\mu)/D$ (curves 1 and 1a), as well as mean free paths $\tilde{\Lambda}^{\lambda\{100\}}_{[100]}(\mu)$ of phonons of different polarizations on the reduced width in germanium films [27]. Solid curves (1, 2, 3, 4, and 5) refer to the Casimir lengths in nanostructure with parameters $W = \mu D$, $D = 50$ nm. Dashed curves (1a, 2a, 3a, 4a, and 5a) refer to the mean free path of phonons in nanostructure with parameters $L = 100\,D$, $D = 50$ nm. Curves 2 and 2a are $\tilde{\Lambda}^{t1\{100\}}_{[100]}(\mu)$, curves 3 and 3a are $\tilde{\Lambda}^{t2\{100\}}_{[100]}(\mu)$, curves 4 and 4a are $\tilde{\Lambda}^{L\{100\}}_{[100]}(\mu)$. Curves 5 and 5a refer to the model of an isotropic medium; curve 5 is Maris' and Tamura's results [25] for the Casimir length. (b) The dependencies of the Casimir lengths $\tilde{\Lambda}^{\{100\}}_{C[100]}(\mu)$ on the reduced width in nanostructures with parameters $W = \mu D$, $D = 50$ nm. Curves: 1 – MgO, 2 – Si, 3 – Ge, 4 – GaAs, 5 – GaN, 6 – diamond, 7 – YAG, 8 – YIG, 9 – NaCl, 10 – CaF_2, 11 – SrF_2, and 12 is for the model of an isotropic medium [25].

As can be seen in Figure 4.1a, the divergence of the mean free path values for both isotropic and monocrystalline films can be eliminated through accounting for their finite length. For a fixed length of $L = 100\,D$ and a film thickness of $D = 50$ nm, the interval of an intensive increase in the phonon mean free paths with increasing the film width $W = \mu D$ is limited by its length. For the values of $W > 10L$, the dependencies of the mean free paths $\tilde{\Lambda}^{\{J\}}_{[I(\psi)]}(\mu)$ reach saturation (see Figure 4.1). From Figure 4.1b it is understood that, for monocrystalline films with the $\{J\} = \{100\}$ orientation and $[I] = [100]$ temperature gradient direction, made of type-I crystals and type-II crystals, the Casimir lengths in the entire range of the μ parameters are greater and lesser, respectively, than those calculated within the model of an isotropic medium.

For square-shaped cross-section nanowires, the Casimir lengths are chiefly controlled by the phonon focusing effect and weakly depend on the orientation of the side faces. Table 4.1 illustrates that they reach their maximum values in focusing directions, exceeding the Casimir lengths of other vibrational modes, as well as the value of $\tilde{\Lambda}_{Ciso} \cong 1.115$ for an isotropic medium. So, for example, for Ge: $\tilde{\Lambda}^{L}_{C[111]} = 1.48$, $\tilde{\Lambda}^{t_1}_{C[110]} = 1.85$, and $\tilde{\Lambda}^{t_2}_{C[100]} = 2.49$. The Casimir lengths become minimal in defocusing directions and turn out to be less than those for isotropic media. For example, for Ge: $\tilde{\Lambda}^{L}_{C[100]} = 0.81$, $\tilde{\Lambda}^{t_1}_{C[111]} = 0.7$, and $\tilde{\Lambda}^{t_2}_{C[110]} = 0.89$.

Table 4.1: The reduced Casimir lengths $\tilde{\Lambda}_c^{\lambda}[I(\psi)]$ and $\tilde{\Lambda}_{c[I(\psi)]}$ in square-shaped cross-section rods for $[I(\psi)]$ symmetrical directions [27].

Compound	k-1	$[I(\psi)]$	$\tilde{\Lambda}^{L}_{C[I(\psi)]}$	$\tilde{\Lambda}^{t1}_{C[I(\psi)]}$	$\tilde{\Lambda}^{t2}_{C[I(\psi)]}$	$\tilde{\Lambda}_{c[I(\psi)]}$	$\tilde{\Lambda}_{c[I(\psi)]}$ [19]
GaN	1.28	[100]	0.76	1.39	1.92	1.59	–
		[110]	1.30	1.46	0.94	1.16	–
		[111]	1.67	0.78	1.12	1.07	–
GaAs	0.90	[100]	0.80	1.34	2.29	1.74	–
		[110]	1.25	1.73	0.89	1.25	–
		[111]	1.52	0.77	1.20	1.09	–
Ge	0.87	[100]	0.81	1.32	2.49	1.80	–
		[110]	1.24	1.85	0.89	1.29	–
		[111]	1.48	0.77	1.24	1.11	–
LiF	0.78	[100]	0.82	1.30	2.64	1.86	1.80
		[110]	1.22	1.93	0.89	1.33	1.28
		[111]	1.44	0.78	1.27	1.11	1.10
MgO	0.70	[100]	0.85	1.25	2.92	1.96	–
		[110]	1.21	2.04	0.90	1.38	–
		[111]	1.40	0.79	1.30	1.13	–
Si	0.67	[100]	0.84	1.30	2.75	1.92	1.86
		[110]	1.22	1.97	0.89	1.34	1.29
		[111]	1.42	0.78	1.26	1.10	1.09
Diamond	0.40	[100]	0.93	1.18	2.11	1.53	1.55
		[110]	1.17	1.32	0.96	1.14	1.11
		[111]	1.27	0.86	1.40	1.17	1.13
YAG	0.03	[100]	1.09	1.12	1.16	1.13	–
		[110]	1.12	1.13	1.08	1.11	–
		[111]	1.13	1.04	1.25	1.15	–
YIG	−0.04	[100]	1.14	1.00	1.10	1.06	–
		[110]	1.10	1.18	1.09	1.13	–
		[111]	1.09	1.05	1.49	1.25	–

Table 4.1 (continued)

Compound	k-1	[I(ψ)]	$\tilde{\Lambda}^{L}_{C[I(\psi)]}$	$\tilde{\Lambda}^{t1}_{C[I(\psi)]}$	$\tilde{\Lambda}^{t2}_{C[I(\psi)]}$	$\tilde{\Lambda}_{C[I(\psi)]}$	$\tilde{\Lambda}_{C[I(\psi)]}$ [19]
SrF_2	−0.20	[100]	1.35	0.82	1.04	0.99	–
		[110]	1.07	2.33	1.00	1.55	–
		[111]	1.00	1.06	1.98	1.48	–
CaF_2	−0.33	[100]	1.59	0.78	0.95	0.97	0.98
		[110]	1.04	1.86	0.98	1.33	1.30
		[111]	0.92	1.13	1.86	1.46	1.45
NaCl	−0.48	[100]	1.91	0.81	0.94	1.03	1.05
		[110]	1.01	1.72	0.99	1.26	1.23
		[111]	0.86	1.19	1.77	1.42	1.43

When going over to nanowires from type-II crystals, the directions of focusing and defocusing of phonons and, accordingly, the minimum and maximum values of the Casimir lengths change places. For example, for NaCl, the minimum values of $\tilde{\Lambda}^{L}_{C[111]} = 0.86$, $\tilde{\Lambda}^{t_1}_{C[100]} = 0.81$, and $\tilde{\Lambda}^{t_2}_{C[100]} = 0.94$ take place for those directions where the maximum values are for Ge and vice versa (see Table 4.1). It should be noted that, for type-I and type-II crystals, the t_1 and t_2 transverse modes are mutually permuted as well [44]. It is worth stressing that, for nanowires with significant anisotropy of elastic energy (GaAs, Ge, LiF, and Si), the minimum and maximum values of the Casimir lengths differ significantly from the value of $\tilde{\Lambda}_{Ciso}$. However, for YAG and YIG crystals with weak anisotropy of elastic energy ($|k-1| < 1$), they are close to the values of $\tilde{\Lambda}_{Ciso}$ (see Table 4.1). Moreover, for each mode, depending on the direction, they can have values both larger and smaller than that of $\tilde{\Lambda}_{Ciso}$. As can be seen from Table 4.1, our results for the Casimir lengths in symmetrical directions are consistent with those obtained in [19] within the error of 1%–4%. For isotropic media, the mean free paths in the boundary scattering regime are completely determined by the geometric parameters of nanostructures. Therefore, they can be used as a convenient comparison system for the Casimir lengths in elastically anisotropic crystals with changing the directions of heat flux or for collating the Casimir lengths in nanostructures fabricated from various materials (see Figures 4.1 and 4.2).

The conducted analysis showed that the Casimir lengths for monocrystalline films are dictated not only by their geometric parameters but markedly depend on the directions of heat flux and orientation of the film plane, that is to say, $\tilde{\Lambda}_C(\mu) \Rightarrow \tilde{\Lambda}^{\{J\}}_{C[I]}(\mu)$. In the limiting case $\mu \gg 1$, the Casimir lengths $\tilde{\Lambda}^{\{J\}}_{C[I]}(\mu)$ can be represented in the form (4.7). In doing so, the A and B coefficients turn into the dependent ones on the orientation parameters $[I(\psi)]$ and $\{J\}$.

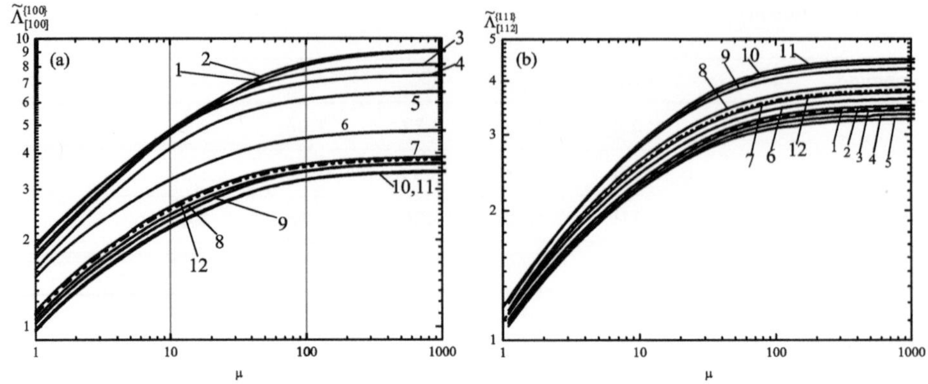

Figure 4.2: The dependencies of the average mean free paths $\tilde{\Lambda}_{[100]}^{\{100\}}(\mu)$ of films with a length of $L = 100\ D$, $D = 50$ nm on the μ parameter: (a) – for the {100} orientation of the film plane and the [100] direction of the temperature gradient; (b) – for the {111} orientation of the film plane and the $[11\bar{2}]$ direction of the temperature gradient Curves: 1 – MgO, 2 – Si, 3 – Ge, 4 – GaAs, 5 – GaN, 6 – diamond, 7 – YAG, 8 – YIG, 9 – NaCl, 10 – CaF_2, and 11 – SrF_2, 12 is for the model of an isotropic medium.

4.2.2 Dependencies of Mean Free Paths of Phonons on Geometric Parameters in Nanostructures with Different Type of Anisotropy of Elastic Energy

Let us examine the dependencies of thermal conductivity and mean free paths of phonons in films made of type-I and type-II crystals on the geometric parameters for various values of the $k-1$ anisotropy coefficient and orientations $\{J\}$ of the planes. For this purpose, we set a certain direction of the temperature gradient and plot the thermal conductivity dependencies on a reduced width of the film, $\mu = W/D$ and $L/D = 100$ $(k_0 = 50)$, for various orientations of the film planes $\{J\}$. Estimates showed that for square-shaped cross-section nanowires $(\mu = 1)$, the average mean free paths for the $\{J\} = \{100\}$ and $\{J\} = \{110\}$ orientations of the side faces differ from each other less than 1.3%. Therefore, their dependencies on thermal conductivity are weak and can be neglected. However, for sufficiently wide monocrystalline films, the phonon mean free path values are chiefly governed by the film orientations. For all the materials, the mean free paths in monocrystalline films increase with increasing their width. As seen in Figure 4.2a, for the $\{J\} = \{100\}$ orientation and $[I] = [100]$ temperature gradient direction, the mean free paths in films made of the type-I and type-II crystals are larger and lesser, respectively, than those for isotropic media. In this case, we have the possibility of tracing how the phonon mean free paths change depending on the values of the $k-1$ anisotropy parameter.

It should be emphasized that the dependence of the phonon mean free paths on the $k-1$ anisotropy parameter is not monotonically increasing, except for small values of this parameter. So, for example, when transitioning from an isotropic medium $(k-1=0)$ to YAG crystals $(k-1=0.03)$ and diamond$(k-1=0.4)$, the mean free path values rise, as the film width remains unchanged (see Figure 4.2a). It would seem that the mean free paths for films made of Si ($k-1=0.67$) and MgO ($k-1=0.69$) crystals should further follow. However, in fact, the next line is occupied by the mean free path values for films made of GaN crystals ($k-1=1.275$) with a dramatically larger anisotropy parameter. They are the maximum ones among the materials in question. Beneath them, the mean free path values for Ge ($k-1=0.87$) and GaAs ($k-1=0.9$) are stationed, whose anisotropy parameters are noticeably greater than those for Si (Figure 4.2a). Thus, such a hierarchy of the mean free paths in the films is caused by the influence of phonon focusing on the density of states and mean free paths of quasi-transverse modes, with latter mainly contributing to the thermal conductivity of the films and, accordingly, to the average mean free paths.

The lack of negative curvature regions in the isoenergetic surfaces for transverse modes is inherent in crystals with small elastic energy anisotropy, diamond, YAG crystals, YIG crystals, and isotropic media. Therefore, changes in the phonon densities of states relatively are scarce and can be described using the enhancement factor (see Section 1.6). For the rest of the type-I crystals with large values of the $k-1$ parameter, negative curvature regions form on the isoenergetic surfaces. In this case, the sector of phonon focusing is assigned by the $2\theta_3$ angles; it collects all states with wave vectors inside the sector $2\theta_4$ (see Section 1.5). The lesser the $2\theta_3$ angle, the more the density of states makes itself felt in the phonon focusing region (see Figure 1.18). The smallest values of the θ_3 angles and the utmost changes in the density of states are observed for Si and MgO with a relatively small value of the $k-1$ parameter (see Table 1.6). For these crystals, the values of the θ_3 angles responsible for the region of phonon focusing increase, and the magnitudes of the density of states diminish as the $k-1$ parameter grows. This results in decreasing the contribution of the mode at hand to the thermal conductivity of the film, and, accordingly, decreasing the mean free path values when going over from Ge ($k-1=0.87$) to GaN ($k-1=1.275$) crystals (see Figure 4.2a). For the type-II crystals, curves of the dependencies of mean free paths on the film width are located beneath the curve that refers to the model of an isotropic medium. For YIG-based films with $k-1=-0.48$, the dependencies of mean free paths are close to those for the model of an isotropic medium and NaCl-based films ($k-1=-0.48$). For films made of CaF_2 crystals and SrF_2, curves 10 and 11 are almost indistinguishable and situated noticeably beneath the curve for the model of an isotropic medium.

From Figure 4.2b, it becomes clear that, for the $\{J\}=\{111\}$ orientation of the film plane and $[I]=[11\bar{2}]$ temperature gradient direction, the scenario concerning the mean free paths in type-I and type-II crystals is profoundly different compared to the case of the {100} orientation of the film plane. For type-II and type-I crystals, they appear to be larger and lesser, respectively, than those in isotropic media. The mean free path values reach their maximum for films made of CaF_2 with the {111} orientation and $[11\bar{2}]$ heat flux direction and become minimum for films fabricated from GaN with the maximum anisotropy parameter among the type-I crystals at hand. Moreover, the given case demonstrates the curve sequence reverse to their values of the $k-1$ anisotropy parameter. The uppermost curve 7 adjacent to the dependence for the model of an isotropic medium pertains to YAG-films; the lowermost curve 5 refers to GaN films.

Let us look into the dependencies of phonon mean free paths in films with the {100} and {111} orientations on the reduced length $k_0 = L/2D$ and compare them for nanowires ($\mu = 1$) and sufficiently wide films($\mu = 100$). As seen in Figure 4.3a, with increasing the nanostructures' length, the magnitudes of the thermal conductivity and phonon mean free paths in films with the {100} orientation and [100] heat flux direction rise faster the larger the value of the μparameter is (see Figure 4.3). For the film lengths exceeding their width by two orders of magnitude, the dependencies of the phonon mean free paths attain saturation. It should be underscored that, for YAG and YIG crystals with weak elastic energy anisotropy ($|k-1| \ll 1$), the dependencies of the mean free paths $\tilde{\Lambda}_{[I]}^{\{J\}}(k_0, \mu)$ on k_0 differ a little from each other and the model of an isotropic medium (see Figure 4.3).

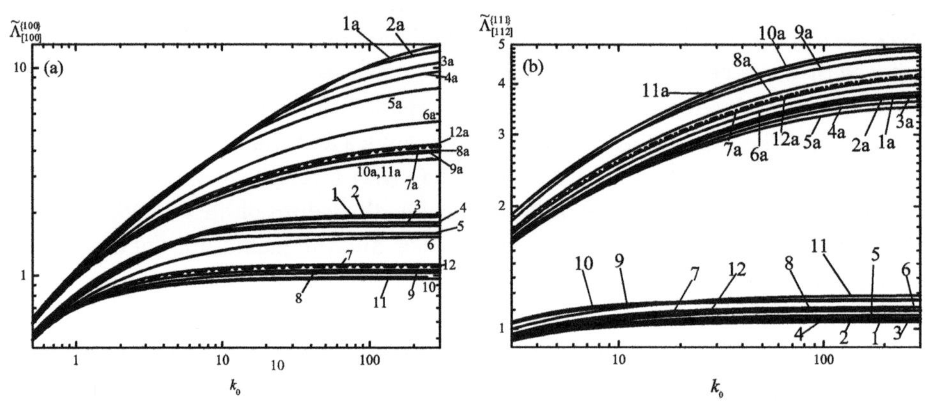

Figure 4.3: The dependencies of the reduced mean free paths on the parameter $k_0 = L/2\,D$ for nanostructures with a thickness of $D = 50$ nm, based on crystals MgO – (1,1a), Si – (2,2a), Ge – (3,3a), GaAs – (4,4a), GaN – (5,5a), diamond – (6,6a), YAG – (7,7a), YIG – (8,8a), NaCl – (9,9a), CaF_2 – (10,10a), SrF_2 – (11,11a), an isotropic medium – (12,12a). Curves 1–12 refer to nanowires ($\mu = 1$), curves 1a–12a refer to nanofilms ($\mu = 100$) for cases: (a) the {100} orientation of the film plane and the temperature gradient is pointing along the [100] direction; (b) the {111} orientation of the film plane and the temperature gradient is pointing along the $[11\bar{2}]$ direction.

It is worth drawing attention to the fact that, as for the {100} and {111] orientations of the films and nanowires, the k_0- dependencies of $\tilde{\Lambda}_{[I]}^{\{J\}}(k_0, \mu)$ are qualitatively different from each other. For both nanofilms and nanowires with the {100} orientation plane and [100] temperature gradient direction, the mean free path values for type-I crystals (GaN, GaAs, Ge, MgO, diamond, YAG) and type-II crystals (NaCl, CaF_2, SrF_2, YIG) turn out to be larger and lesser, respectively, than those for isotropic media (see Figure 4.3).

Moreover, the sequence of curves (the dependencies of the mean free paths on the reduced film length) depicted in Figure 4.3a coincides with the sequence of curves (the dependencies of the mean free paths on the reduced film width) in Figure 4.2a. Among the materials in question, both dependencies demonstrate the maximum values of the mean free paths for films made of type-I Si and MgO crystals, whereas the minimum values refer to films made of type-II CaF_2 crystals.

Figure 4.3b clarifies that, for the films with the $\{J\} = \{111\}$ plane orientations and $[11\bar{2}]$ temperature gradient direction, the sequence of the mean free paths becomes considerably reverse relative to the films with the {100} orientations: for type-II and type-I crystals, they are larger and lesser, respectively, than those for isotropic media. Analogously to the above dependencies of mean free paths on the film width (see Figure 4.2b), the maximum values of the mean free paths on the reduced length for the films with the {111} and $[11\bar{2}]$ heat flux direction are characteristic of type-I CaF_2 crystals, whereas GaN films exhibit the minimum values. The sequence of curves pictured in Figures 4.3b on the reduced film length coincides with the sequence of curves pictured in Figure 4.2b for the dependencies of mean free paths on the reduced film width.

4.2.3 Dependencies of Thermal Conductivity Anisotropy on the Lengths of monocrystalline Films

Let us analyse the change in anisotropy of phonon transport in the film plane, depending on its length for various compounds. For this, we set the film width as $W = 100\ D$ ($\mu = 100$, $D = 50$ nm) and plot the dependencies of the phonon mean free paths on the reduced length $k_0 = L/2D$ for the directions of heat flux to provide focusing and defocusing of the acoustic modes. It was shown in Section 2.3 (see also [20]) that the mean free paths in bulk materials reach their maximum and minimum values in the phonon focusing and defocusing directions, respectively. Therefore, if the mean free paths calculated for films with a specific ratio of geometric parameters coincide for these directions, their dependencies are expected to be isotropic in the film plane for this region of parameters. If they differ, the mean free paths in the film plane will be accordingly anisotropic. It should be stressed that in films made of type-I crystals and with the {100} orientation, the slow transverse mode is focused and defocused in the [100] and [110] directions; on the contrary, for films fabricated from type-II crystals, the fast transverse mode is defocused and focused in these directions.

For the fast and slow transverse modes in films with the {111} plane orientation, local maxima and minima of focusing are realized in the $[11\bar{2}]$ and [110] directions. Therefore, the dependencies of the mean free paths on the reduced length, $\tilde{\Lambda}^{\lambda\{J\}}_{[I(\psi)]}(k_0) = \Lambda^{\lambda\{J\}}_{[I(\psi)]}(k_0)/D$, normalized to the film thickness, should be computed for these directions (see [70]). Figure 4.4 show that the mean free path dependencies for different vibrational modes in films with the {100} and {111} orientations for all materials are increasing functions of the sample length. With an increase in the film length of $L > 100W$, they attain saturation. For short films with the {100} and {111} orientations, the length of which is less than or equal to its width ($L \le W$), the dependencies of the mean free paths in the directions of focusing and defocusing of phonons for all acoustic modes differ insignificantly. In any case, when $k_0 = 50$ ($L = W$), these dependencies intersect, and it should be expected that the angular dependencies of the thermal conductivity are isotropic for square-shaped films with the {100} and {111} orientations. As to short films with $L < W$, their angular dependencies of the mean free paths and thermal conductivity are expected to be almost isotropic.

However, an increase in the length of films with the {100} and {111} orientations when $L > W(k_0 > 50)$ leads to the fact that the mean free paths in focusing directions become noticeably larger than those in phonon defocusing directions. In this case, the anisotropy of the mean free paths lengths increases, and when $L > 10^2 W$ it reaches its maximum values (see Figure 4.4). So, for example, the

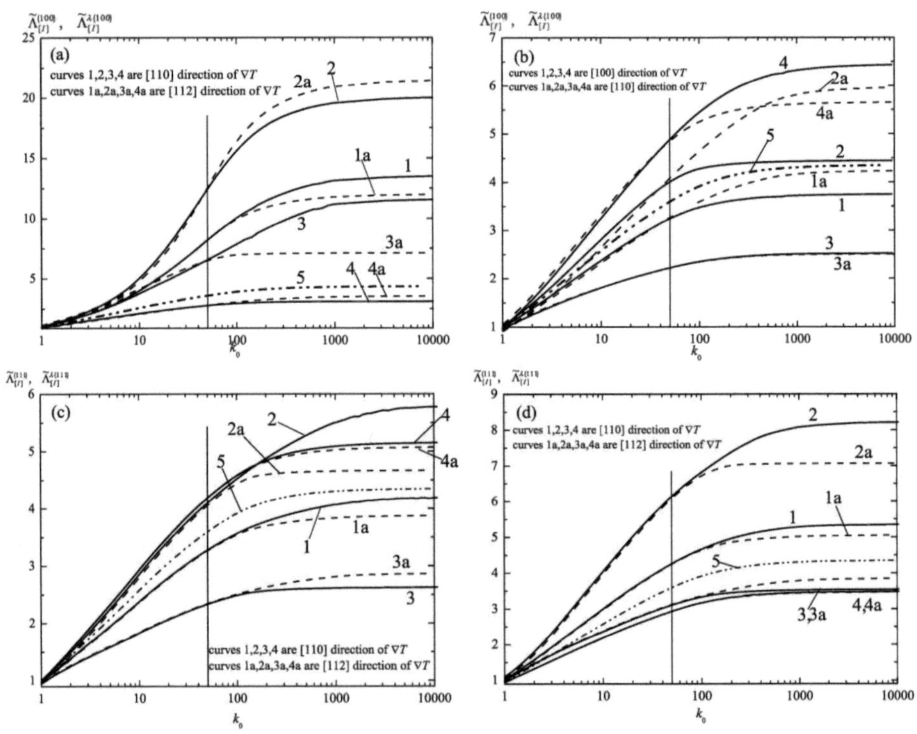

Figure 4.4: The dependencies of the reduced mean free paths $\tilde{\Lambda}_{[I]}^{\{100\}}(k_0)$ (curves 1 and 1a), as well as the mean free paths $\tilde{\Lambda}_{[I]}^{\lambda\{100\}}(k_0)$of phonons of different polarizations in the Si (a, c) and CaF_2 (b, d) films with the {100} (a, b) and {111} (c, d) orientation of the film plane on the reduced length k_0 [30]. Curves (2, 2a) refer to a fast transverse mode, curves (3, 3a) refer to the t_2 mode, curves (4, 4a) are for longitudinal phonons, and curve 5 are for the model of an isotropic medium.

mean free paths of the slow transverse mode for Ge, Si films with the {100} orientation, as well as of the fast transverse mode for CaF_2 films turn out to be 1.65, 1.62, and 1.34 times larger in the focusing direction than those in the defocusing direction, respectively. As to type-I crystals, the anisotropy of the mean free paths of fast transverse modes in films with the {100} orientation when $L > 100W$ is small: it ranges about 2%–3% for Ge and GaAs crystals and rises to 9% for InSb (see Table 4.3). An exception is MgO films: the mean free paths of the t_1 mode in the [100] directions are 1.33 times smaller than those in the [110] direction (see Table 4.2). For the t_1 mode travelling in CaF_2 and NaCl films, the situation with anisotropy of the mean free paths $\Lambda_{[I]}^{t_1\{100\}}$ is similar to one with MgO films: the mean free paths in the [110] direction turn out to be 1.34 and 1.30 times longer than those in the [100] direction (see Table 4.2). When going over to the films with the {111} plane, the anisotropy of the full thermal conductivity drops, not exceeding 10%. This result is caused by focusing and defocusing of the slow and fast transverse modes. The latter make a dominant contribution to the full thermal conductivity and average mean free path. The fact of the matter is that the fast transverse mode is focused and defocused in the [110] direction. In the $[11\bar{2}]$ direction, it has a local maximum of focusing. Therefore, due to the mutual compensation of these contributions, the resulting anisotropy of the full thermal conductivity is small. For GaAs-, InSb-, LiF-, diamond-, CaF_2-, and NaCl-based films, the dependencies of the phonon mean free paths on geometric parameters are similar to those calculated for Si and CaF_2 films (see Table 4.2). When $L \le W$, these dependencies for each vibrational mode in the directions of focusing and defocusing of phonons almost coincide. When $L > 100W$, they become anisotropic. The anisotropy parameters of the mean free paths for the studied films are given in Table 4.2.

Table 4.2: Anisotropy of mean free paths in films with the parameters $L = 100W$, $W = 100D$, and $D = 50nm$.

$\Lambda_{[I_1]}^{\lambda\{J\}}/\Lambda_{[I_2]}^{\lambda\{J\}}$	GaAs	Ge	InSb	GaN	LiF	MgO	Si	diamond	CaF_2	NaCl
$\Lambda_{[110]}^{L\{100\}}/\Lambda_{[100]}^{L\{100\}}$	1.18	1.16	1.18	1.21	1.15	1.13	1.13	1.08	0.88	0.82
$\Lambda_{[110]}^{L\{111\}}/\Lambda_{[11\bar{2}]}^{L\{111\}}$	1.04	1.02	1.03	1.05	1.02	1.01	1.02	1.00	1.00	1.02
	1.02	1.03	1.09	1.04	1.06	1.33	1.07	1.01	1.34	1.30
$\Lambda_{[110]}^{t1\{111\}}/\Lambda_{[11\bar{2}]}^{t1\{111\}}$	1.27	1.23	1.18	1.15	1.27	1.16	1.27	1.09	1.16	1.12
$\Lambda_{[100]}^{t2\{100\}}/\Lambda_{[110]}^{t2\{100\}}$	1.52	1.65	1.33	1.22	1.67	1.67	1.62	1.24	1.01	1.01
$\Lambda_{[11\bar{2}]}^{t2\{111\}}/\Lambda_{[110]}^{t2\{111\}}$	1.13	1.09	1.09	1.05	1.09	1.11	1.09	1.07	1.09	1.07
$\Lambda_{[100]}^{\{100\}}/\Lambda_{[110]}^{\{100\}}$	1.21	1.21	1.16	1.13	1.16	1.05	1.12	1.08	0.89	0.93
$\Lambda_{[110]}^{\{111\}}/\Lambda_{[11\bar{2}]}^{\{111\}}$	1.08	1.08	1.06	1.05	1.09	1.05	1.1	1.02	1.06	1.04

4.3 The Influence of Focusing on the Density of Phonon States and Phonon Mean Free Paths in Nanowires with Different Types of Elastic Energy Anisotropy

Let us consider phonon transport in monocrystalline nanowires at low temperatures when the thermal resistance is due to diffuse phonon scattering at the boundaries. Since the anisotropy of the mean free paths and density of phonon states are governed by focusing of phonons, we examine the correlation of these quantities for transverse modes in nanowires with different types of anisotropy of elastic energy. For this, we calculate the angular dependencies of the phonon mean free paths for cases when heat flux rotates in the YZ $\{J\} = \{100\}$ plane of the cube face or in the $\{J\} = \{110\}$ diagonal plane. The next step is to compare the predicted angular dependencies with the angular dependencies of the densities of phonon states for transverse modes [71, 109]. As can be seen in Figure 4.5, the angular dependencies of the mean free paths and densities of phonon states for fast and slow transverse modes in the {100} and {110} planes correlate well with each other: the regions of maximum and minimum values of both quantities for all crystals are determined by phonon focusing and defocusing sectors, respectively (see Section 1.5). Moreover, their maximum values are limited by the $\pm\theta_3^{\lambda\{J\}}$angles relative to the focusing direction. These angles are responsible for the directions of group velocities at zero curvature points in the isoenergetic surface. For thermal conductivity anisotropy and phonon mean free paths of Ge-, Si-, GaSb-, LiF-, GaAs-, and GaN-based (type-I crystals) nanowires, we have obtained close results. The phonon mean free paths take the maximum and minimum values in the phonon focusing and defocusing sectors, respectively. Moreover, their values for each vibrational mode in the focusing regions exceed those both of other vibrational modes and of deduced within the model of an isotropic medium (see dotted curves 5 in Figure 4.5). At the same time, in the defocusing sectors, their values turn out to be less than those for the model of an isotropic medium. In all nanowires made of crystals of the first type, the maximum values of the thermal conductivity are achieved in or close to the [100] directions and provided by a slow transverse mode. So, for example, for wave vectors in the cube face plane, the slow transverse mode is focused in the [100] directions, and its mean free path for Ge and GaAs is 3.0 and 2.8 times longer than that for longitudinal phonons and 2.2 and 2.0 times larger than that in the model of an isotropic medium. For nanostructures with large anisotropy parameter values (GaN, Ge, GaSb, LiF, and GaAs), the mean free paths, as well as the density of phonon states (DPS) of the quasi-transverse mode t_2, reach their maxima at the angles $\psi = n\pi/2 \pm \theta$. The latter regulate the directions of group velocities at zero curvature points. In the [100] focusing directions, the mean free paths have a local minimum. This minimum is most clearly manifested for GaN nanowires with a maximum

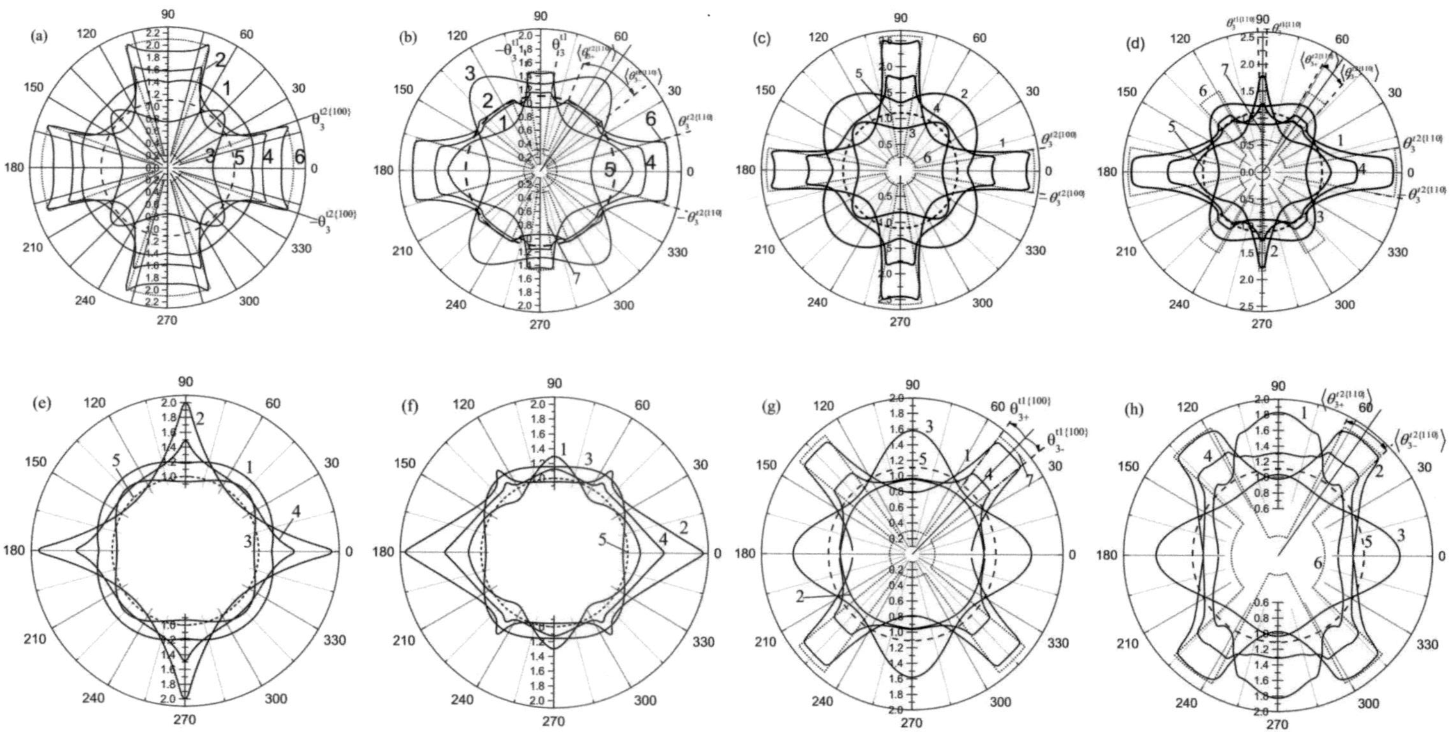

Figure 4.5: The angular dependencies of the reduced phonon mean free paths $\tilde{\Lambda}^{\lambda\{J\}}_{[I(\psi)]} = \Lambda^{\lambda\{J\}}_{[I(\psi)]}/D$ and $\tilde{\Lambda}^{\{J\}}_{[I(\psi)]} = \Lambda^{\{J\}}_{[I(\psi)]}/D$ in GaN (a,b), Ge (c,d), diamond (e,f), and CaF_2 (g,h) for square-shaped cross-section samples with a length of $L = 100\ D$ and $D = 50$ nm in cases when the temperature gradient rotates in the cube face plane (a, c, e, g) and in the diagonal plane (b, d, e, h). 1 – a fast transverse mode, 2 – a slow transverse mode, 3 – a longitudinal mode, 4 – the average mean free path [71]. Dashed curves 5 refer to an isotropic medium ($\Lambda_{iso} = 1.11D$). Dotted lines 6 and 7 illustrate the angular distribution of the density of states for slow and fast transverse modes in the {100} and {110} planes.

value of the $k-1$ anisotropy parameter: the mean free path of the t_2 mode at the minimum is 11% shorter than at the $\pm\theta_3$ angles (see Figure 4.5). Against the aforementioned crystals, in diamond-based nanowires with a smaller value of the anisotropy parameter, the mean free paths for the slow mode reach a sharp maximum exactly in the [100] directions (see Figure 4.5). For diamond, the mean free path turns out to be 1.8 times longer than that for an isotropic medium and 1.7 and 2.2 times longer than that for the fast transverse mode and longitudinal phonons, respectively. Thus, for all crystals having regions of negative curvature in the isoenergetic surfaces, the angular dependencies of the DPS and mean free paths in nanowires for fast and slow transverse modes in the {100} and {110} planes correlate with each other: their maximum and minimum values are observed in ones and the same angle intervals. These angle ranges are assigned by the directions of group velocities at zero curvature points (see Figure 4.5).

The maximum anisotropy of the lattice thermal conductivity (71%) takes place for Si, and the minimum one (31%) is for diamond nanowires (see Table 4.3). For nanowires made of Ge, GaAs, GaS, LiF crystals, the maximum thermal conductivity is 60% higher than the minimum one in the [111] directions. However, for the most anisotropic material such as GaN, the anisotropy of thermal conductivity is noticeably less; it is 47%. The thermal conductivity maxima for all the materials are furnished by the slow quasi-transverse mode. Although the maximum and minimum values of the $k-1$ anisotropy parameter (among crystals with a region of negative curvature on the isoenergetic surface) are attributed to GaN and Si crystals, respectively, the maximum anisotropy of thermal conductivity, however, is observed for Si crystals. In Si, the isoenergetic surface for the t_2 mode has a minimum concavity and, accordingly, minimum values of the $\theta_3^{t_2\{100\}}$ angle (see Table 1.5). Therefore, the t_2 mode has the largest values of the ratios of the density of phonon states for the regions of phonon focusing and defocusing in the {100} plane, among the above-specified type-I crystals (see Table 1.5).

It should be stressed that the predominant contribution to the thermal conductivity of the nanostructures considered is provided by quasi-transverse modes (see Table 4.4).

For nanowires made of type-I crystals, the contribution of the quasi-transverse modes in the [100] heat flux direction amounts to 93%–94% of the full thermal conductivity, except for diamond nanowires – it diminishes to 88%. In this case, the contribution of the t_2 mode to the thermal conductivity of nanowires fabricated from type-I crystals with a region of negative curvature on the isoenergetic surface in the [100] focusing direction is 68% for Si and decreases to 64% in GaN. For nanowires made of diamond, this contribution goes down to 58% (see Table 4.4). As can be noticed from Table 4.4, the t_1 mode makes the dominant contribution to the thermal conductivity in the [110] directions, whereas the t_2 mode makes the dominant

Table 4.3: Ratios of mean free paths in symmetrical directions for square-shaped cross-section nanowires with a length of $L = 100\ D$ and $D = 50$ nm for Ge, Si, GaSb, LiF, GaAs, and diamond [71].

Ratios of mean free path	Ge	Si	GaSb	GaN	LiF	GaAs	Diamond
$\Lambda_{[100]}^{\{110\}}:\Lambda_{[110]}^{\{110\}}:\Lambda_{[111]}^{\{110\}}$	1.61:1.16:1	1.71:1.18:1	1.61:1.15:1	1.47:1.08:1	1.65:1.17:1	1.60:1.20:1	1.31:0.98:1
$\Lambda_{[100]}^{t2\{110\}}:\Lambda_{[111]}^{t2\{110\}}:\Lambda_{[110]}^{t2\{110\}}$	2.76:1.38:1	3.01:1.40:1	2.58:1.33:1	2.02:1.18:1	2.92:1.41:1	2.56:1.33:1	2.14:1.41:1
$\Lambda_{[100]}^{t1\{110\}}:\Lambda_{[111]}^{t1\{110\}}:\Lambda_{[110]}^{t1\{110\}}$	2.31:1.65:1	2.34:1.61:1	2.21:1.69:1	1.84:1.75:1	2.34:1.62:1	2.19:1.74:1	1.50:1.35:1
$\Lambda_{[111]}^{L\{110\}}:\Lambda_{[110]}^{L\{110\}}:\Lambda_{[100]}^{L\{110\}}$	1.79:1.51:1	1.65:1.43:1	1.83:1.53:1	2.16:1.70:1	1.72:1.47:1	1.86:1.55:1	1.34:1.24:1

Table 4.4: The contributions of vibrational modes to thermal conductivity of nanowires made of type-I crystals with a length of $L = 100\ D$ and $D = 50$ nm in symmetric directions (%) [71].

Temperature gradient direction	Si			Ge			GaSb			GaN			LiF			GaAs			Diamond		
	L	t_1	t_2	L	t_1	t_2	L	t_1	t_2	L	t_1	t_2	L	t_1	t_2	L	t_1	t_2	L	t_1	t_2
[100]	7	25	68	7	27	66	6	28	66	6	30	64	7	26	67	7	28	65	12	30	58
[110]	14	53	33	15	52	33	14	50	36	14	43	43	15	53	32	14	50	36	19	45	36
[111]	19	27	54	20	26	54	19	26	55	20	25	55	20	26	54	20	26	54	21	29	50

contribution to the thermal conductivity in the [001] directions. The contribution of this mode in the [111] directions exceeds 50%. The contribution of longitudinal phonons is small: for nanowires made of anisotropic crystals in the [100] direction of defocusing is 6%–7%, and in the [111] direction of focusing, it increases by 3 times and makes up 19%–20%. For diamond, it increases to 21%.

For nanowires made of type-II crystals, the ratios of mean free paths in symmetrical directions are given in Table 4.5. The thermal conductivity values reach their maximum in the [111] directions due to the slow quasi-transverse mode. The minimum values are ensured in the [100] direction. As can be seen in Figure 4.5d, the dependencies of the DPS and the mean free paths of the slow mode correlate well with each other.

Table 4.5: Ratios of mean free paths in symmetrical directions for nanowires made of type-II square-shaped cross-section crystals with a length of $L = 100\ D$ and $D = 50$ nm [71].

Ratios of mean free paths	SrF_2	CaF_2	PbS
$\Lambda_{[111]}^{\{110\}}:\Lambda_{[110]}^{\{110\}}:\Lambda_{[100]}^{\{110\}}$	1.47:1.44:1	1.49:1.35:1	1.39:1.20:1
$\Lambda_{[111]}^{t2\{110\}}:\Lambda_{[100]}^{t2\{110\}}:\Lambda_{[110]}^{t2\{110\}}$	1.93:1.04:1	1.88:1.00:1	1.71:0.95:1
$\Lambda_{[110]}^{t1\{110\}}:\Lambda_{[111]}^{t1\{110\}}:\Lambda_{[100]}^{t1\{110\}}$	2.45:1.30:1	2.35:1.46:1	1.99:1.53:1
$\Lambda_{[100]}^{L\{110\}}:\Lambda_{[110]}^{L\{110\}}:\Lambda_{[111]}^{L\{110\}}$	1.34:1.07:1	1.72:1.12:1	2.26:1.18:1

The maximum anisotropy of the lattice thermal conductivity of nanowires amounts to 49% for CaF_2, and the minimum is 39% for PbS (see Table 4.5). The fast transverse mode focusing itself in the [110] directions plays a smaller role in the thermal conductivity of nanostructures made of type-II crystals. The maximum values of the DPS and mean free paths in the angle intervals $[n\pi/4 - \theta_3^{t_1\{100\}}, n\pi/4 + \theta_3^{t_1\{100\}}]$ owe to this mode (see Figure 4.5). Its mean free path in nanowires made of CaF_2 and SrF_2 crystals is 1.8 and 1.9 times longer than that for longitudinal phonons and 1.7 and 1.8 times longer than that computed within the model of an isotropic medium.

For type-II crystals, the slow transverse mode makes the dominant contribution to thermal conductivity. In the [111] focusing direction, its contribution exceeds 60% for nanowires extracted from CaF_2 and SrF_2 crystals, and it decreases to 59% for PbS (see Table 4.6). In the [110] direction, the fast transverse mode makes the dominant contribution to the thermal conductivity.

Its contribution comes up to 58% for nanowires fabricated from SrF_2 crystals and decreases to 49% for PbS nanowires (see Table 4.6). In nanowires made of type-II crystals, longitudinal phonons are focused and defocused in the [100] and [111] directions, respectively. Their contribution to thermal conductivity in the [100] directions varies from 17% to 24%, and it declines to 8%–9% in the [111] directions

Table 4.6: Contributions of vibrational modes to thermal conductivity of SrF_2, CaF_2, and PbS square-shaped cross-section rods with a length of $L/D = 100$ in symmetrical directions, (%).

Temperature gradient direction	SrF_2			CaF_2			PbS		
	L	t_1	t_2	L	t_1	t_2	L	t_1	t_2
[100]	17	34	49	20	31	49	24	30	46
[110]	10	58	32	10	54	36	11	49	40
[111]	9	30	61	8	30	62	8	33	59

(see Table 4.6). Thus, a correlation has been established between the angular dependencies of the densities of phonon states (DPS) and the mean free paths of quasi-transverse modes. It has been shown that for all the nanowires, the regions of maxima of the mean free paths and DPS for fast and slow transverse modes in the {100} and {110} planes coincide and caused by the phonon focusing.

4.4 Change in Thermal Conductivity Anisotropy When Transitioning from Nanowires to Square-Shaped Films with Different Plane Orientation

The present section deals with the angular dependencies of thermal conductivity $k^{\{J\}}_{[I(\psi)]}(T)$ and average mean free paths $\tilde{\Lambda}^{\{J\}}_{[I(\psi)]}(\mu)$ of phonons in nanostructure as the temperature gradient rotates along the planes with different orientations. With a temperature fixed, the angular dependencies of the thermal conductivity and mean free paths of phonons differ only in a constant multiplier (see Section 4.1). Therefore, we first compare the changes in anisotropy of the average phonon mean free paths for square-shaped cross-section nanowires and sufficiently wide films. The subsequent step is to analyse the changes in anisotropy of the mean free paths for various orientations of the film planes and directions of heat flux and to match with the outcomes derived within the model of an isotropic medium. We also specify the orientations of the planes of monocrystalline films and heat flux directions, which provide the maximum and minimum values of the phonon mean free paths and, accordingly, the thermal conductivities.

Figures 4.6–4.8 indicate that the angular dependencies of the phonon mean free paths in monocrystalline nanostructures as the heat flux rotates along the {100}, {110}, and {111} planes differ dramatically both from each other and from those predicted by the model of an isotropic medium. For crystals of the same type, the angular dependencies of the mean free paths for all directions of the heat flux and the above orientations are qualitatively similar, distinguishing themselves only

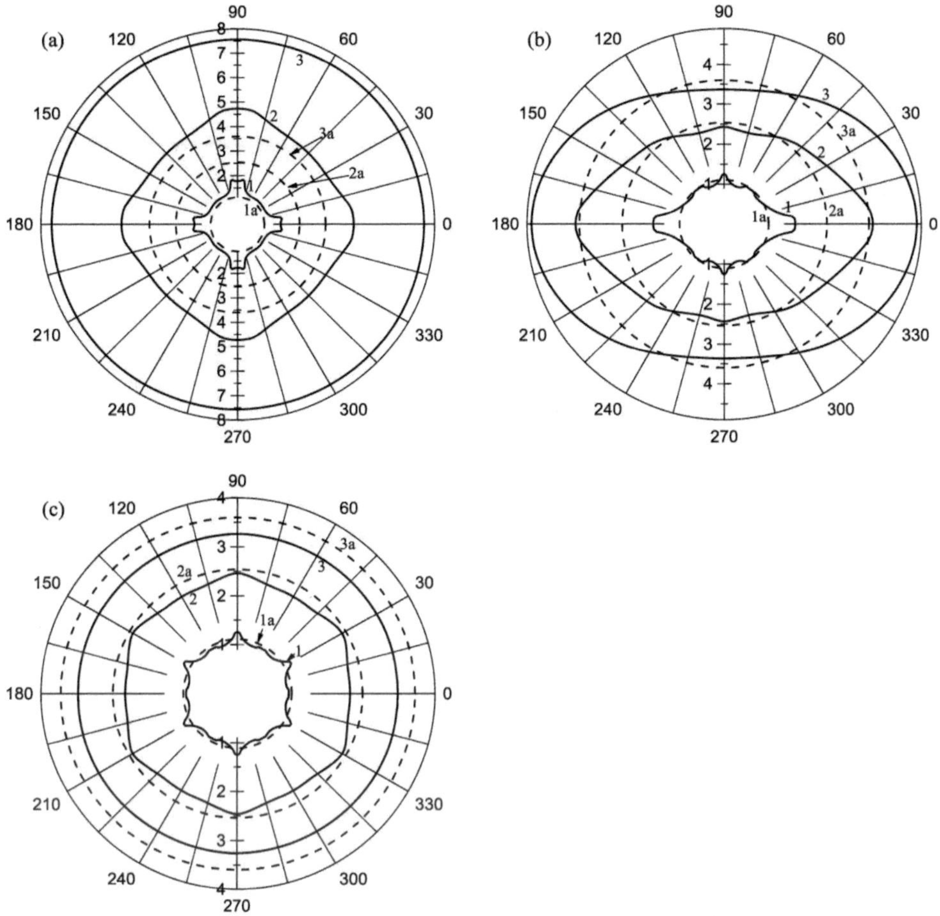

Figure 4.6: The angular dependencies of mean free paths normalized to the film thickness of $D = 50$ nm for nanostructures with a length of $L = 100\ D$, based on Ge crystals for cases where the temperature gradient rotates along the film plane: (a) – {100}, (b) – {110}, and (c) – {111}. The curves (1.1a), (2.2a), and (3.3a) are plotted for the parameter values $\mu = 1$, 10, and 100, respectively. The curves (1,2,3) illustrate the anisotropic continuum model, the dashed curves (1a,2a,3a) are for the model of an isotropic medium (see [27]).

by a higher or lesser degree of anisotropy. Therefore, it would be reasonable to plot the angular dependencies only for two type-I crystals (Ge and diamond) and a type-II crystal (CaF_2) (see Figures 4.6–4.8) [27].

Let us discuss anisotropy of phonon transport in nanowires when rotating the heat flux along the {100}, {110}, and {111} planes and compare with the results obtained for films. As already noted in the previous section, for type-I crystals (LiF, GaAs, Ge, Si, diamond, and YAG), the values of the nanowires' thermal conductivity reach their maximum in the [001] directions owing to a slow transverse mode t_2

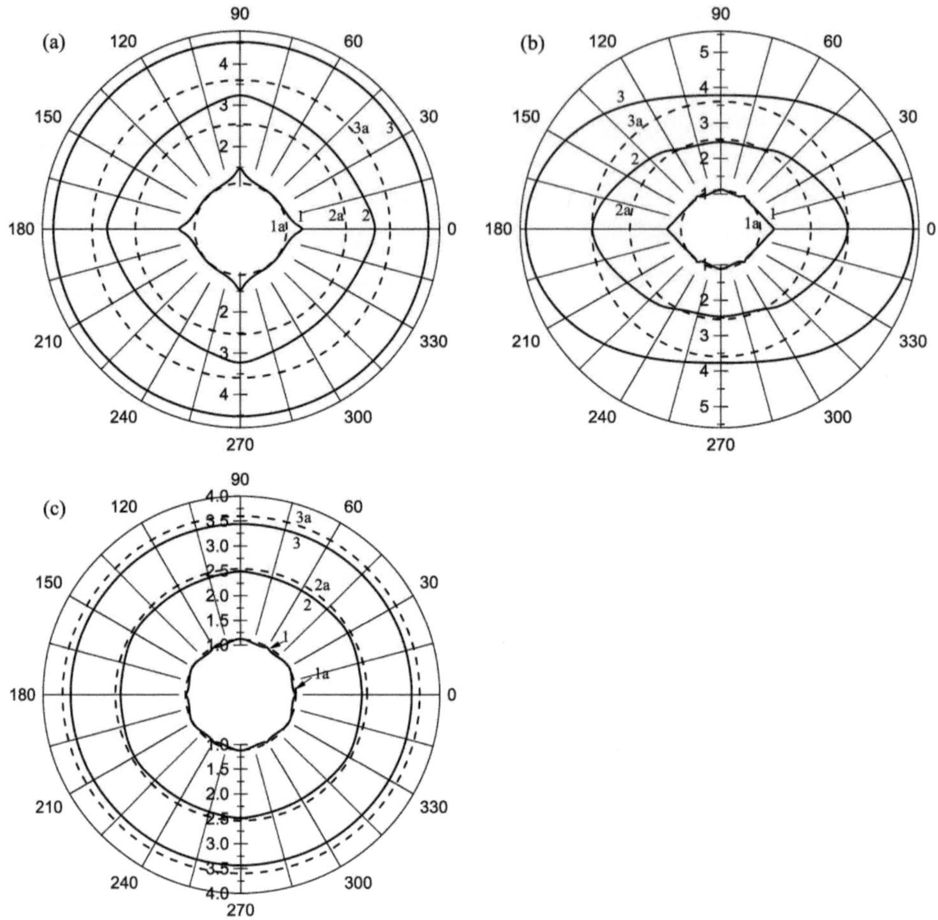

Figure 4.7: The same as in Figure 4.6 but for diamond nanostructures [27].

(see Figures 4.5 and 4.6). In the [110] and [111] directions, the angular dependencies of the thermal conductivity and average mean free paths have local maxima caused by focusing of fast transverse and longitudinal modes, respectively (see Figure 4.6b).

As the heat flux rotates along the {100} plane, the phonon mean free paths $\tilde{\Lambda}^{\{100\}}_{[I(\psi)]}(1)$ in nanowires made of type-I crystals for all directions are longer than $\tilde{\Lambda}_{iso}(1)$ for the model of an isotropic medium. However, in the {110} plane, the mean free paths $\tilde{\Lambda}^{\{100\}}_{[I(\psi)]}(1)$ exceed the values of $\tilde{\Lambda}_{iso}(1)$ only in the directions close to [100] and [110]. When the heat flux rotates in the {111} plane, the phonon mean free paths $\tilde{\Lambda}^{\{111\}}_{[I(\psi)]}(1)$ in a wide interval of the angles ψ turn out to be less than in an isotropic medium (see Figures 4.6c and 4.7c). Only in the directions close to [110], these values become larger than those for isotropic media.

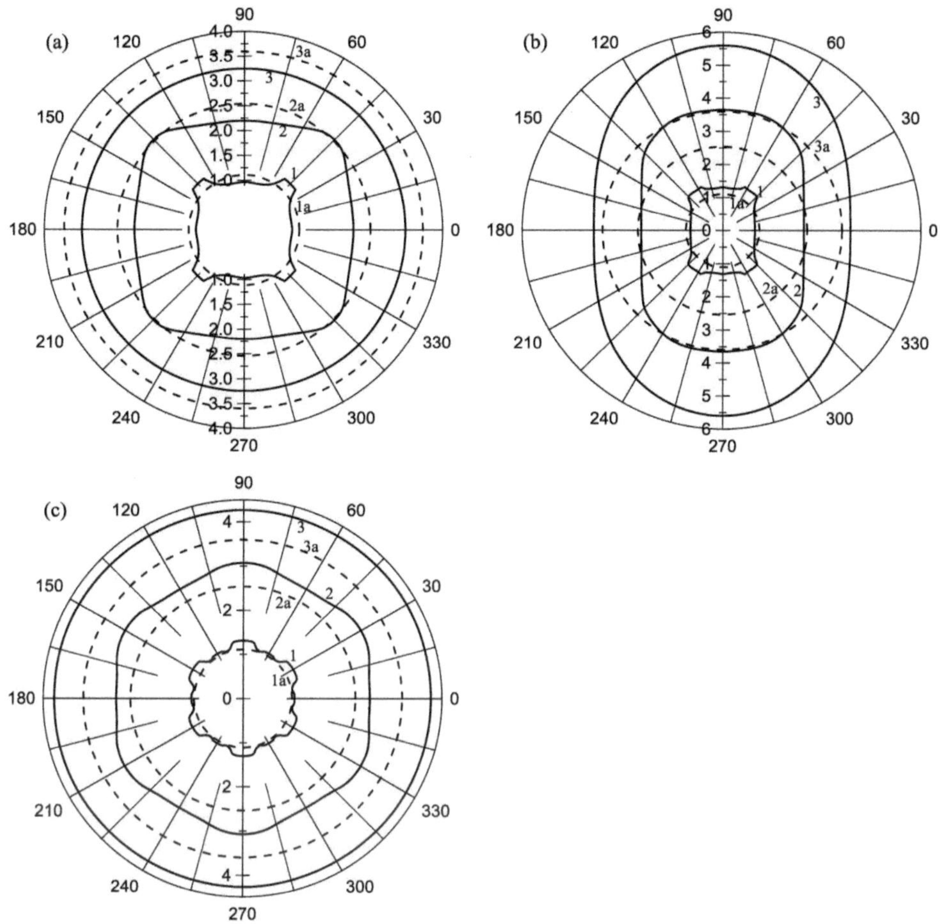

Figure 4.8: The same as in Figure 4.6 but for CaF_2-based nanostructures [27].

In contrast to the above, for nanowires made of type-II crystals (CaF_2, NaCl, and YIG), the maximum thermal conductivity values are attained in the [111] direction and provided by a slow transverse mode that is focused in this direction (see Figures 4.5 and 4.6). In the [110] and [100] directions, the angular dependencies and mean free paths $\tilde{\Lambda}^{\{J\}}_{[I(\psi)]}(1)$ have local maxima produced by focusing of fast transverse and longitudinal modes, respectively. With the heat flux rotating along the {100} plane, the phonon mean free paths $\tilde{\Lambda}^{\{100\}}_{[I(\psi)]}(1)$ for nanowires made of type-II crystals are less than $\tilde{\Lambda}_{iso}(1)$ for isotropic media, except the directions close to [110] (see Figure 4.6). The {111} plane contributes to an increase in the mean free paths $\tilde{\Lambda}^{\{111\}}_{[I(\psi)]}(1)$ inside a wide-angle range compared to those in isotropic media, except the directions close to [110].

As is seen in Figures 4.6–4.8, rising the film width causes an increase in the mean free paths and qualitatively changing in their angular dependencies,

$\tilde{\Lambda}^{\{J\}}_{[I(\psi)]}(\mu)$, compared with nanowires. For both types of crystals, they become almost isotropic in the {100} and {111} planes when $\mu = 10$, although small anisotropy takes place (see Figures 4.6a, 4.6c, 4.7a, 4.8c). Moreover, for films oriented along the {100} plane, made of type-I crystals, the mean free paths $\tilde{\Lambda}^{\{100\}}_{[I(\psi)]}(100)$ are larger than those for isotropic media. However, for the {111} orientation, the situation alters: for all directions, the mean free paths $\tilde{\Lambda}^{\{100\}}_{[I(\psi)]}(100)$ are less than those for isotropic media. For films made of type-II crystals, the angular dependencies of the mean free paths $\tilde{\Lambda}^{\{J\}}_{[I(\psi)]}(\mu)$ are qualitatively different. For them, the reverse inequalities hold: for the {100} orientation when $\mu = 100$, the mean free paths $\tilde{\Lambda}^{\{100\}}_{[I(\psi)]}(100)$ in the entire ψ-angle range are less and their magnitudes for the {111} orientation $\tilde{\Lambda}^{\{100\}}_{[I(\psi)]}(100)$ are larger compared with isotropic media. For sufficient wide films oriented along the {110} plane, the angular dependencies $\tilde{\Lambda}^{\{J\}}_{[I(\psi)]}(\mu)$ take an ellipsoidal shape. For type-I crystals, the long axis of the ellipsoid lies along the [100] direction. For type-II crystals, it is observed along the [100] direction (see Figures 4.6b–4.8b). For LiF, GaAs, Ge, Si, diamond, and YAG type-I crystals when $\mu = 100$, the ratios of the mean free paths, $\tilde{\Lambda}^{\{110\}}_{[100]}(\mu)/\tilde{\Lambda}^{\{110\}}_{[110]}(\mu)$, in films with the {110} plane are 1.45, 1.39, 1.42, 1.50, 1.44, and 1.03, respectively.

Their mean free paths in the {110} plane turn out to be less than those in the {100} plane at all ψ angles. However, in the directions close to [100], diamond films demonstrate larger values of the mean free paths $\tilde{\Lambda}^{\{100\}}_{[I(\psi)]}(100)$ than $\tilde{\Lambda}^{\{100\}}_{[I(\psi)]}(100)$ while the situation is reversed for the direction close to [110]. Moreover, as to diamond films with the {J} = {100} orientation, the mean free paths $\tilde{\Lambda}^{\{100\}}_{[I(\psi)]}(100)$ are larger than those for isotropic media at all angles ψ (see Figure 4.7b). By contrast to this, the inequality $\tilde{\Lambda}^{\{100\}}_{[I(\psi)]}(100) > \tilde{\Lambda}_{iso}(100)$ holds for LiF, GaAs, Ge, and Si crystals in the angle range of $-(\pi/3) < \psi < (\pi/3)$, whereas the inequality $\tilde{\Lambda}^{\{110\}}_{[I(\psi)]}(100) < \tilde{\Lambda}_{iso}(100)$ is fulfilled beyond the above interval. It should be underscored that the thermal conductivity values for LiF, GaAs, Ge, Si, and YAG (crystals of the first type) reach their maximum when the film is oriented along the {100} plane and $\mu = 100$. The minima values can be found along the {111} orientation (see Figure 4.6b). As to films made of CaF_2, NaCl, and YIG (crystals of the second type), the mean free paths have minimum values along the {100} plane. The maximum values of the thermal conductivity can be revealed when the film is set along the {110} plane and [110] heat flux direction, and when $\mu = 100$. In the process, the ratios of the mean free paths, $\tilde{\Lambda}^{\{110\}}_{[100]}(100)/\tilde{\Lambda}^{\{110\}}_{[110]}(100)$, amount to 0.69, 0.77, and 0.94, respectively.

So, we have shown: firstly, not only the spectrum and polarization vectors [44], but also anisotropy of mean free paths of phonons in nanostructures made of crystals with positive $(k-1>0)$ and negative $(k-1<0)$ anisotropy of second-order elastic moduli are qualitatively different. Secondly, the angular dependencies of the thermal conductivity and phonon mean free path for crystals of both types change qualitatively upon transition from nanowires to sufficiently wide films. Thirdly, the maximum values of the thermal conductivity of LiF-, GaAs-, Ge-, Si-, and YAG-based films with parameters $L = W = 100\,D$ and $D = 50$ nm are realized in the {100} orientation and the minimum ones are implemented in the {111} orientation. The thermal conductivity of diamond films when $\mu = 100$ reaches its maximum along the {110} plane and in the [100] direction, and the minimum values can be observed along the {111} plane. The maximum values of thermal conductivity in films with parameters $L = W = 100\ D$ and $D = 50$ nm, made of type-II CaF_2, NaCl, and YIG crystals when $\mu = 100$, can be recorded for the {110} plane orientation and the [110] heat flux direction, while the minimum values are fixed for the {100} plane orientation.

4.5 Conclusion

The influence of various types of elastic energy anisotropy on phonon transport in monocrystalline nanostructures has been considered. The dependencies of the lattice thermal conductivity on the nanostructures' geometric parameters at low temperatures, when diffuse phonon scattering at the boundaries dominates, are analysed. The main results on the influence of phonon focusing on phonon transport in monocrystalline films made of type-I (GaN, GaAs, Ge, MgO, Si, diamond, and YAG) and type-II (NaCl, CaF_2, SrF_2, and YIG) crystals can be formulated as follows:

1. For isotropic media, the mean free paths of phonons of different polarizations are the same and equal to the average mean free path. They do not depend on elastic moduli but are completely determined by geometric parameters. Therefore, they are a convenient comparison system for elastically anisotropic nanostructures.
2. In elastically anisotropic nanostructures, the mean free paths of phonons of different polarizations are different and depend substantially not only on their geometric parameters but also on the directions of heat flux and the orientation of the side faces of the nanostructures. For square-shaped cross-section nanowires, the values of thermal conductivity and phonon mean free paths are mainly determined by the heat flux direction, and their dependence on the orientation of the side faces is weak. However, for sufficiently wide films, the thermal conductivity magnitudes are largely controlled by the film plane orientation.
3. Accounting for finite length of the films eliminates the logarithmic divergence of the Casimir lengths when the film width tends to infinity in both isotropic media and elastically anisotropic materials. The length of the films limits the interval of an intensive increase in phonon mean free paths with rising film width. When the width of the films is 10 times greater than their length, the dependencies of the phonon mean free paths come to saturation.
4. For both nanofilms and nanowires made of type-I and type-II crystals, the anisotropy of the thermal conductivity and mean free paths of phonons differs qualitatively. For crystals of the same type, the angular dependencies of the mean free paths for all directions of heat flux and orientations of the film planes are qualitatively similar, distinguishing themselves only by a greater or lesser degree of anisotropy.

5. A correlation has been established between the angular dependencies of the densities of phonon states and the mean free paths of quasi-transverse modes. It has been shown that, for all nanowires, the maximum regions of DPS and mean free paths for fast and slow transverse modes in the {100} and {110} planes coincide and are due to phonon focusing.
6. The analysis of the dependencies of the mean free paths of phonons in films with the {100} orientation and [*I*] = [100] temperature gradient direction on the geometric parameters has shown that their values for films fabricated from type-I and type-II crystals turn out to be larger and smaller, respectively, than those for isotropic media. In films with the {*J*} = {111} plane orientations and $[11\bar{2}]$ temperature gradient direction, the sequence of the dependencies of the mean free paths in materials with different types of elastic energy anisotropy becomes inverse with respect to films with the {100} orientation. Their values in films made of type-I and type-II are larger and smaller, respectively, than those calculated for isotropic media.
7. The orientations of the planes of monocrystalline films and the directions of heat flux, which provide the maximum or minimum heat removal from the elements of semiconductor microcircuits, have been identified. For square-shaped films based on type-I crystals (GaN, GaAs, MgO, LiF, GaAs, Ge, Si, and YAG), the thermal conductivity values reach their maximum in the {100} plane orientation, while their minimum is observed for the {111} orientations. For square-shaped films based on type-II crystals (CaF_2, NaCl, and YIG), the minimum thermal conductivity values can be revealed for the {111} plane orientations, while the maximum values are typical of the {110} plane orientation and [110] heat flux direction.

Chapter 5
Phonon Propagation and Phonon Transport in Films with Different Types of Elastic Energy Anisotropy

The present section covers the influence of focusing on phonon propagation and phonon transport in monocrystalline films with various types of anisotropy of elastic energy. As a first step, we calculate the angular dependencies of thermal conductivity and mean free paths ($\tilde{\Lambda}^{\{J\}\lambda}_{[I(\psi)]}$) of phonons of different polarizations upon rotation of the temperature gradient in the plane of the films. For this, we address square-shaped films ($L = W$) oriented in various planes. The subsequent step is to analyse the physical reasons for changing in the anisotropy of the phonon mean free paths for long films ($L \gg W$). After that, we compare the angular dependencies of the mean free paths $\tilde{\Lambda}^{\{J\}\lambda}_{[I(\psi)]}$ in elastically anisotropic nanostructures with the results derived within the model of an isotropic medium. As noted earlier, in the isotropic media, the mean free paths of phonons of different polarizations are equal and completely controlled by geometric parameters. This is a convenient system to match the mean free paths in elastically anisotropic nanostructures when varying the film plane's orientations or the directions of heat flux. The construction of this chapter is based on the papers [70, 109, 117].

5.1 Thermal Conductivity Anisotropy of Square-Shaped Films with Different Plane Orientation

Let us look into the features of phonon transport in square-shaped films with different types of anisotropy of elastic energy. As an example, we may take GaAs and CaF_2 crystals. When comparing the angular dependencies of the mean free paths of phonons in the films with different orientations of planes, the following features attract attention (see Figure 5.1). Firstly, the thermal conductivity and average phonon mean free paths in square-shaped films ($L = W$) in films oriented along the {100} and {111} planes are isotropic for crystals of both types. However, for type-I and type-II crystalline films with the {110} orientation, the aforementioned dependencies have an ellipsoidal shape, with the long axes of the ellipsoids pointing along the [100] and [110] directions, respectively. Secondly, for all films with the orientations $\{J\}$ = {001} and fabricated from type-I and type-II crystals, the average phonon mean free paths are longer and shorter, respectively, than those offered by the model of an isotropic medium, (see Table 5.1). For example, for GaAs and CaF_2 films with $\{J\}$ = {100}, the average mean free paths turn out to be 1.95 times larger and by 10% less, respectively, than those predicted by the model of an isotropic medium. Thirdly, the opposite is true for films with the {111} plane's orientations:

https://doi.org/10.1515/9783110670509-006

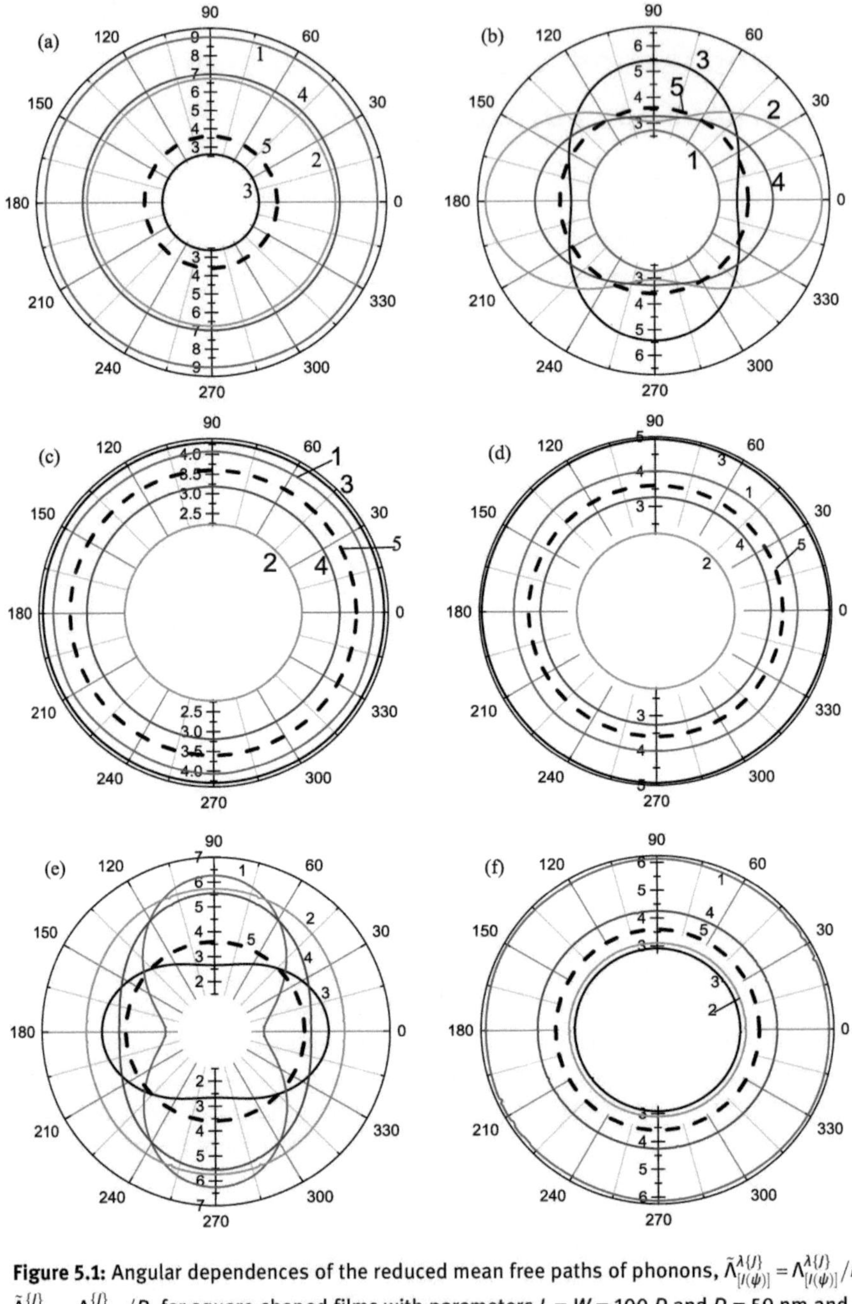

Figure 5.1: Angular dependences of the reduced mean free paths of phonons, $\tilde{\Lambda}^{\lambda\{J\}}_{[I(\psi)]} = \Lambda^{\lambda\{J\}}_{[I(\psi)]}/D$ and $\tilde{\Lambda}^{\{J\}}_{[I(\psi)]} = \Lambda^{\{J\}}_{[I(\psi)]}/D$, for square-shaped films with parameters $L = W = 100\ D$ and $D = 50$ nm and orientations of {100} (a,d), {110} (b,e), and {111} (c,f), for GaAs (a,b,c) and CaF_2 (d,e,f) crystals. Curves 1 – a fast transverse mode, 2 – a slow transverse mode, 3 – a longitudinal mode, 4 – the average mean free path, 5 – the mean free paths calculated within the model of an isotropic medium.

Table 5.1: Ratios of mean free paths in symmetrical directions for square-shaped films with the parameters $L = W = 100\ D$ and $D = 50$ nm.

Compound	$\Lambda_{[I(\psi)]}^{t1\{100\}}:\Lambda_{[I(\psi)]}^{t2\{100\}}:\Lambda_{[I(\psi)]}^{L\{100\}}:\Lambda_{iso}$	$\Lambda_{[I(\psi)]}^{t1\{111\}}:\Lambda_{[I(\psi)]}^{t2\{111\}}:\Lambda_{[I(\psi)]}^{L\{111\}}:\Lambda_{iso}$	$\Lambda_{[100]}^{\{100\}}:\Lambda_{[100]}^{\{110\}}:\Lambda_{[110]}^{\{110\}}:\Lambda_{[111]}^{\{111\}}:\Lambda_{iso}$
GaAs	2.50:1.87:0.73:1	1.12:0.62:1.19:1	1.95:1.27:0.91:0.88:1
Ge	3.00:1.85:0.74:1	1.13: 0.64:1.18:1	2.10:1.33:0.93:0.91:1
InSb	2.23:1.89:0.73:1	1.10:0.61:1.19:1	1.86:1.22:0.88:0.86:1
LiF	3.36:1.84:0.76:1	1.13:0.65:1.18:1	2.23:1.38:0.95:0.91:1
MgO	3.40:1.81:0.77:1	1.12:0.67:1.16:1	2.24:1.46:0.97:0.92:1
Si	3.46:1.84:0.77:1	1.1:0.65:1.2:1	2.29:1.40:0.94:0.91:1
diamond	1.48:1.23:0.85:1	1.07:0.79:1.10:1	1.26:1.51:1.05:0.96:1
CaF_2	1.12:0.62:1.37:1	1.69:0.87:0.81:1	0.90:1.07:1.55:1.19:1
NaCl	1.20:0.60:1.57:1	1.59:0.88:0.74:1	0.96:1.06:1.38:1.12:1

type-I and type-II crystals demonstrate shorter and longer average mean free paths, respectively, than the model of an isotropic medium (see Table 5.1). For example, for GaAs films with $\{J\} = \{111\}$, the average mean free paths turn out to be 12% less than those for the model of an isotropic medium; for CaF_2 films these values are 19% larger compared to the aforementioned model. For other materials, see Table 5.1. Next, we disclose the physical causes that are responsible for the dependencies of the thermal conductivity and phonon mean free paths on the film plane's orientations.

As noted already, the Casimir–McCurdy theory [13, 19] suggests that all phonons are absorbed by the surface when collided with it and then re-emitted isotropically into half-space towards the sample.

Therefore, phonons of different polarizations are diffusely scattered at each point of the surface regardless of the latter's orientation. Consequently, it would seem that the thermal conductivity should be independent on the orientation of the planes. However, we will show that these dependencies are caused by the influence of elastic anisotropy on the propagation of acoustic modes throughout the films having different orientation.

To understand how focusing affects the phonon propagation in films, it is mandatory to analyse the distribution of the phonon mean free paths both over the angles $\Phi(\theta, \varphi)$ in the films with various orientation planes and over the angles $\Theta(\theta, \varphi)$ of the transverse cross-sections, along which heat flux moves. For this purpose, we define a coordinate system related to the film as follows. The axis Z is set perpendicular to the film plane. Then, the Φ and Θ angles assign the distribution of the heat flux along the film plane and in the transverse cross-section of the film, respectively. For films with $\{J\} = \{100\}$, we have $\Phi = \varphi$ and $\Theta = \theta$. In the new coordinate system, the distribution of the phonon mean free paths over the angles Φ and Θ can be represented as [70]:

$$\Lambda_{[I]}^{\lambda\{J\}}(\Theta) = \frac{1}{2\pi}\int_0^{2\pi} d\Phi \Lambda_{[I(\psi)]}^{\lambda\{J\}}(\Theta, \Phi), \quad \Lambda_{[I]}^{\lambda\{J\}}(\Phi) = \frac{1}{2}\int_{-1}^{1} d(\cos\Theta)\Lambda_{[I(\psi)]}^{\lambda\{J\}}(\Theta, \Phi). \tag{5.1}$$

Let us find the relevant interval of angles, which mainly contributes to averaging the mean free paths over the Φ angles in the planes of films with different orientations. For this, we perform the averaging over the transverse cross-section of the film in formula (5.1) and construct the dependencies of the mean free paths $\Lambda_{[I(\psi)]}^{\lambda\{J\}}(\Phi)$ on the Φ angle. The analysis shows that, for square-shaped films with the {100} and {111} planes, this interval of angles is about equal to $\Delta\Phi \cong \pi/2$ but exceeds this value for some modes.

For films with the {100} plane orientation, the slow quasi-transverse modes t_2 are focused and defocused in the [100] and [110] directions. Longitudinal phonons are defocused in the [100] direction, having a local maximum of focusing in the [110] direction. The angle between these directions amounts to $\Delta\varphi = \pi/4$. For films oriented along the {111} plane, the angle between the [110] focusing and $[11\bar{2}]$ defocusing directions for the t_2 mode is equal to $\Delta\varphi = \pi/6$. From this, it follows that with an arbitrary direction of the heat flux, the averaging simultaneously captures the phonon focusing and defocusing directions when it comes to films with the {100} and {111} orientations and made of crystals of both types. It is not surprising therefore that the mean free paths become isotropic after the averaging over the Φ angles (see Figure 5.1). In this case, for type-I crystalline films, the ratios of the phonon mean free paths vary from 1.3 for diamond to 2.5 for Si. For type-II crystalline films, these ratios are less than unity (see Table 5.2). Thus, the anisotropy of thermal conductivity and phonon mean free paths in square-shaped films with the {100} and {111} planes in the boundary scattering regime is chiefly determined by the film plane's orientation.

Let us elucidate the physical reasons for emerging an ellipsoidal shape of the dependencies of mean free paths (thermal conductivity) in square-shaped films with the $\{J\} = \{110\}$ plane orientation. As is clear from Figure 5.1, these dependencies are due to the slow transverse mode t_2 and fast transverse mode t_1 for type-I and type-II crystals, respectively. In these films, the previously mentioned modes are focused and defocused in the [100] and [110] directions, the angle between which is $\Delta\Phi = 90°$. It turns out to be too large to fully blur the focusing effect when averaging over the Φ angle. For an arbitrary direction of the temperature gradient, the averaging over angles in the {110} film plane cannot simultaneously cover both directions. Consequently, the angular dependencies of the phonon mean free paths remain anisotropic. Therefore, the anisotropy of the contributions of the phonon

Table 5.2: Ratios of mean free paths in symmetrical directions for square-shaped films with the parameters $W = 100\ D$ and $D = 50$ nm.

Compound	$\Lambda_{[110]}^{L\{110\}}:\Lambda_{[110]}^{L\{111\}}:\Lambda_{[100]}^{L\{110\}}:\Lambda_{[100]}^{L\{100\}}$	$\Lambda_{[100]}^{t1\{100\}}:\Lambda_{[110]}^{t1\{111\}}:\Lambda_{[110]}^{t1\{110\}}:\Lambda_{[100]}^{t1\{110\}}$	$\Lambda_{[100]}^{t2\{100\}}:\Lambda_{[100]}^{t2\{110\}}:\Lambda_{[110]}^{t2\{110\}}:\Lambda_{[110]}^{t2\{111\}}$	$\Lambda_{[I(\psi)]}^{\{100\}}:\Lambda_{[100]}^{\{110\}}:\Lambda_{[110]}^{\{110\}}:\Lambda_{[I(\psi)]}^{\{111\}}$
GaN	2.5:1.78:1.3:1	2.6:1.5:1.05:1	2.89:2.37:1.24:1	1.97:1.35:1.04:1
GaAs	2.06:1.63:1.21:1	3.64:1.62:1.09:1	3.01:2.90:1.37:1	2.20:1.44:1.03:1
Ge	1.97:1.60:1.19:1	4.29:1.62:1.10:1	2.90:3.06:1.41:1	2.33:1.47:1.04:1
InSb	2.06:1.62:1.20:1	3.20:1.59:1.09:1	3.09:2.75:1.33:1	2.17:1.42:1.02:1
LiF	1.88:1.56:1.17:1	4.85:1.65:1.11:1	2.82:3.21:1.43:1	2.44:1.51:1.04:1
MgO	1.76:1.50:1.15:1	4.78:1.59:1.10:1	2.70:3.40:1.46:1	2.43:1.58:1.05:1
Si	1.79:1.51:1.15:1	4.93:1.63:1.10:1	2.83:3.27:1.43:1	2.52:1.54:1.03:1
diamond	1.40:1.29:1.08:1	1.90:1.37:1.07:1	1.57:3.11:1.51:1	1.32:1.58:1.10:1
CaF_2	0.54:0.59:0.92:1	2.03:3.10:3.16:1	0.72:1.66:1.85:1	0.77:0.91:1.31:1
NaCl	0.42:0.47:0.91:1	2.21:2.96:2.63:1	1.69:1.57:1.79:1	0.86:0.95:1.23:1

modes to the thermal conductivity is not averaged, and the latter in the $\{J\} = \{110\}$ plane is described by ellipsoid (see Figure 5.1). In type-I crystals, the long axis of the ellipsoid is pointing along the [100] direction, in which the mode t_2 focuses. In this direction, its mean free path Λ_{iso} in GaAs crystals exceeds Λ_{iso} more than 1.8 times (see Table 5.1). The short ellipsoid axis corresponds to the [110] defocusing direction, in which the mean free path turns out to be 17% less than Λ_{iso}. For longitudinal phonons, we have the reversed situation: the focusing and defocusing directions swap around compared to the slow transverse mode, t_2. The angle between these directions is equal to $\Delta\Phi = 90°$. The angular dependence of the mean free paths also has an ellipsoidal shape, but the long and short axes of the ellipsoid coincide with the [110] and [100] directions, respectively. The mean free paths of longitudinal phonons in Ge and Si films in the [110] direction turn out to be 1.5 and 1.4 times longer and 13% and 12% less in the [100] direction, respectively than those within the model of an isotropic medium (see Table 5.2). However, the heat capacity of the longitudinal phonons is small in comparison with the transverse phonons. Therefore, the contribution of the former is insufficient to drastically change the shape of full thermal conductivity of the materials at hand (see Figure 5.1). Thus, in films with the {110} orientation and made of crystals of the first type, the maximum and minimum values of the thermal conductivity are reached in the [001] and [110] directions, respectively. Figure 5.1 shows that, in films fabricated from crystals of the second type, the ellipsoidal shape of the thermal conductivity dependence is due to the fast transverse mode t_1. The latter is focused and defocused in the directions [110] and [001], respectively. Therefore, the long axis of the ellipsoid lies along [110], and the short axis resides in the [100] defocusing direction. In type-II crystal-based films, the mean free paths of longitudinal phonons have maximum and minimum values in the [001] and [110] directions, respectively.

For CaF_2 films, their ratio is 1.7. However, their heat capacity values are small as compared with those of transverse phonons (see Table 5.3). Therefore, the contribution of the longitudinal phonons is insufficient to change the form of full thermal conductivity of the materials under consideration (see Figure 5.1). In this regard, in films with the $\{J\} = \{110\}$ orientation and made of type-I crystals, the maximum and minimum values of the thermal conductivity are observed in the [110] and [100] directions, respectively. For CaF_2 films, their ratio is equal to 1.46 (see Table 5.2). Thus, the main contribution to the thermal conductivity of films with the {100} orientations and made of type-I crystals such as GaAs, InSb, and GaN is made by the slow transverse mode, while the fast transverse mode dominantly contributes to the thermal conductivity of Ge, LiF, MgO, Si, and diamond crystals. The fast transverse mode also makes the main contribution to the thermal conductivity of films with the {111} plane and made of crystals of both types (see Table 5.3).

Table 5.3: Contributions of vibrational modes to thermal conductivity of square-shaped films with the parameters $L = W = 100\ D$ and $D = 50$ nm, based on crystals of the first and second groups for different directions of heat flux (%).

{*J*} [*I*]		{100}	{110} [100]	{110} [110]	{111}
GaAs	L	5	10	23	19
	t_1	47	20	30	46
	t_2	48	70	47	35
Ge	L	5	10	24	20
	t_1	53	19	30	46
	t_2	42	71	46	34
InSb	L	5	9	22	17
	t_1	43	20	31	46
	t_2	52	71	47	37
GaN	L	5	10	25	18
	t_1	37	21	28	42
	t_2	58	69	47	40
LiF	L	5	10	24	20
	t_1	56	19	30	46
	t_2	39	71	46	34
MgO	L	6	10	23	20
	t_1	57	18	30	46
	t_2	37	72	47	34
Si	L	5	9	22	19
	t_1	57	19	31	47
	t_2	38	72	47	34
diamond	L	12	11	21	22
	t_1	46	20	31	43
	t_2	42	69	48	35
CaF_2	L	18	14	6	8
	t_1	48	20	44	56
	t_2	34	66	50	36

Table 5.3 (continued)

{*J*} [*I*]		{100}	{110} [100]	{110} [110]	{111}
NaCl	*L*	24	19	7	9
	t_1	46	19	38	53
	t_2	30	62	55	38

5.2 Peculiarities of Thermal Conductivity of Long Films with Different Types of Elastic Energy Anisotropy

The analysis performed using formula (5.1) shows that the intervals of angles bringing the main contribution when averaging the mean free paths are strongly narrowed as the length of films oriented in the {100} and {111} planes increases. Therefore, for long films ($L \gg W$), averaging over the Φ angles in the planes of these films is already insufficient to fully blur the effect of phonon focusing. As a result, the mean free paths in films with the {100} and {111} planes and made of crystals of both types, become anisotropic (see Figure 5.2). As Figure 5.2 indicates, the main contribution to the anisotropy of thermal conductivity and average phonon mean free paths in films produced from type-I crystals is made by the slow transverse mode, which is focused in the [100] directions.

However, in these directions, the mean free paths $\Lambda_{[100]}^{t_2\{100\}}$ for films with sufficiently large anisotropy parameters (GaN, GaAs, LiF, and Ge) have local minima, For example, for GaAs films, their values are 18% smaller than those at angles $\psi = n\pi/2 \pm \theta_3$ (see Figure 5.2a). In the [100] directions, the mean free path $\Lambda_{[100]}^{t_2\{100\}}$ of the slow mode in GaAs films is 53% longer than that in the [110] directions.

Anisotropy of the thermal conductivity and average mean free paths in GaAs films with $L \gg W$ is noticeably less due to the contribution of the fast transverse mode: In long GaAs films, the average phonon mean free paths are longer and shorter for the {100} and {111} orientations, respectively, than those predicted by the model of an isotropic medium.

It should be stressed that the regions of maxima of the phonon mean free paths for films with the {100} and {111} orientations are limited by angles defining the phonon focusing regions $\pm\theta_3^{t_1}$ and $\pm\theta_3^{t_2}$. When $L \gg W$, the main contribution to the anisotropy of thermal conductivity of films with the {100} plane and fabricated from type-II crystals is made by the fast transverse mode, which is focused in the [110] direction (see Figure 5.2). So, for example, for CaF_2 films, the mean

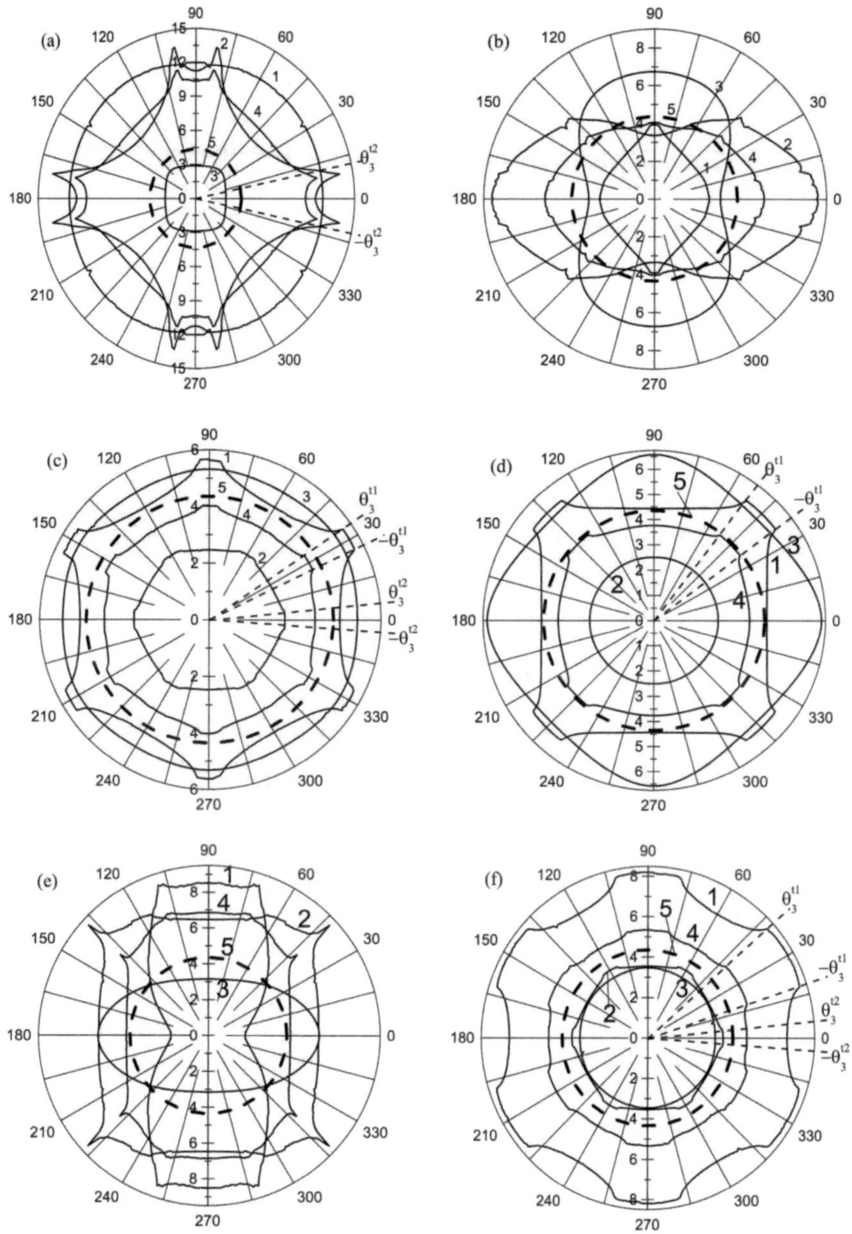

Figure 5.2: Angular dependencies of the reduced mean free paths of phonons $\tilde{\Lambda}^{\lambda\{J\}}_{[I(\psi)]} = \Lambda^{\lambda\{J\}}_{[I(\psi)]}/D$ and $\tilde{\Lambda}^{\{J\}}_{[I(\psi)]} = \Lambda^{\{J\}}_{[I(\psi)]}/D$ for films with the parameters $L = 2000\ W$, $W = 100\ D$, and $D = 50$ nm and orientations of {100} (a,d), {110} (b,e), and {111} planes (c,f), based on GaAs (a,b,c) and CaF_2 (d,e) crystals. Curves 1 – a fast transverse mode, 2 – a slow transverse mode, 3 – a longitudinal mode, 4 – the average mean free path, 5 – the mean free path in the model of an isotropic medium.

free path $\Lambda_{[I]}^{t_1\{100\}}$ in the [110] direction is larger by 35% than that in the [001] direction. However, the anisotropy of the full thermal conductivity of CaF_2 films with the {100} orientation declines to 13%.

For films with the $\{J\}$ = {111} orientation and made of crystals of both types, the thermal conductivity anisotropy is produced by the fast transverse mode focused in the [110] direction. In this direction, this mode ensures the maximum values of the average mean free path (see Figure 5.2). For long GaAs and CaF_2 films, these values are larger by 24% and 16% than those in the $[11\bar{2}]$ directions. The contribution of the slow transverse mode to thermal conductivity is small, and its mean free path in films made of crystals of both types is shorter than that in the model of an isotropic medium. Let us compare the angular dependencies of the mean free paths in elastically anisotropic films when $L \gg W$ with those constructed within the model of an isotropic medium. As can be understood from Figure 5.2, we get the same results as for square-shaped films. For type-I and type-II crystals, the mean free paths in films with the {100} orientations turn out to be longer and shorter, respectively, than those offered by the model of an isotropic medium. For GaAs and CaF_2 films, their ratios can be written in the following way:$\Lambda_{[100]}^{\{100\}}:\Lambda_{[110]}^{\{100\}}:\Lambda_{iso} = 2.29:1.98:1$ and $\Lambda_{[100]}^{\{100\}}:\Lambda_{[100]}^{\{100\}}:\Lambda_{iso} \cong 0.87:0.98:1$, respectively. For films with the {111} plane, the situation is the opposite: the average mean free paths in the films made of type-I and type-II crystals are less and larger, respectively, than those in the model of an isotropic medium. For GaAs and CaF_2, their ratios can be represented as: $\Lambda_{[100]}^{\{100\}}:\Lambda_{[110]}^{\{100\}}:\Lambda_{iso} = 0.92:0.85:1$ and $\Lambda_{[100]}^{\{100\}}:\Lambda_{[110]}^{\{100\}}:\Lambda_{iso} = 1.23:1.16:1$, respectively. Thus, for long films with the {100} and {111} orientations and fabricated from crystals of the first type, the main contribution to the thermal conductivity anisotropy is made by the slow and fast transverse modes, respectively.

For films with the {100} plane and made of crystals of the second type, the thermal conductivity anisotropy is mainly caused by the fast transverse mode. When going over to long films with the $\{J\}$ = {110} orientation, an ellipsoidal shape of the angular dependencies and the characteristic features of the mean free paths of phonons of different polarizations are preserved (see Figure 5.2). For type-I crystalline films, the average mean phonon free path of the slow transverse mode takes the maximum and minimum values in the [100] focusing and [110] defocusing directions, respectively. However, for films made of type-I crystals, the long axis of the ellipsoid lies along the [110] focusing direction of the fast transverse mode. The short axis is along its defocusing direction [100]. Additional features such as the form of local maxima in the dependence of the mean free path in CaF_2 films are due to focusing of the transverse modes (see Figure 5.2).

5.3 Influence of Focusing on the Propagation of Phonon Modes in Square-Shaped Films with the {100} and {111} Plane's Orientations

To begin with, we demonstrate how focusing affects the propagation of phonon modes in both-type crystalline square-shaped films ($L = W$) and how much the values of the latter's thermal conductivity and, accordingly, average mean free paths are distinguished depending on their orientation. It will be shown that type-I crystal-based films with the {100} orientation have larger and type-II crystal-based films with the $\{J\} = \{100\}$ orientation have smaller values of the thermal conductivity than films with the {111} plane. It is well known that the main contribution to thermal resistance is made by phonon scattering at the planes of films, while the predominant contribution to thermal conductivity is made by phonons propagating almost parallel to the plane of the film (see, for example, [24, 25, 31, 32]). Therefore, the object of the present section is to explore the distribution of the phonon flux throughout the transverse cross-section of the films, as well as the influence of phonon focusing on this flux. For this purpose, we average the phonon mean free paths over the Φ angles resting in the film planes, using formula (5.1). Further, we plot their dependencies on the Θ angles lying in the transverse cross-section, through which the heat flux flows (see Figure 5.3).

As can be inferred from Figure 5.3, the phonons of the slow mode in type-I crystalline films with the {100} orientation are focused and defocused in the [100] and [110] directions, respectively. As for films made of type-II crystals, the focusing and defocusing directions change places in this plane (see Figure 5.3). In films with the {110} orientation and made of type-I crystals, the slow and fast transverse modes are focused in the [100] and [110] directions, respectively.

In type-II crystal-based films with the {J} = {110} orientation, the fast mode is focused at the angle (n is an integer). As can be seen in Figure 5.4, transverse phonons in films oriented along the {100} plane and made of crystals of the first type are dominantly focused; the mean free paths and averaged over the Φ angle, are larger than for all angles Θ (see Figure 5.4a). At the same time, the situation is opposite for longitudinal phonons: their mean free paths turn out to be less than. In contrast to the above, transverse phonons in type-II crystalline films with the {100} orientation are chiefly defocused. Their mean free paths $\Lambda_{[001]}^{t1\{100\}}(\Theta)$ and $\Lambda_{[001]}^{t1\{100\}}(\Theta)$, averaged over the Φ angle, become shorter than $\Lambda_{iso}(\Theta)$, for all angles Θ (see Figure 5.4b). However, the mean free paths for longitudinal phonons in films produced from crystals of the second type are longer than $\Lambda_{iso}(\Theta)$, at all angles. As was noted before, the contribution of longitudinal phonons to the thermal conductivity and average mean free paths is small compared with that of transverse phonons (see Table 5.3). Thus, the thermal conductivity and mean phonon mean free paths for square-shaped films with the {100} plane and made of crystals of the first type and

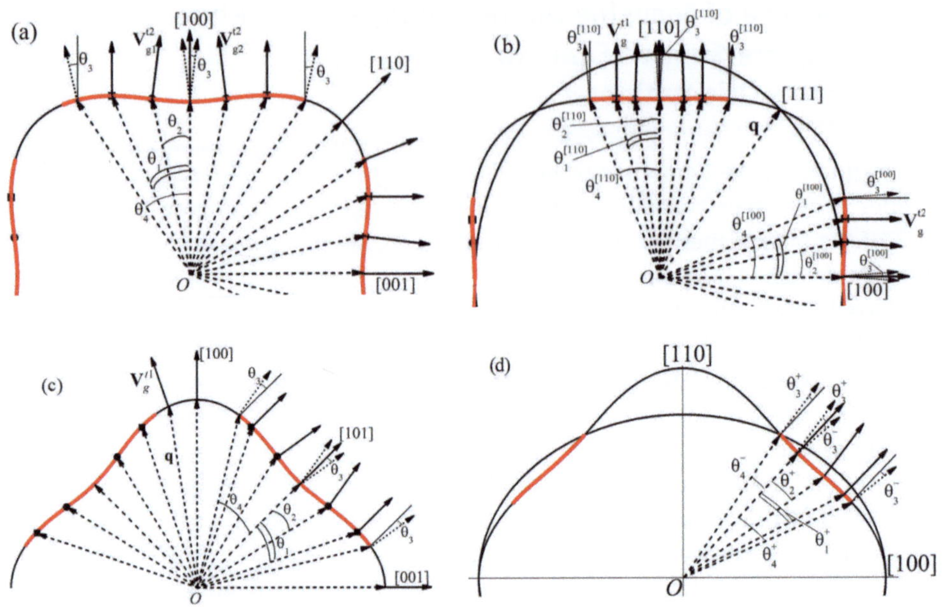

Figure 5.3: Scheme that illustrates focusing of slow and fast transverse modes in GaAs (a,b) and CaF_2 (c,d) crystals for cross-sections of the isoenergetic surface by the XZ plane for the t_2 mode (a,c) and by the diagonal plane for the t_1 and t_2 modes (b,d). The arrows show the wave vectors inside the surface and the corresponding group velocities of phonons beyond it.

are larger than those for films fabricated from crystals of the second type. This may be explained by the prevalence of the phonon focusing and defocusing effect for the former and the latter, respectively.

It should be emphasized that the distribution of the mean free paths over the transverse cross-section of the film, depicted in Figures 5.4, is built in space of wave vectors. For the slow transverse mode in films made of type-I crystals, the side peaks of $\Lambda^{t_2\{100\}}_{[I(\psi)]}(\Theta)$ correspond to the wave vectors close to the $\pm\theta_1^{t_2}$ angles in the coordinate space (or space of group velocities) (see Section 1.5, Table 1.5). Due to the focusing effect, the phonons of this mode propagate in the [100] direction (see Figure 5.3); their mean free paths may be limited by either the length of the sample or bulk relaxation mechanisms. Thus, the side maxima are shifted to the central peak that corresponds to the temperature gradient direction (see Figures 5.3 and 5.4). The side maxima in the curve of $\tilde{\Lambda}^{t_1\{110\}}_{[100]}(\Theta)$ for the fast transverse mode focused in the [110] direction for the wave vector lying in the {110} plane can be interpreted in the analogous way. As becomes clear from Figure 5.3, the main contribution to the mean free paths of transverse modes in the coordinate presentation is made by regions of the angles limited by the values of $\pi/2\pm\theta_3^{t_2}$ and $\pi/2\pm\theta_3^{t_1}$ for slow and fast transverse modes, respectively. Thus, the distribution of mean free paths in films made of type-I crystals for these modes in the coordinate space has a single-peaked shape (as in an isotropic medium) with the characteristic widths $2\theta_3^{t_1}$ and $2\theta_3^{t_2}$, respectively.

For square-shaped films with {111} plane, the dependencies of the mean free paths for all acoustic modes qualitatively change as compared with the dependencies for films with the {100} orientation (see Figure 5.4). The central maximum is preserved only for the model of an isotropic medium ($\Lambda_{iso}(\theta)$). At the same time, the mean free paths $\Lambda_{[I(\psi)]}^{\lambda\{111\}}(\Theta)$ of all acoustic modes in films fabricated from both types of crystals have smooth minima in the vicinity of the heat flux direction (the axis of the films) (see Figure 5.4). In contrast to films with the $\{J\} = \{100\}$ plane orientation, this stands for the prevalence of the defocusing effect for all modes in the directions of wave vectors, close to the heat flux direction.

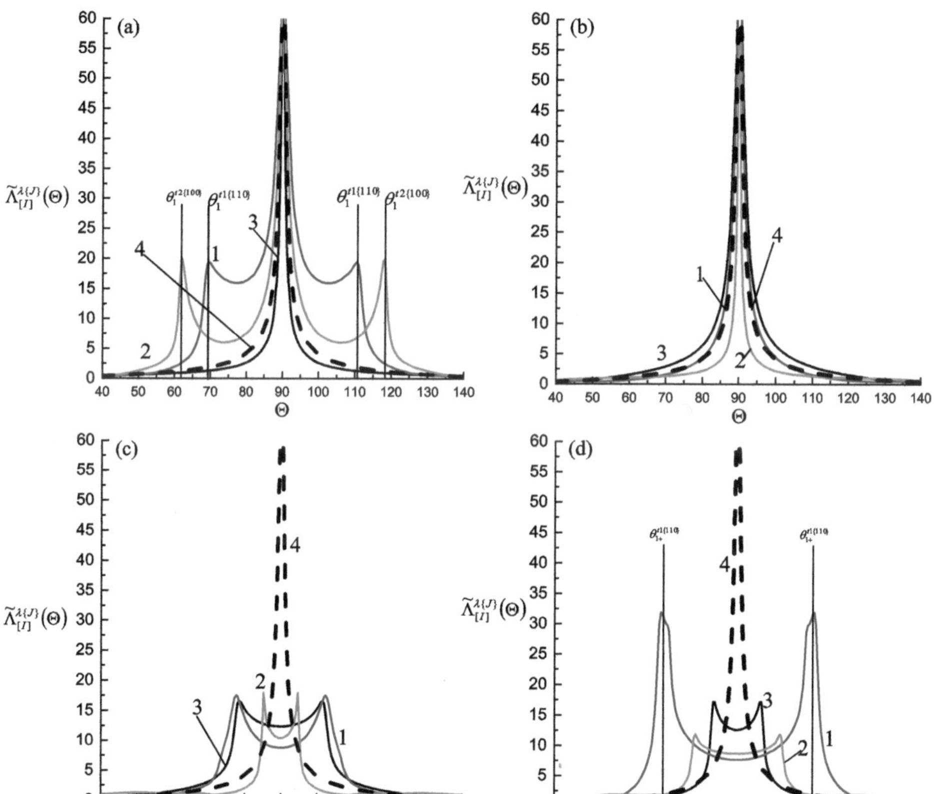

Figure 5.4: Distribution of the mean free paths $\tilde{\Lambda}_{[I(\psi)]}^{\lambda\{J\}}(\Theta)$ over the transverse cross-section for films made of GaAs (a,b) and (c,d) CaF_2 crystals and oriented along the {100} and {111} planes with the parameters $L = W = 100\ D$, $D = 50$ nm: curves 1 – for the t_1 mode, curves 2 – for the t_2, mode, curves 3 – for longitudinal phonons, and curves 4 – for the model of an isotropic medium.

Moreover, the mean free paths $\Lambda_{[I]}^{t_2\{111\}}(\Theta)$ are shorter than $\Lambda_{iso}(\Theta)$ within an almost entire interval of the Θangles except the narrow ranges such as $84^\circ < \Theta < 87^\circ$ and $93^\circ < \Theta < 96^\circ$ for type-I GaAs crystals and $69^\circ < \Theta < 85^\circ$ and $95^\circ < \Theta < 111^\circ$ for type-II CaF_2 crystals where they have local maxima.

In square-shaped films with the {111} orientation and made of crystals of the second type, the defocusing effects also prevail for slow transverse and longitudinal phonons. Their mean free paths in the region of the central peak have a smooth minimum. Nevertheless, they can exceed $\Lambda_{iso}(\Theta)$ but only in narrow intervals of angles, where they have local maxima (see Figure 5.4). However, the mean free path $\Lambda_{[I(\psi)]}^{t_1\{111\}}(\Theta)$ of the fast transverse mode has two side maxima where $\Lambda_{[I(\psi)]}^{t_1\{111\}}(\Theta) > \Lambda_{iso}(\Theta)$. Since these maxima are significantly higher than the value of $\Lambda_{iso}(\Theta)$, the average value for the fast transverse mode exceeds $\Lambda_{iso}(\Theta)$ (see Figure 5.2(f), curve 1). Finally, the mean free paths for films with the {111} plane and fabricated from crystals of the second type turn out to be longer than both $\Lambda_{iso}(\Theta)$ and those for films made of crystals of the first type (see Figure 5.2). For example, for CaF_2 films with the {111} plane, the average mean free paths are 1.3 times longer than those for GaAs films.

The papers of Fuchs and Sondheimer [30, 31] state that, at sufficiently low temperatures, the thermal resistance of films is mainly determined by phonon scattering at the film planes. Some research works utilize the term 'scattering power' of the planes of films having different orientations to characterize the latter's thermal conductivity during diffuse phonon scattering at the film boundaries (see, for example, [28]). It should be underscored that the term used for characterizing the influence of phonon focusing on the thermal conductivity of films is physically incorrect. The fact is that the Casimir theory suggests that every point of the surface, regardless of its orientation, diffusely scatters phonons of all polarizations. When collided with the surface, the phonons are absorbed and then re-emitted isotropically into half-space towards inside the sample. Thus, the scattering power or scattering intensity of films with different plane orientations is the same. The influence of phonon focusing on the thermal conductivity of films is associated with the difference in directions of the wave vector and propagation of phonons. Besides, it is caused by the elastic anisotropy of the crystals. If the heat flux is pointing along the axis of the film and parallel to the focusing direction of one of the modes, the directions of phonon propagation of this mode deviate from the film planes to its axis. In addition, the mean free path of these phonons increases, as is the case for slow transverse phonons in films oriented along the {100} plane and when the heat flux flows in the [100] direction (see Figure 5.5). Moreover, phonons of the t_2 mode with the $\mathbf{q}_1$ wave vector and moving at the angle $\theta = \pm\theta_1$ will propagate in the [100] direction (see Figure 5.5). Phonons with the $\mathbf{q}_2$ wave vector and travelling at the angle $\theta = -\theta_2^{t2}$ will propagate at the angle $\theta = \pm\theta_3^{t2}$ (see Figures 5.3a and 5.5). Since, for GaAs crystals, we have these angles equal to $\theta_3^{t2} = 12.0^\circ$ and $\theta_2^{t2} = 13.8^\circ$, the mean free path $\Lambda_{[100]}^{t_2\{100\}}(\theta_3^{t2})$ is shorter than

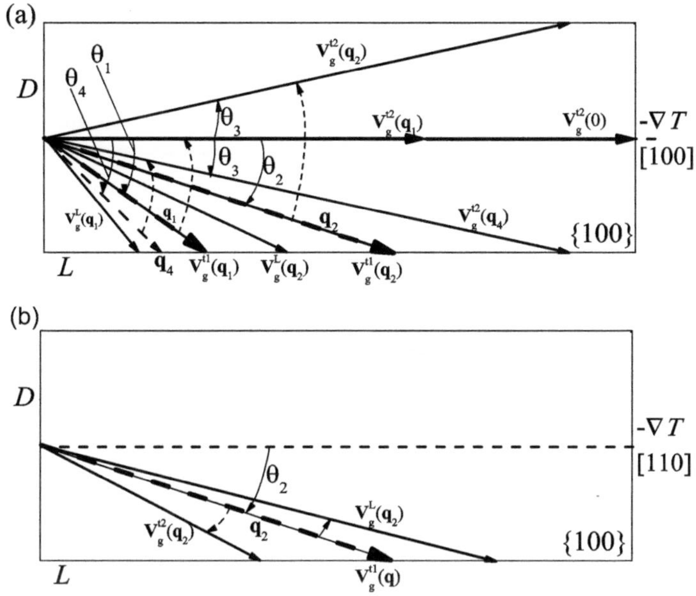

Figure 5.5: Scheme that illustrates the influence of focusing on the phonon propagation in the axial cross-section of a GaAs film with the {100} plane orientation for the temperature gradient in the directions [100] (a) and [110] (b) and wave vectors at angles $\theta_1^{t2} = 28.6$, $\theta_2^{t2} = 13.8°$, $\theta_3^{t2} = 12.0°$, and $\theta_4^{t2} = 34.4°$ to the direction [100]. L is the length of the film, D is its thickness.

$\Lambda_{iso}(\theta_2)$ within the model of an isotropic medium (see Figure 5.5). ***With diffuse boundary scattering, the mean free path of a phonon with the q wave vector is determined by the distance travelled by the phonon until it collides with the sample surface.*** Figures 5.5 outline the length of the group velocity vector as the phonon mean free path for each direction. This vector is directed from the point of emission and ends at the point of the collision with the surface. For a fast transverse mode in the plane of the cube face, the phonon spectrum is isotropic and there is no focusing effect. Therefore, the directions of the $\mathbf{q}_2$ wave vector and group velocity $V_g^{t_1}(\mathbf{q}_2)$ coincide. Consequently, the mean free path is equal to $\Lambda_{[100]}^{t_1\{100\}}(\theta_2) = \Lambda_{iso}(\theta_2)$ (see Figure 5.5). If the heat flux is pointing along the axis of the film and parallel to the defocusing direction of one of the modes, the directions of the phonon propagation of this mode deviate from the axis of the film to its planes. In this case, the mean free path of these phonons diminishes. This takes place for longitudinal phonons with the $\mathbf{q}_1$ wave vector, with the propagation angle determined by the group velocity direction becoming larger than the θ_1 angle. Then, the mean free path $\Lambda_{[100]}^{L\{100\}}(\theta_1)$ turns out to be shorter than $\Lambda_{iso}(\theta_1)$ (see Figure 5.5a).

Thus, in the heat flux flowing along [100], the phonons of the slow transverse mode, having wave vectors in the sector $-\theta_4 \le \theta \le \theta_4$, will deviate from the planes of the film with {100} plane to its axis. The value of their mean free path can dramatically exceed that for an isotropic medium. However, the [110] heat flux direction for films with the {100} orientation corresponds to the direction of the defocusing of slow transverse modes. Therefore, the latter's mean free path is shorter than $\Lambda_{iso}(\theta_1)$ (see Figure 5.5).

The phonons of the fast transverse mode with wave vectors in the {110} plane are focused in the [110] direction. Therefore, they will deviate from the planes of the film toward the [110] temperature gradient direction. The value of their mean free path can significantly exceed the value for an isotropic medium (see Figure 5.6). Figure 5.6 shows that in GaAs films with the $\{J\} = \{110\}$ plane orientation, the phonons of the t_1 mode with wave vectors in the sector $-\theta_4 \leq \theta \leq \theta_4$ (the angle is $\theta_4^{t1} = 25.4°$ for the mode t_1) will deviate from the side faces to the axis of the rod and propagate in the sector $-\theta_3 \leq \theta \leq \theta_3$ (the angle is $\theta_3^{t1} = 3.6°$ for GaAs). Thus, the density of states for the t_1 mode is significantly higher in the [110] direction and much lower for the defocusing directions than that in the model of an isotropic medium (see Section 1.5).

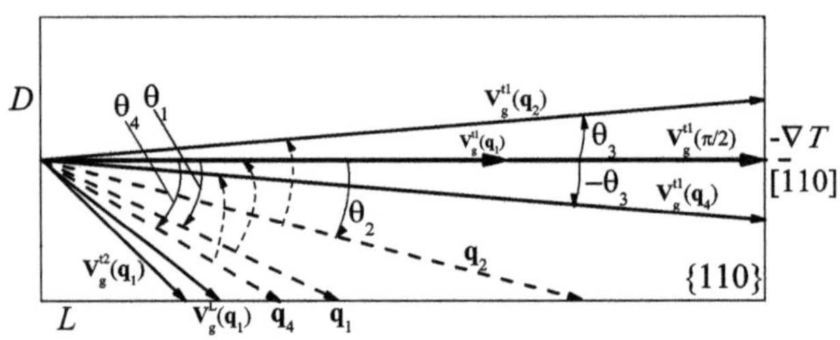

Figure 5.6: Diagram that illustrates the influence of focusing on the propagation of phonons in the cross-section of a GaAs film with the {110} orientation for wave vectors at the angles $\theta_1^{t1} = 21.4°$, $\theta_2^{t1} = 11.4°$, $\theta_3^{t1} = 3.6°$, and $\theta_4^{t1} = 25.4°$ to the [110] temperature gradient direction.

In this case, phonons of the t_1 mode in the {110} plane with the $\mathbf{q}_1$ wave vector and θ_1 angle (the angle is $\theta_1^{t1} = 21.4°$ for GaAs) will propagate in the [110] heat flux direction (see Figure 5.6), while phonons with the wave vector $\mathbf{q}_2$ and θ_2 angle (the angle is equal to $\theta_2^{t1} = 11.4°$ for the t_1 mode) will propagate at the angle $-\theta_3(\theta_3^{t1} = 3.6°)$ (see Figure 5.6). The [110] direction is the direction of defocusing of the slow transverse mode, and its mean free path is shorter than that in an isotropic medium (see Figure 5.6). Longitudinal phonons are focused in the [111] direction, and the [110] direction corresponds to the defocusing direction for wave vectors in the {110} plane. Therefore, they will deviate from the temperature gradient direction towards the wide face of the sample. Their mean free path is shorter than that for an isotropic medium. As noted earlier, the role of longitudinal phonons in phonon transport is small due to their low heat capacity.

Let us consider the propagation of phonons in films with the {100} orientation and made of crystals of the second type. In the process, we use CaF_2 crystals as an example. Suppose the temperature gradient is set along the [001] direction. For a slow transverse mode in the plane of the cube face, the phonon spectrum is isotropic and there is no focusing effect. Therefore, as can be clear from Figure 5.7, the direction of the q_2 wave vector coincides with the direction of the group velocity $V_g^{t1}(\mathbf{q}_2)$. Consequently, the mean free path is equal to $\Lambda_{[100]}^{t_2\{100\}}(\theta_2) = \Lambda_{iso}(\theta_2)$. For wave vectors in the {100} plane, the phonons of the fast transverse mode are defocused in the [001] direction. Therefore, they will deviate from the [001] temperature gradient direction to the planes of the film. Their mean free path is far much shorter than that for an isotropic medium (see Figure 5.7a). Longitudinal phonons are focused in the [001] direction and their mean free paths in films made of type-II crystals are larger than Λ_{iso}.

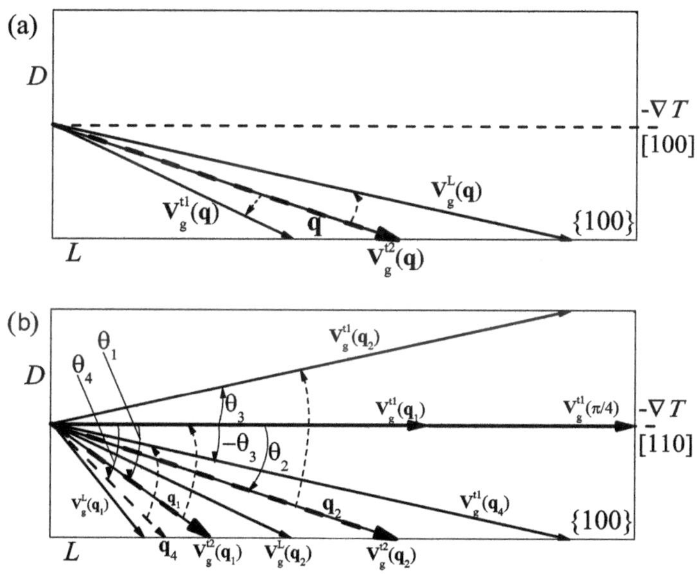

Figure 5.7: Scheme that illustrates the influence of focusing on the propagation of phonons in the cross-section of a CaF_2 film with the {100} orientation: (a) for [100] the temperature gradient direction and (b) for wave vectors at the angles $\theta_1^{t1} = 25.9°$, $\theta_2^{t1} = 13.4°$, $\theta_3^{t1} = 7.8°$, and $\theta_4^{t1} = 31.0°$ to the [110] temperature gradient direction.

The opposite situation develops for films with the {100} orientation and made of type-II crystals for the [011] temperature gradient direction. Phonons of the fast transverse mode are focused in this direction. Therefore, they will deviate from the planes of the film to the direction of the heat flux. Their mean free path can be significantly longer than Λ_{iso} (see Figure 5.7a, b). Figure 5.7b shows that, in CaF_2 films with the {100} plane, phonons of the t_1 mode with wave vectors in the sector $-\theta_4 \le \theta \le \theta_4$ (the angle is $\theta_4^{t_1} = 31°$ for the t_1 mode) will deviate from the side faces to the axis of the film and propagate in the sector $-\theta_3 \le \theta \le \theta_3$ (the angle is $\theta_3^{t_1} = 7.8°$ for GaAs). Thus, the density of states for the t_1 mode is significantly higher in the [110] direction and much lower for the defocusing directions than that in the model of an isotropic medium (see Section 1.5). In this case, phonons of the t_1 mode in the {110} plane with the q_1 wave vector and θ_1 angle (the angle is $\theta_1^{t_1} = 25.9°$ for GaAs) will propagate in the [110] heat flux direction; their mean free path are limited by either the length of the film or bulk relaxation mechanisms of phonon relaxation (see Figure 5.7). Phonons with the q_2 wave vector and θ_2 angle (the angle is $\theta_2^{t_1} = 13.4°$for the t_1 mode) will propagate at the angle $-\theta_3(\theta_3^{t_1} = 7.8°)$ (see Figure 5.7b).

For a slow transverse mode in the plane of the cube face, the phonon spectrum is isotropic and there is no focusing effect for CaF_2 films. Therefore, the direction of the group velocity $\boldsymbol{V}_g^{t_1}(\mathbf{q}_2)$ coincides with the direction of the $\mathbf{q}_2$ wave vector. Consequently, the mean free path is equal to $\Lambda_{[100]}^{t_2\{100\}}(\theta_2) = \Lambda_{iso}(\theta_2)$ (see Figure 5.7b). Longitudinal phonons are defocused in the [011] direction. Therefore, the propagation angle determined by the direction of the group velocity $V_g^{t_1}(q_2)$ becomes larger than the θ_2 angle and, accordingly, the mean free path $\Lambda_{[100]}^{L\{\bar{1}10\}}(\theta_2)$ becomes less than $\Lambda_{iso}(\theta_1)$ (see Figure 5.7b).

5.4 Conclusion

The fifth chapter explores the influence of anisotropy of elastic energy on phonon transport in monocrystalline films in the Knudsen phonon gas flow regime. The influence of phonon focusing on the propagation of phonon modes and phonon transport in monocrystalline films with different types of elastic energy anisotropy has been analysed. The main results of studies of phonon transport in films made of type-I (GaN, GaAs, Ge, MgO, Si, diamond, and YAG) and type-II crystals (NaCl, CaF_2, SrF_2, and YIG) can be formulated as follows:

1. The analysis of the distribution of phonon mean free paths over angles in the planes of the films makes it possible to determine physical causes leading to the isotropic dependence of thermal conductivity for square-shaped films with the {100} and {111} orientations. It is shown that, for an arbitrary direction of the heat flux, the region of averaging the mean free paths over angles in the film plane simultaneously captures the phonon focusing and defocusing directions. Therefore, the thermal conductivity, as well as the mean free paths of phonons of different polarizations for crystals of both types, become isotropic. The anisotropy of their thermal conductivity in the boundary scattering regime is fully dependent on the film plane's orientation. In films oriented along the {100} plane, the average mean free paths turn out to be longer and shorter for type-I and

type-II crystals, respectively, than those within the model of an isotropic medium. However, the opposite situation is realized in films with the {111} orientations: the average mean free paths in films made of type-I and type-II crystals are less and larger, respectively, than those calculated within the model of an isotropic medium.

2. In square-shaped films with the {110} orientation, the angle between the focusing and defocusing directions doubles as compared with the {100} and {111} orientations. Therefore, for an arbitrary direction of the heat flux, averaging the mean free paths over angles in the film plane cannot simultaneously cover both directions. In this regard, the thermal conductivity in the {110} film plane becomes anisotropic, and so do the contributions of acoustic modes. The values of the thermal conductivity for type-I crystalline films reach their maximum in the [001] directions, whereas the minimum values are observed in the [110] directions. For films made of crystals of the second type, the situation is quite the opposite.
3. With an increase in the length of the films, the intervals of angles that make the main contribution when averaging the mean free paths in the plane of the films with {100} and {111} are significantly narrowed. Therefore, in the limiting case of long films ($L \gg W$), the averaging over the angles is already insufficient to blur the phonon focusing effect. In this case, the mean free paths in films with the {100} and {111} orientations and made of crystals of both types become anisotropic.
4. The analysis of the influence of focusing on the propagation of phonon modes in films of crystals of the first type shows that the focusing effect prevails for fast and slow transverse phonons in square-shaped films with the {100} orientation. At the same time, in films oriented along the {111} plane, the defocusing effect dominates for the slow transverse mode. Therefore, the thermal conductivity and average mean free paths in films with the {100} orientation and made of type-I crystals turn out to be 2–2.4 times greater than those in films with the {111} planes.
5. The analysis of the distribution of mean free paths in the transverse cross-section of square-shaped films with the {100} orientation shows that the defocusing effect predominates for transverse phonons travelling in films made of type-II crystals. Therefore, the average phonon mean free paths in films oriented along the {100} plane and fabricated from type-II and type-I crystals turn out to be shorter and longer, respectively, than those predicted by the model of an isotropic medium. The situation is opposite for films with the {111} orientation: the focusing effect is dominant for the fast transverse mode in type-II crystalline films, while the defocusing effect is prevalent for transverse phonons in films produced from crystals of the first type. Thus, the average mean free paths in films with the {111} plane orientation and fabricated from crystals of the second and first types turn out to be longer and shorter, respectively than those within the model of an isotropic medium.

Chapter 6
Anisotropy and Temperature Dependencies of Thermal Conductivity of Silicon Films and Nanowires

As far as the temperature dependencies of thermal conductivity of monocrystalline silicon nanowires and films are concerned, the research works [4–7] took the measurements at temperatures ranged from 20 K to 300 K and from 17 K to 350 K, respectively. The former's diameter varied from 22 nm to 115 nm, whereas the latter's thicknesses changed from 0.02 μm to 1.6 μm. The findings secured were interpreted, as a rule, within the model of an isotropic medium [4–7, 25, 110]. In the process, either the Casimir theory [13] or the theory of Fuchs and Sondheimer [30, 31] was used. The effects due to focusing of phonons were excluded. The Fuchs and Sondheimer theory [30, 31] suggests that the thermal conductivity of the films depends on only one parameter, namely, their thickness. This approximation is incorrect. It was shown in [24, 27, 106] (see also Chapter 4) that the mean free paths in silicon films substantially depend on their geometric parameters. Moreover, the Casimir lengths logarithmically diverge as the film width tends to infinity. Earlier, the papers [25, 32] have noted this divergence for the thermal conductivity of films within the model of an isotropic medium. An analysis of the thermal conductivity of monocrystalline films, carried out in [27, 106], shows that the Casimir lengths for them also diverge logarithmically when the film width tends to infinity. For phonons of different polarizations, their values differ significantly as compared to an isotropic medium. As was demonstrated in the previous chapter, the thermal conductivity of the films largely depends on the orientation of their plane. It should be emphasized that the reviews [1–3], as well as the papers [4–7], ignore the directions of heat flux and planes' orientation relative to crystallographic axes.

In contrast to the preceding papers, the present chapter accounts for focusing of phonons and the phonon relaxation rates at the sample boundaries, calculated in [20, 21], to describe the temperature dependencies of the thermal conductivity of silicon films and nanowires (see also formulas (3.15) and (3.16)). Bearing in mind the phonon focusing, we will show that a change in the orientation of the silicon film can lead to a change in the thermal conductivity by more than 2 times.

The lack of information about the film plane's orientation raises certain difficulties in interpreting the experimental results [5–7] (see Section 6.2). Section 6.1 evaluates the temperature dependencies of the thermal conductivity of silicon nanowires for various directions of heat flux relative to crystallographic axes. Section 6.2 analyses the anisotropy and temperature dependencies of the thermal conductivity of silicon films having different plane orientations. The content of chapter VI is based on the papers [23, 24, 113].

https://doi.org/10.1515/9783110670509-007

6.1 Phonon Transport in Silicon Nanowires

The paper [4] were measured the temperature dependencies of the thermal conductivity of monocrystalline silicon nanowires with diameters from 22 nm to 115 nm in the temperature range from 20 K to 300 K. The axis of the grown-up nanowires lies in the [111] direction. It was shown that, for nanowires with diameters of 56 nm and 115 nm in the temperature range from 20 K to 60 K, the thermal conductivity follows the dependence $\kappa(T){\sim}T^3$, and so does the specific heat of bulk samples within the Debye theory. Therefore, the phonon spectrum for bulk Si crystals in calculating the thermal conductivity of these nanowires may be utilized. Obviously, analysing the thermal conductivity of nanoscale samples in the temperature range of 20 K–300 K requires accounting for the dispersion of the spectrum. For this purpose, as in Chapter 3, an approximation spectrum of thermal phonons in Si is used (see Section 1.2). This approximation preserves cubic symmetry and makes it possible to explore a change in the phonon focusing when going over from the long-wavelength $(x \ll 1)$ to short-wavelength limit $(x \approx 1)$. In the limit of long waves $(x \ll 1)$, this approximation is applied within the anisotropic continuum model [63].

6.1.1 Temperature Dependencies of Thermal Conductivity of Silicon Nanowires

For analysing the lattice thermal conductivity of silicon nanowires, as well as bulk silicon crystals, the three-mode Callaway model is used. In doing so, the contributions of the diffusion and drift motion of phonons should be extracted (see Section 3.1). According to (3.10) and (3.11), this model offers the total thermal conductivity as an additive sum of these contributions:

$$\kappa_{dif[I(\psi)]} = \frac{k_B q_T^3}{4(2\pi)^3} \sum_\lambda \int_{-1}^{1} d(\cos\theta) \int_0^{2\pi} d\varphi y^3 \int_0^1 \frac{\left(V_{g3}^\lambda\right)^2 Z_\lambda^2 x^2}{\nu_{[I(\psi)]}^\lambda (sh(Z_\lambda/2))^2} dx,$$
$$\kappa_{dr[I(\psi)]}^{(1)} = \frac{k_B q_T^3}{12(2\pi)^3} \sum_\lambda B_{[I(\psi)]}^{(1)}(T) \Psi_{N[I(\psi)]}^\lambda. \tag{6.1}$$

The designations are given in Section 3.1. Against square-shaped and rectangular cross-section samples, circular cross-section nanowires exhibit the thermal conductivity as dependent on only one orientation parameter, namely, the direction of heat flux relative to the crystal axes, $[I(\psi)]$, with the ψ angle controlling the heat flux direction. The total phonon relaxation rate includes the contributions of resistive $(\nu_{[I(\psi)]R}^\lambda)$ and normal $(\nu_N^\lambda(q))$ processes of phonon relaxation. Normal processes are responsible for the drift contribution to the thermal conductivity (see Section 3.1). Phonon scattering by isotopic disorder $(\nu_{iso}^\lambda(q))$ and at the sample's boundaries $(\nu_{[I(\psi)]B}^\lambda)$, and the Umklapp processes $(\nu_U^\lambda(q))$ are regarded as resistive scattering processes. The phonon relaxation rates $\nu_{iso}^{\lambda}(q)$, $\nu_{U}^{\lambda}(q)$, and $\nu_{N}^{\lambda}(q)$ in bulk relaxation mechanisms are defined by formulas (3.5), (3.6), and (3.17)–(3.19).

The phonon relaxation rate during diffuse boundary scattering in cylinder-shaped samples is calculated using piecewise-smooth functions for various intervals of angles. The borderline between the angles is governed by the parameter $\Delta^{\lambda}(\psi, x, \theta, \varphi) = (L/4\mathrm{R})V_{g\perp}^{\lambda}/|V_{g3}^{\lambda}|$. The latter's magnitude depends on the ratio between the group velocity components parallel (V_{g3}^{λ}) and perpendicular ($V_{g\perp}^{\lambda}$) to the temperature gradient and the geometric parameter $k_{\mathrm{O}} = L/2D$ and $D = 2R$. If the condition $\Delta^{\lambda}(\psi, x, \theta, \varphi) \leq 1$ holds, the relaxation rate has the form (see Section 2.2):

$$\nu_{B[I(\psi)]}^{\lambda}(x, \theta, \varphi) = \frac{3\pi V_{g\perp}^{\lambda}}{4\mathrm{D}}\left\{1 - \frac{3}{2}\left[(1 + (\Delta^{\lambda}))^2\sqrt{1 - (\Delta^{\lambda})^2} - \frac{1}{3}\left(\sqrt{1 - (\Delta^{\lambda})^2}\right)^3 - \Delta^{\lambda}\left(\arccos \Delta^{\lambda} + \Delta^{\lambda}\left(\sqrt{1 - (\Delta^{\lambda})^2}\right)\right)\right]\right\}^{-1}, \tag{6.2}$$

If the opposite inequality is fulfilled, we have:

$$\nu_{B[I(\psi)]}^{\lambda}(x, \theta, \varphi) = \left(3\pi V_{g\perp}^{\lambda}\right)/4D, \quad V_{g\perp}^{\lambda} = \sqrt{(V_{g}^{\lambda})^2 - (V_{g3}^{\lambda})^2}. \tag{6.3}$$

A further analysis of the thermal conductivity of nanowires implies the choice of the cube edges as a coordinate system to determine the spectrum and group velocities of phonons. Consider the rotation of the heat flux (the axis of the sample) along the diagonal plane. This plane captures all symmetric directions, including the [111] direction of the nanowire's axis. It is worth recalling that this direction was involved in exploring phonon transport in [4]. Let the ψ angle set a deviation of the heat flux in the diagonal plane from Z-axis pointing along the cube edge. Then, the group velocity component V_{g3}^{λ} parallel to the temperature gradient can be expressed through:

$$\begin{gathered}\nabla_r T = \{\nabla_x T, \nabla_y T, \nabla_z T\} = |\nabla_r T|\left\{-\sin\psi/\sqrt{2}, \sin\psi/\sqrt{2}, \cos\psi\right\}, \\ V_{g3}^{\lambda} = \left(-V_{gx}^{\lambda} + V_{gy}^{\lambda}\right)\sin\psi/\sqrt{2} + V_{gz}^{\lambda}\cos\psi.\end{gathered} \tag{6.4}$$

Accounting for the dispersion of phonons gives rise to the dependence of the group velocity, $\Delta^{\lambda}(\psi, x, \theta, \varphi)$ parameter, and relaxation rates $\nu_{B[I(\psi)]}^{\lambda}(x, \theta, \varphi)$ in formulas (6.2) and (6.3) on not only the angles ψ,θ, and φ but the reduced phonon wave vector x (see Section 3.2). As is clear from formulas (6.1)–(6.3), the $[I(\psi)]$-dependencies of the thermal conductivity on the temperature gradient direction relative to the crystal axes are determined by the magnitudes of the phonon group velocities components $V_{g3}^{\lambda}(\psi)$ and $V_{g\perp}^{\lambda}$, which are directly contained in the expressions for the thermal conductivity and relaxation rate $\nu_{B[I(\psi)]}^{\lambda}(x, \theta, \varphi)$.

It should be noted that when measuring the temperature dependencies of thermal conductivity [19], the bulk Si samples were preliminarily processed with sandpaper to furnish diffuse phonon-boundary scattering. The diffuse nature of scattering is realized if the geometric parameters of the sample's boundary roughness are larger or comparable with the wavelength of a phonon [8, 9, 14, 15]. Such a procedure is impossible to create in practice. Therefore, a partial specularity of the reflection of phonons from the nanowire's boundaries needs to be taken into account. Its accounting is carried out in the usual way [8, 9, 14, 15, 35]:

$$\tilde{\nu}^{\lambda}_{B[I(\psi)]}(\theta, \varphi) = \frac{(1-P)}{(1+P)} \nu^{\lambda}_{B[I(\psi)]}(\theta, \varphi). \qquad (6.5)$$

Figure 6.1 illustrates the influence of various relaxation mechanisms on the temperature dependencies of the thermal conductivity of silicon nanowires in different temperature ranges. The calculated data on the thermal conductivity when switching various phonon scattering mechanisms are given. In the low temperature region, when phonon scattering at the boundaries dominates, the thermal conductivity follows the dependence $\kappa(T) \sim T^3$. However, fully diffuse boundary scattering of phonons corresponds to a theoretical curve (curve 6) that goes 26% below the experimental data. Accounting for the specularity of the reflection ($P = 0.15$) allows one to reconcile the predicted values of the thermal conductivity with the experimental data. For nanowires with diameters of 115 nm and 56 nm, accounting for boundary and isotopic

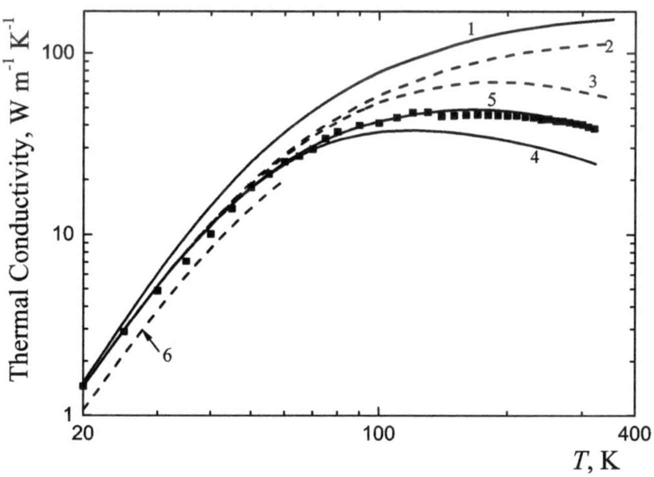

Figure 6.1: The temperature dependencies of thermal conductivity of silicon nanowires ($D = 115$ nm, $L = 3$ μm, $P = 0.15$) for the [111] temperature gradient direction and various phonon scattering mechanisms: curve 1 – the boundary scattering regime ($\nu = \tilde{\nu}^{\lambda}_{B[111]}$); curve 2 – $\nu = \tilde{\nu}^{\lambda}_{B[111]} + \nu_{iso}$; curve 3 – $\nu = \tilde{\nu}^{\lambda}_{B[111]} + \nu_{iso} + \nu_U$; curve 4 – the diffusion contribution ($\kappa_{dif\,[111]}(T)$); curve 5 – the total thermal conductivity; curve 6 – the boundary scattering regime when $P = 0 (\nu = \nu^{\lambda}_{B[111]})$. Symbols ■ are the experimental data of [4].

scattering makes it possible to match the calculated and experimental data in the temperature range from 20 K to 60 K with an accuracy of 4% without resorting to additional fitting parameters (see Figure 6.1, curve 2). The parameter $P = 0.15$ is further used for all calculations shown in Figure 6.1. As can be understood from Figure 6.1, at temperatures above 60 K, anharmonic scattering processes play a significant role in heat resistance. In calculating the temperature dependencies of the thermal conductivity of nanowires, the parameters of anharmonic scattering processes, found earlier in [22] for bulk Si crystals, should be accepted as the basis (see also Table 6.1). These parameters give the possibility of reconciling the computed temperature dependencies of the thermal conductivity of bulk Si rectangular and square-shaped cross-section crystals with the experimental data in [19] for various directions of the temperature gradient and side faces of the samples in the temperature range of 3 K–40 K. It should be underscored that this interval actually covers only the region of maximum thermal conductivity in bulk Si crystals. It is too narrow to evaluate the anharmonic scattering parameters. Therefore, analysing the thermal conductivity of silicon nanowires in the range of 20 K–300 K requires the specification of the anharmonicity parameters. To better agree with the calculated results of [4], the $A_U^{t_1}$ parameter that characterizes the phonon relaxation frequency for the fast transverse mode in the Umklapp processes should be increased (see Tables 3.1 and 6.1).

Table 6.1: Parameters of anharmonic phonon–phonon scattering and isotopic disorder scattering for Si_{nat} crystals.

A_N^l, $K^{-5}sec^{-1}$	A_N^l, $K^{-5}sec^{-1}$	A_U^l, $K^{-3}sec^{-1}$	$A_N^{t_1}$, $K^{-3}sec^{-1}$	$A_U^{t_2}$, $K^{-3}sec^{-1}$	C_N^l, K	$C_U^{t_l}$, K
0.8	$2.0 \cdot 10^{-3}$	$2.0 \cdot 10^3$	$1.80 \cdot 10^3$	$0.70 \cdot 10^3$	310	98

It should be noted that the values of thermal conductivity of nanoscale samples in the temperature range of 20 K–300 K are much weaker depending on the anharmonicity parameters because of the more significant (than in bulk materials) role of boundary scattering (see below).

Let us examine the contribution of the drift motion of phonons to the thermal conductivity of silicon nanowires in the temperature range from 20 K to 300 K. The performed analysis shows that, over the entire temperature range, the drift contribution of transverse phonons is small compared to the diffusion one and it can be neglected (see Figure 6.2). At $T < 60$ K, the drift contribution of longitudinal phonons to the thermal conductivity, $\kappa_{[111]dr}^{(1)L}(T)$, is much smaller than the diffusion one. However, at higher temperatures of $100\text{K} < T < 300\text{K}$, normal phonon-phonon scattering processes become dominant, and the drift contribution is vastly superior to the contribution of diffusion motion (see Figure 6.2). As for nanowires with diameters of 115 nm and 56 nm, the $\kappa_{[111]dr}^{(1)L}(T)$ contribution at $T = 100$ K is 13% and 9%

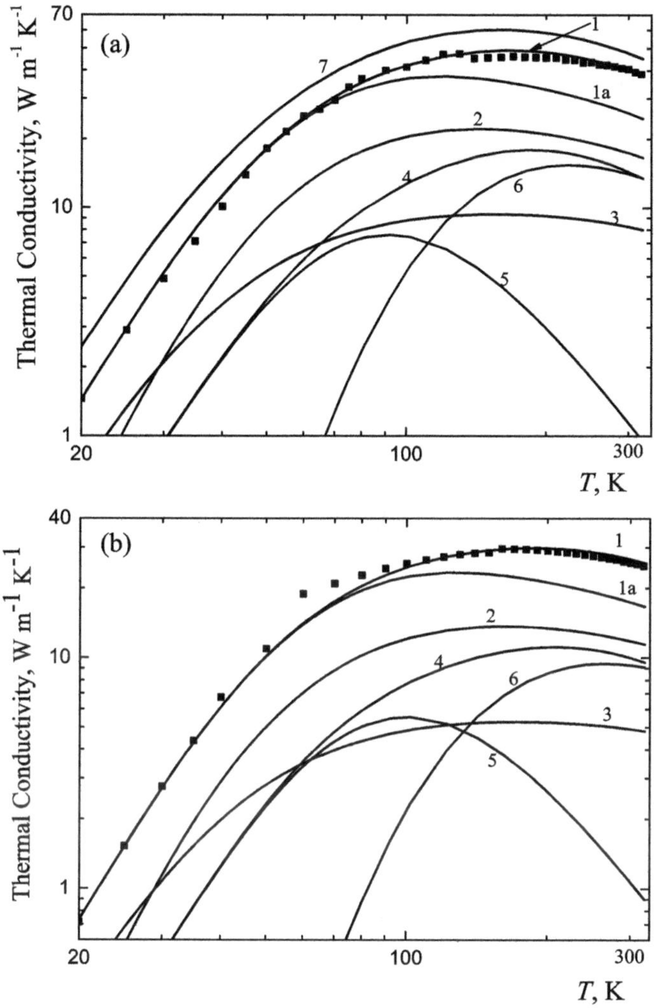

Figure 6.2: The temperature dependencies of thermal conductivity of silicon nanowires with a diameter of (a) $D = 115$ nm and (b) $D = 56$ nm in the [111] temperature gradient direction: curves 1 – for the total thermal conductivity; curves 1a – for the contributions of the diffusion motion; curves 2, 3, and 4 – for the contributions of fast, slow transverse and longitudinal phonons, respectively; curves 5, 6 – for the diffusion and drift contributions of longitudinal phonons; curve 7 refers to the total thermal conductivity in the [100] direction. Symbols ■ are the experimental data of [4].

of the total thermal conductivity, respectively. At $T = 300$ K, it reaches 35%. Figure 6.2 indicates that the accounting for the drift motion of longitudinal phonons allows one to match the calculated results with the experimental data of [4] over the entire temperature range studied.

To estimate the thermal conductivity anisotropy of silicon nanowires, we should plot the temperature dependencies $\kappa_{[I(\psi)]}(T)$ for the [111] and [100] temperature gradient directions (see Figure 6.2a, curves 1 and 7). As can be inferred from the calculation, the ratio of thermal conductivities $\kappa_{[100]}/\kappa_{[111]}$ is 1.7 for 115 and 56 nm diameter nanowires at $T = 20$ K, while at $T = 300$ K it declines to 1.2. It can be stressed that, for bulk Si rods investigated at $T = 3$ K, this ratio is equal to 1.50 and tends to unite above the maximum temperature (see Chapter 3).

The role of various branches of the phonon spectrum in the thermal conductivity of nanowires varies significantly with increasing temperature. These changes are due to both the dispersion of thermal phonons and the drift motion of longitudinal phonons. Figure 6.2 shows that, at low temperatures $T < 30$ K, the slow transverse mode $\kappa^{t_2}(T)$ dominantly contributes to the thermal conductivity. At $T = 20$ K, its contribution is 49% and twice the contribution of the fast transverse mode. However, as the temperature grows, the contribution of the fast transverse mode $\kappa^{t_1}(T)$ increases much faster than the contribution of the slow transverse mode $\kappa^{t_2}(T)$. At $T > 30$ K, the values of $\kappa^{t_1}(T)$ become larger than those of $\kappa^{t_2}(T)$ (see Figure 6.2). Therefore, in the entire temperature range from 30 K to 300 K, the fast transverse mode makes the dominant contribution to the thermal conductivity. The growth slowdown in the contribution of the slow transverse mode $\kappa^{t_2}(T)$ with increasing temperature is associated with a strong dispersion of the t_2 mode.

When $q_{max}/2 < q < q_{max}$ (see Figure 1.5), the long flat segments in the short-wavelength spectrum of the slow transverse mode provoke abnormally low values of the group velocity and, accordingly, a drastic decline in the contribution of $\kappa^{t_2}(T)$ with increasing the temperature. At $T = 20$ K, the contribution of longitudinal phonons to the thermal conductivity is much less than that of transverse phonons and amounts to 19%. However, with increasing the temperature, this contribution rapidly increases due to drift motion and, at $T > 65$ K, the contribution of longitudinal phonons becomes larger than that of the slow transverse mode. At $T = 190$ K, the contribution of $\kappa^{L}(T)$ reaches a maximum and amounts to 36% of the total thermal conductivity.

Figure 6.3 displays the temperature dependencies of the total thermal conductivity in the regime of the boundary $(\nu = \nu_B)$ and bulk $(\nu = \nu_{iso} + \nu_U + \nu_N)$ phonon relaxation mechanisms.

The points of intersection of these curves give the transition temperatures (T_{BV}) from the boundary scattering to the bulk relaxation mechanisms. At these temperatures, the mean free paths for the boundary scattering and for bulk mechanisms turn out to be equal, $\Lambda_{B[I(\psi)]} = \Lambda_V$, and the boundary scattering provides 50% of the total heat resistance. A numerical analysis shows that for 115 nm and 56 nm diameter Si nanowires, the transition temperatures T_{BV} are 265 K and 353 K, respectively. As can be concluded from Figure 6.3, the transition temperatures T_{BV} for bulk materials are an order of magnitude lower. At room temperature, for nanowires with diameters of 56 nm and 115 nm, the contribution of boundary scattering to the thermal resistance is 60% and 42%, respectively.

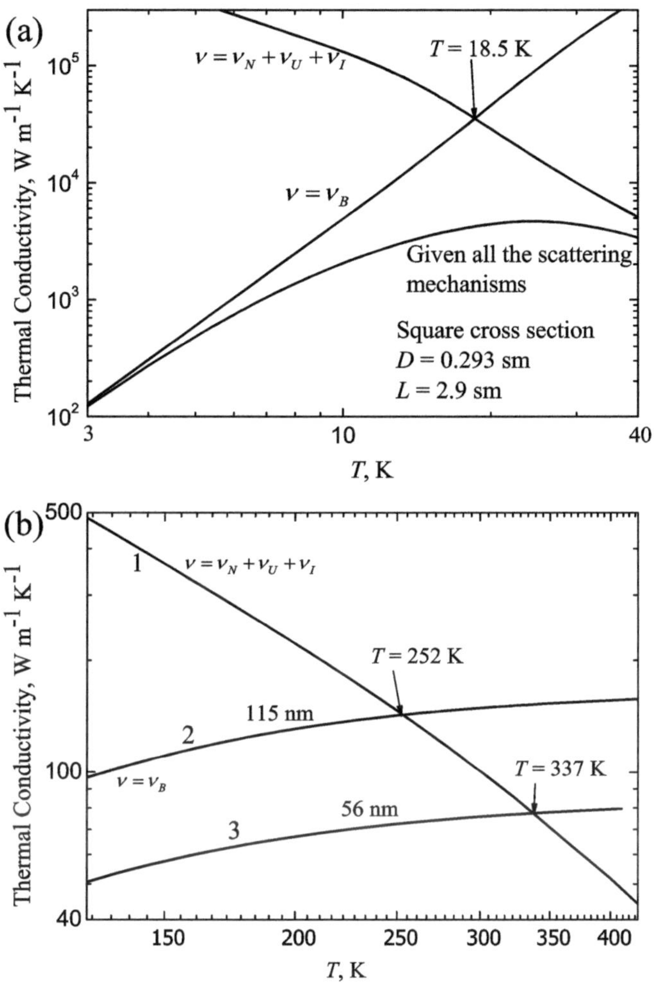

Figure 6.3: The temperature dependencies of thermal conductivity (a) of bulk samples and (b) nanowires in the [111] direction: curves 1 – for bulk phonon relaxation mechanisms, curves 2 and 3 – for the phonon-boundary scattering regime in nanowires with diameters $D = 115$ nm and $D = 56$ nm, respectively.

6.1.2 Anisotropy of the Thermal Conductivity of Silicon Nanowires

Let us look into the change in the anisotropy of the total thermal conductivity and the contributions to it from various branches of the phonon spectrum with increasing temperature. To do this, we analyse the change in the angular dependencies of the thermal conductivity of silicon nanowires when the temperature gradient (the sample's axis) rotates in the diagonal plane and temperature is altered. The group velocity components V_{g3}^{λ} and $V_{g\perp}^{\lambda}$ necessary for this analysis are determined by expression (6.4), with the ψ angle defining the deviation of the heat flux from the Z-axis pointing along the cube edge.

The calculated findings of the angular dependencies of thermal conductivity of silicon nanowires are shown in Figure 6.4. As can be seen from the figures, at T = 20 K, the slow transverse phonons make the dominant contribution to the thermal conductivity (49%). At this temperature, phonon boundary scattering provides 94% of the total heat resistance; therefore, its anisotropy is maximal. Maximum values of the thermal conductivity are realized in the [001] directions

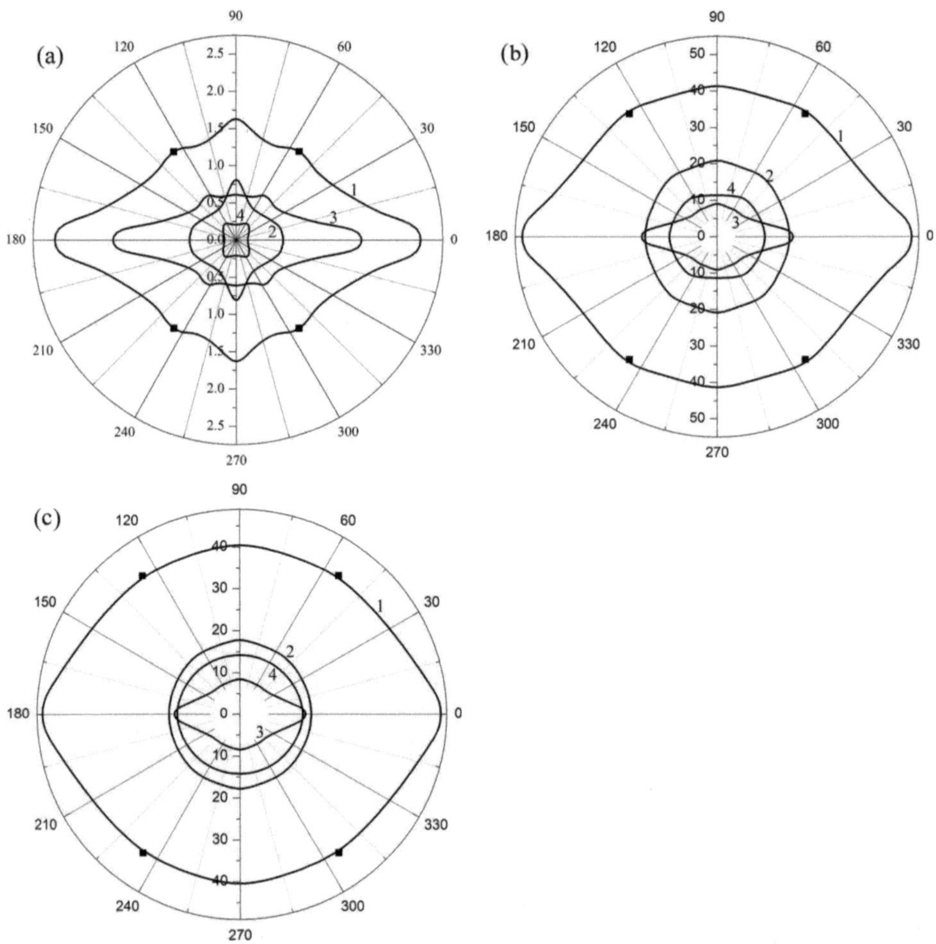

Figure 6.4: The $\kappa(\psi)$ angular dependencies of the total thermal conductivity, $W \bullet m^{-1} \bullet K^{-1}$, (curve 1) and contributions of different branches of the phonon spectrum to it for nanowires with D = 115 nm in the case of rotation of the nanowire axis in the diagonal plane for temperatures: (a) 20 K, (b) 100 K, (c) 300 K. Curves 2, 3, and 4 are the contributions of fast, slow transverse modes, and longitudinal phonons, respectively. Symbols ■ refer to the experimental data of [4].

due to focusing of slow transverse phonons and minimum ones are reached at $\psi_{min} \approx 68°$ (see Figure 6.4a). At this temperature, the ratio $\kappa_{[001]}/\kappa_{min} \cong 1.8$ specifies the maximum anisotropy of the thermal conductivity. In the [011] and [111] directions, local maximums of the thermal conductivity are seen. They are due to focusing of fast transverse and longitudinal phonons, respectively (see Figure 6.4a).

In the symmetric directions for nanowires with D = 115 nm, the ratio of thermal conductivity magnitudes is $\kappa_{[001]}:\kappa_{[011]}:\kappa_{[111]} = 1.69:1.12:1$ For nanowires with D = 56 nm, the anisotropy is slightly higher: $\kappa_{[001]}:\kappa_{[011]}:\kappa_{[111]} = 1.74:1.13:1$.

As is clear from Figure 6.4, the thermal conductivity anisotropy of silicon nanowires is mainly provided by the slow transverse mode, for which the ratio $\kappa^{t_2}_{[100]}/\kappa^{t_2}_{[110]}$ is equal to 2.7, 2.3, and 1.85 at temperatures of 20 K, 100 K, and 300 K, respectively. This stands for the fact that the decrease in the thermal conductivity anisotropy with increasing temperature is mainly due to the decrease in the contribution of the slow mode (see Figure 6.4). At T = 20 K, the contribution of the fast transverse mode $\kappa^{t_1}_{[100]}$ in the [001] direction is 2.7 times less than the contribution of $\kappa^{t_2}_{[100]}$. However, in the [011] direction, it exceeds the contribution of the slow mode by 1.3 times and provides a local maximum of the thermal conductivity at T = 20 K. With increasing the temperature, the $\kappa^{t_2}_{[111]}$ contribution of slow transverse phonons drops from 49% at T = 20 K to 21% at T = 100 K. On the other hand, the contributions of fast transverse and longitudinal phonons to the thermal conductivity in this direction increase, reaching 49% and 30% at T = 100 K, respectively. At T = 100 K, these contributions become almost isotropic. Therefore, the anisotropy of the total thermal conductivity decreases to 26% (see Figure 6.4b). At T = 100 K, the ratio of magnitudes of the thermal conductivities for symmetrical directions is $\kappa_{[001]}:\kappa_{[111]}:\kappa_{[011]} = 1.3:1:1$. The minimum values of the thermal conductivity are achieved at $\psi_{min} \approx 74°$; they are 3% lower than the values of $\kappa_{[011]}$. A further increase in temperature up to T = 300 K increases the contribution of longitudinal phonons to the thermal conductivity in the [111] directions to 36%. This occurs due to the drift motion of phonons. At the same time, the contributions of fast and slow transverse phonons to the thermal conductivity go down to 43% and 21%, respectively. The anisotropy of the total thermal conductivity diminishes. In the symmetric directions, for 115 nm diameter nanowires, the ratio of the thermal conductivity magnitudes is $\kappa_{[001]}:\kappa_{[111]}:\kappa_{[011]} = 1.2:1:1$ at T = 300 K. In spite of dominating at T = 300 K, the bulk relaxation mechanisms $(T_{BV} = 265K)$ and boundary scattering $(T_{BV} = 353K)$ in 115 nm and 56 nm diameter nanowires, respectively, have a weak influence on the thermal conductivity anisotropy. At T = 300 K, the ratio of thermal conductivities of 56 nm diameter nanowires for symmetrical directions does not increase significantly: $\kappa_{[001]}:\kappa_{[111]}:\kappa_{[011]} = 1.2:1:1$.

Thus, the conducted analysis shows that the anisotropy of thermal conductivity in silicon nanowires is chiefly provided by the slow transverse mode, which is focused in the [100] direction and provides maximum thermal conductivity in this direction. The decrease in the anisotropy of thermal conductivity from 70% at T = 20 K to 20% at T = 300 K is mainly explained by a decrease in the contribution of the slow transverse mode to the thermal conductivity with rising temperature and dispersion of thermal phonons.

6.2 Phonon Transport in Silicon Films

The present section deals with phonon transport in rectangular cross-section films with a thickness D, width $W = \mu D$, and length L. Let us compare the predicted results with the data of experimental studies [5–7], in which the thermal conductivity of silicon films with thicknesses of D = 1.6, 0.83, 0.42, 0.10, and 0.02 μm was explored in the temperature range from 17 K to 350 K. In the range from 17 K to 40 K, the thermal conductivity of these films follows the $\kappa(T) \sim T^3$ dependence [6, 7] and so does the heat capacity of bulk samples within the Debye theory. Therefore, in calculating the thermal conductivity of the films, a three-dimensional phonon spectrum was used. To take into account the dispersion of thermal phonons, the approximation spectrum was used [39, 63] (see also Section 1.2).

6.2.1 Temperature Dependencies of Thermal Conductivity of Silicon Films

To analyse the thermal conductivity of nanostructures such as films and nanowires, we resort to the three-mode Callaway model [91, 92, 94]. The latter offers the lattice thermal conductivity as the additive sum of the diffuse ($\kappa^{\{J\}}_{dif[I(\psi)]}$) and drift ($\kappa^{\{J\}}_{dr[I(\psi)]}$) contributions. Against circular cross-section nanowires, films demonstrate the thermal conductivity as dependent on not only the direction $[I(\psi)]$ of heat flux but on the film plane's orientation $\{J\}$ (see Section 3.1):

$$\kappa^{\{J\}}_{dif[I(\psi)]} = \frac{k_B q_T^3}{4(2\pi)^3} \sum_\lambda \int_{-1}^{1} d(\cos\theta) \int_0^{2\pi} d\varphi y^3 \int_0^1 \frac{\left(V^\lambda_{g3}\right)^2 Z_\lambda^2 x^2}{\nu^{\{J\}\lambda}_{[I(\psi)]} (sh(Z_\lambda/2))^2} dx,$$

$$\kappa^{\{J\}(1)}_{dr[I(\psi)]} = \frac{k_B q_T^3}{12(2\pi)^3} \sum_\lambda B^{\{J\}(1)}_{[I(\psi)]}(T) \Psi^{\{J\}\lambda}_{N[I(\psi)]}. \qquad (6.6)$$

The total phonon relaxation rate $\nu^{\{J\}\lambda}_{[I(\psi)]}$ consists of the contributions of resistive ($\nu^{\{J\}\lambda}_{R[I(\psi)]}$) and normal ($\nu^\lambda_N(q)$) relaxation processes. The latter involve the drift contribution to the thermal conductivity (see Section 3.1). As resistive scattering processes, phonon scattering by isotopic disorder ($\nu^\lambda_{iso}(q)$), at the boundaries of the samples ($\nu^{\lambda\{J\}}_{B[I(\psi)]}$), and the Umklapp processes ($\nu^\lambda_U(q)$) are considered. The phonon relaxation rates $\nu^\lambda_{iso}(q)$, $\nu^\lambda_U(q)$, and $\nu^\lambda_N(q)$ are given by formulas (3.5), (3.6), (3.17)–(3.19). The phonon relaxation rates $\nu^{\lambda\{J\}}_{B[I(\psi)]}(q)$ at the boundaries of the films are defined by the same functions as for bulk materials with rectangular cross-section. If the inequalities

$\mu|V_{g1}^{\lambda}| > |V_{g2}^{\lambda}|$ and $|V_{g3}^{\lambda}/V_{g1}^{\lambda}| \geq k_0$ or $\mu|V_{g1}^{\lambda}| < |V_{g2}^{\lambda}|$ and $|V_{g3}^{\lambda}/V_{g2}^{\lambda}| \geq k_0/\mu$ are satisfied, they have the form (see Section 3.2):

$$\nu_{B[I(\psi)]}^{\lambda\{J\}}(x,\theta,\varphi) = \frac{|V_{g3}^{\lambda}|}{k_0 D}\left\{1 - \frac{k_0}{2}\frac{\left(|V_{g2}^{\lambda}| + \mu|V_{g1}^{\lambda}|\right)}{\mu|V_{g3}^{\lambda}|} + \frac{(k_0)^2}{3}\frac{|V_{g1}^{\lambda}||V_{g2}^{\lambda}|}{\mu\left(V_{g3}^{\lambda}\right)^2}\right\} \tag{6.7}$$

When the opposite inequalities hold, we arrive at:

$$\nu_{B[I(\psi)]}^{\lambda\{J\}}(x,\theta,\varphi) = \frac{6}{D}\frac{\mu\left(V_{g1}^{\lambda}\right)^2}{\left(3\mu|V_{g1}^{\lambda}| - |V_{g2}^{\lambda}|\right)}, \mathit{if} \quad \mu|V_{g1}^{\lambda}| > |V_{g2}^{\lambda}| \quad \mathit{and} \quad |V_{g3}^{\lambda}/V_{g1}^{\lambda}| < k_0,$$

$$\nu_{B[I(\psi)]}^{\lambda\{J\}}(x,\theta,\varphi) = \frac{6}{\mu D}\frac{\left(V_{g2}^{\lambda}\right)^2}{\left(3|V_{g2}^{\lambda}| - \mu|V_{g1}^{\lambda}|\right)}, \quad \mathit{if} \quad \mu|V_{g1}^{\lambda}| > |V_{g2}^{\lambda}| \quad \mathit{and} \quad |V_{g3}^{\lambda}/V_{g2}^{\lambda}| < k_0/\mu. \tag{6.8}$$

It should be stressed that accounting for the dispersion brings in dependence both the relaxation rates in formulas (6.7) and (6.8) and the group velocity components V_{g1}^{λ}, V_{g2}^{λ}, and V_{g3}^{λ} on not only the angles ψ, θ and φ but the reduced phonon wave vector x. The case when $\mu = 1$ corresponds to square-shaped transverse cross-section nanowires. This section is aimed at calculating the temperature dependencies of the thermal conductivity of films of different orientations {J}. The next step is to analyse the $\kappa_{[I(\psi)]}^{\{J\}}(T)$ angular dependencies of thermal conductivity at fixed temperatures when heat flux rotates along the {100}, {110}, and {111} film plane's orientations. This will allow one to study the change in the anisotropy of thermal conductivity with temperature, as well as to determine the film plane's orientations and the directions of the heat flux, which provide the maximum or minimum values of the thermal conductivity of monocrystalline silicon films.

As in Section 4.1, the group velocity components V_{g1}^{λ}, V_{g2}^{λ}, and V_{g3}^{λ} for various film plane's orientations {J} are found in a coordinate system related to the films (see formulas (4.4)). Let the coordinate system be assigned as follows. Axis 1 (the axis of rotation of heat flux) defines the plane orientation {J] and is perpendicular to the latter. Axis 3 is pointing along the heat flux $[I(\psi)]$. The ψ angle that controls the heat flux direction in the plane {J} is measured from Z-axis set along the cube edge.

In contrast to bulk crystals, for thin films and nanowires, the diffuse nature of phonon boundary scattering cannot be fully implemented in practice. Therefore, partial specular reflection of phonons from the film boundaries is given by the standard way [8, 9, 14, 15, 35]:

$$\tilde{\nu}_{B[I(\psi)]}^{\lambda\{J\}}(P,x,\theta,\varphi) = \frac{(1-P)}{(1+P)}\nu_{B[I(\psi)]}^{\lambda\{J\}}(x,\theta,\varphi), \tag{6.9}$$

where P is the specularity factor that can be derived from the matching between the predicted and experimental data. As for Si circular cross-section nanowires, accounting for the specularity factor equal to $P = 0.15$ aids in well agreeing the calculated outcomes with the experimental data of [4] in the entire temperature range. Unfortunately, the papers [5–7] indicate no directions of heat flux and film plane's orientations relative to the crystallographic axes when interpreting the experimental data. The SOI technology (silicon on insulator) is usually used for films with the {100} or {111} orientations (see, for example [111]) to cut. The present section states that the thermal conductivity in the plane of sufficiently wide films ($L >> D$ and $W >> D, L \sim W$) with the above orientations almost does not depend on the direction of heat flux (see Section 6.2.2). However, the dependence of their thermal conductivity on the film plane's orientation is essential: when going over from the {111} to {100} orientation, the thermal conductivity may increase more than two times. Since the papers [5–7] leave aside information about the orientation of the film planes, we have calculated the temperature dependencies of thermal conductivity for films with three orientations such as $\{J\} = \{100\}$, {110}, and {111}. Besides, we have matched the predicted and experimental data using the specularity factor P as a fitting parameter. The results of such adjusting are shown in Figure 6.5.

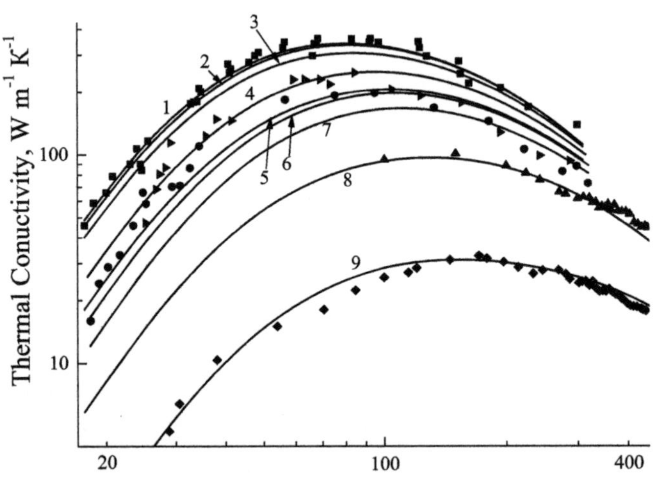

Figure 6.5: The temperature dependencies of thermal conductivity of silicon films with various thicknesses D, specularity parameters P, and film lengths L: curves 1, 2, 3 – $D = 1.6$ μm, $P = 0.48$, $L = 8$ μm; curve 4 – $D = 0.83$ μm, $P = 0.29$, $L = 8$ μm; curves 5, 6, 7– $D = 0.42$ μm, $P = 0.29$, $L = 8$ μm; curve 8- $D = 0.10$ μm, $P = 0.14$, $L = 100$ D; curve 9 – $D = 0.02$ μm, $P = 0$, $L = 100$ D. Curves 1, 4, 5, 8, 9 are for the {100} film orientations; curves 2, 6 – for the {110} orientation, curves 3, 7- for the {111} orientation. The width of films is $W = 100$ D. Symbols correspond to the experimental data of [5–7]: ■ – for $D = 1.6$ μm, ►- for $D = 0.83$ μm, • – for $D = 0.42$ μm, ▲- for $D = 0.10$ μm, and ♦ – for $D = 0.02$ μm.

For films with the {100} orientation and D = 1.6 μm, the values of the thermal conductivity are in good agreement with the experimental data of [6, 7] when $P_{\{100\}} = 0.48$. It is worth noting that when $P = 0.48$ and T = 20 K, they turn out to be 7% and 21% less for films with the {110} and {111} orientations, respectively, than those for films with the {100} orientation. The ratios of the thermal conductivity values are $\kappa_{[100]}^{\{100\}}:\kappa_{[100]}^{\{110\}}:\kappa_{[110]}^{\{111\}} = 1.2:1.1:1$. When $P_{\{110\}} = 0.51$ $P_{\{111\}} = 0.56$, the thermal conductivity values of $\kappa_{[100]}^{\{110\}}(T)$ and $\kappa_{[110]}^{\{111\}}(T)$ are also in good agreement with the experimental data and close to the values of $\kappa_{[100]}^{\{100\}}(T)$ when $P_{\{100\}} = 0.48$. For films with the {100} orientation and D = 0.83 and 0.42 μm, the calculated findings are matched with the experimental data when $P_{\{100\}} = 0.29$. However, for these films with the {110} and {111} orientations, the agreement is observed when $P_{\{110\}} = 0.37$ and $P_{\{111\}} = 0.50$ (see Figure 6.5). The ratios of the thermal conductivity values for films with different plane orientations and D = 0.42 μm appear as $\kappa_{[100]}^{\{100\}}:\kappa_{[100]}^{\{110\}}:\kappa_{[110]}^{\{111\}} = 1.2:1.1:1$ when $P_{\{100\}} = 0.29$ and T = 20 K. As Figure 6.5 displays, the predicted values of thermal conductivity for nanofilms with the {100} plane orientation and D = 100 and 20 nm are also in satisfactory accordance with the results obtained in [5]. The decrease in the film thickness leads to the weighty role of boundary scattering and, accordingly, an increase in the anisotropy of the thermal conductivity. The ratios of the thermal conductivity values for films with different plane orientations and D = 100 nm amount to $\kappa_{[100]}^{\{100\}}:\kappa_{[100]}^{\{110\}}:\kappa_{[110]}^{\{111\}} = 2.2:1.5:1$ when $P_{\{100\}} = 0.14$ and T = 20 K (see Figure 6.6). For 20 nm thickness films, the specular parameter P appears to be

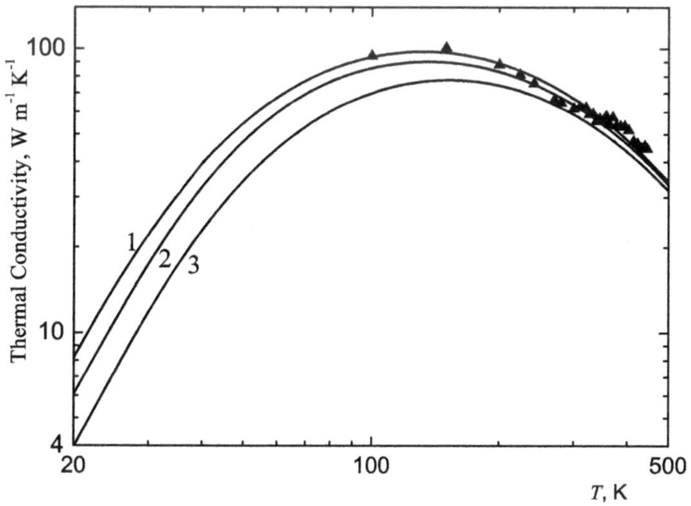

Figure 6.6: The temperature dependencies of thermal conductivity of silicon films with a thickness of D = 100 nm, a width of W = 100 D, and a length of L = W for various orientations $\{J\}$ and heat flux directions $[I\ (\psi)]$: curves 1 – {100} [100]; 2 – {110} [100]; 3 – {111} [112]. The specularity parameter is P = 0.14. Symbols are the experimental data of [5].

close to zero because the temperature dependence of thermal conductivity is satisfactorily described when $P = 0$. Furthermore, as Figure 6.6 implies, the role of the bulk scattering processes goes up with increasing the temperature, and the thermal conductivity anisotropy diminishes.

In this case, the dependence of the thermal conductivity on the film plane's orientation weakens as well. When $\mu = 100$ and $T = 20$ K, the ratio of the thermal conductivity values for films with the {100} and {111} orientations is equal to 2.2. However, it drops to 1.2 at $T = 320$ K. Thus, with diffuse scattering of phonons, films with the {100} orientation exhibit maximum thermal conductivity, whereas minimum thermal conductivity is inherent in films with the {111} orientation.

The analysis of the dependence of thermal conductivity of films on the geometric parameters shows that, for unchanged magnitudes of D and L, an increase in the film width initiates an increase in the thermal conductivity (see Figure 6.7). Then, the region of its intensive growth is limited by the values of $\mu < 20\,(L/D)$ or $W < 20L$. When $\mu > 20\,(L/D)$, the dependencies of the thermal conductivity reach saturation (see Figure 6.7). When $\mu = 20\,(L/D)$ (e.g. for films with $D = 1.6$ μm and $\mu = 100$), the thermal conductivity value is 0.5% less than the limited value.

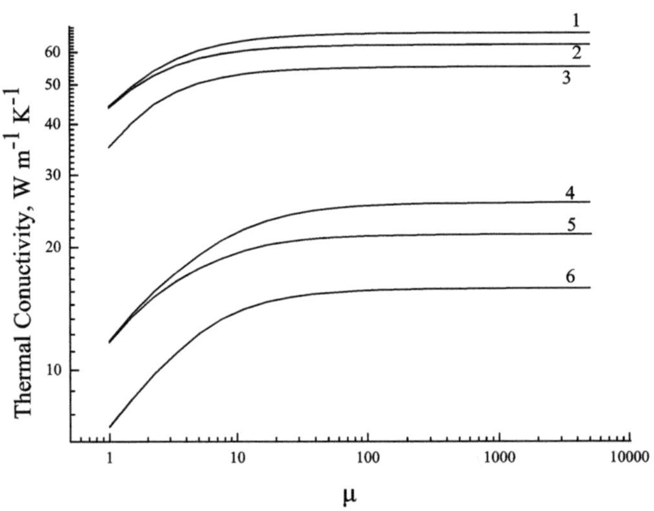

Figure 6.7: Dependencies of thermal conductivity of silicon films on the parameter μ when $L = 8$ μm and film thickness D: curves 1, 2, 3 correspond to $D = 1.6$ μm, curves 4, 5, 6 are for $D = 0.42$ μm; and film plane's orientations {J}: curves 1, 4 refer to {J} = {100}, curves 2, 5 are for {J} = {110}, curves 3, 6 are for {J} = {111}.

As for the dependence of thermal conductivity on the film length, the region of its intensive growth is limited by the values of $L < \mu D$ when D and W are fixed. When $L > 20\,\mu D$, it reaches saturation. It should be emphasized that a huge number of publications (see, e.g. [5–7, 28, 29]) take into account phonon boundary scattering in sufficiently thin films ($D << L$ and $D << W$) in an analogous way as it was done in the works of Fuchs [30] and Sondheimer [31] when analysing the conductivity of thin metallic films.

In [5–7, 28, 29], mean free paths and phonon relaxation times in the films are assumed to depend only on their thickness. The papers [5–7, 25, 110] claim that the model of an isotropic medium regards boundary scattering as not leading to the anisotropy of thermal conductivity A direct generalization of the results of [30, 31] to elastically anisotropic crystals gives an expression for the boundary relaxation rate, with the latter depending only on the thickness D and the group velocity component V^{λ}_{g1} perpendicular to the film plane [28]:

$$v^{\lambda}_{B}(\theta,\varphi)=\frac{(1-P)}{(1+P)}\frac{V^{\lambda}_{g1}}{D}. \tag{6.10}$$

The method of determining the specularity factor P and its relation to surface roughness does not change the essence of the problem [8, 14, 28, 29, 30, 31, 112]. Accounting for boundary scattering in the form of (6.10) results in incorrectly treating the dependence of thermal conductivity on the geometric parameters and orientations of the film planes. According to the analysis performed in [27, 106], the thermal conductivity of films and phonon mean free paths substantially depends not only on the film width but also on the other geometric dimensions. Moreover, the papers [25, 27, 32, 106] show that the Casimir lengths in the model of an isotropic medium not only significantly depend on the film width but also logarithmically diverge when its width tends to infinity. As noted in [25, 32], this divergence is due to phonons propagating almost parallel to the film plane. As demonstrated in [27, 106] (see also Section 4.2.1), taking the finite length of films into account eliminates this divergence.

According to expression (6.10), when calculating the temperature dependencies, the thermal conductivity of silicon films in [28] reaches its maximum and minimum values for the {110} and {100} orientations, respectively (see [28], Figure 4a). The authors conclude that in the case of phonon diffuse scattering at the film boundaries, the plane with the {110} orientation possesses the least scattering power (and the maximum thermal conductivity), whereas the plane with the {100} orientation has the maximum scattering power (and the minimum thermal conductivity). These results contradict the experimental data of [19] and the outcomes of [27, 106]. As shown in [19], two identical rectangular cross-section Si samples set in the [110] temperature gradient direction exhibit different thermal conductivity values for the {100} and {110} wide faces. In the first case (for the {100} wide faces), it turned out to be 33% higher than for the {110} wide faces. This result is the opposite of the conclusion made in [28]. It should be underscored that the predicted values of the thermal conductivity of bulk Si samples using expressions (6.7) and (6.8) are in good agreement with experimental data of [19]. They quantitatively describe the dependencies of the thermal conductivity of Si, both on the heat flux direction and on the orientation of the side faces of the samples (see [22]).

In this regard, we have recalculated the temperature dependencies of the thermal conductivity of a silicon film with a thickness of D = 100 nm and a specularity parameter $P_{\{100\}} = 0.14$ using formula (6.10) for three orientations of its plane (see dashed lines 1a, 2a, and 3a in Figure 6.8) and compared with our results (solid lines 1, 2, and 3 in Figures 6.7 and 6.8).

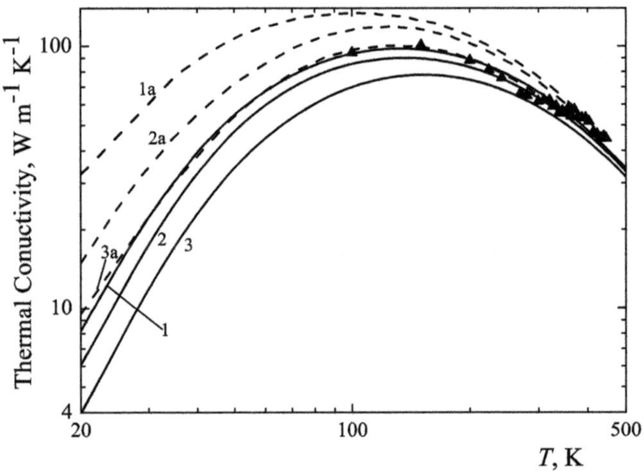

Figure 6.8: Temperature dependencies of thermal conductivity in silicon films with a thickness of D = 100 nm, width of W = 100 D, and length of L = W for various orientations of the film plane and temperature gradient: curves 1, 1a are for {100}, [100]; curves 2, 2a refer to {110}, [100]; curves 3, 3a are for {111}, [112]. Curves 1, 2, 3 relate to the calculation using expressions (6.7) and (6.8) for relaxation time, curves 1a, 2a, 3a are for the calculation of the relaxation times (6.10). The specularity parameter is P = 0.14. Symbols are the experiment of [5].

As can be seen in the figure, the results obtained (curves 1a, 2a, and 3a) are in qualitative agreement with our results. That is to say, the planes with the {100} and {111} orientations have maximum and minimum values of the thermal conductivity, respectively. Obviously, the authors of [28] made a number of numerical errors in calculating the temperature dependencies of the thermal conductivity of silicon films for various orientations. However, the temperature dependencies computed by using expressions (6.7) and (6.8) for the phonon-boundary relaxation rates quantitatively differ significantly from the results deduced using formula (6.10) (see Figure 6.8). At a temperature of T = 20 K, the above expressions for the relaxation rate of phonons at the film boundaries for various orientations of the film plane yield the ratio $\kappa_{[100]}^{\{100\}}:\kappa_{[100]}^{\{110\}}:\kappa_{[110]}^{\{111\}} = 2.2{:}1.5{:}1$. At the same time, using formula

(6.10), we have: $\kappa_{[100]}^{\{100\}}:\kappa_{[100]}^{\{110\}}:\kappa_{[110]}^{\{111\}} = 3.4{:}1.6{:}1$. Thus, with diffuse scattering at the boundaries, the maximum thermal conductivity is typical of Si films with the {100} orientation, while films with the {111} orientation possess the minimum thermal conductivity. The detailed discussion of the orientation dependence of thermal conductivity, presented here, speaks volumes about how important this problem is for silicon microelectronics.

As noted earlier, the use of the term "scattering power" of planes of films with different orientations for characterizing the influence of phonon focusing on their thermal conductivity is physically incorrect. Casimir's theory assumes that all phonons upon collision with the surface are absorbed and then re-emitted isotropically into half-space towards inside the sample. Therefore, whatever the orientation is, the surface in each point diffusely scatters all phonons of all polarizations. Hence, the scattering intensity of films with different plane's orientations is the same. The analysis carried out in Chapter 5 shows that the anisotropy of the film thermal conductivity is due to the influence of focusing on the propagation of phonon modes in the films. The geometric constructions presented in Chapter 4 show (see Figures 4.4 and 4.5) that, when the directions of focusing and temperature gradient coincide, the phonon flux from the heater to the cold sink deviates towards the film axis. Consequently, the mean free paths for these phonons become longer than those in an isotropic medium.

To illustrate the influence of various relaxation mechanisms on the temperature dependencies of the thermal conductivity of silicon films in different temperature ranges, Figure 6.9 sketches the calculated outcomes for the films with $D = 0.42$ μm. In the low-temperature region, boundary and isotopic scattering are the main phonon relaxation mechanisms for films with $D = 1.6$ and 0.42 μm. In the temperature range from 17 to 40 K, accounting for these mechanisms makes it possible to agree the computed results with the experimental data of [6, 7]. The contributions of isotopic scattering at $T = 20$ K are 33% and 21% for films with $D = 1.6$ and 0.42 μm, respectively. It is worth stressing that in the case of diffuse phonon scattering at the boundaries, the theoretical curves for films oriented along the {100} planes are 57% and 36% below the experimental curves for films with $D = 1.6$ and 0.42 μm, respectively (see Figure 6.9, curve 6).

Further, when analysing the temperature dependencies of the thermal conductivity of films with $D = 1.6$ and 0.42 μm, we set the film plane's orientation as $\{J\} = \{100\}$ and the specularity parameters equal to $P_{\{100\}} = 0.48$ and $P_{\{100\}} = 0.29$, respectively. As can be seen in Figure 6.9, at temperatures above 50 K, anharmonic scattering processes play a significant role in heat resistance. When analysing the thermal conductivity of silicon films, we use the same parameters of the anharmonic scattering processes as it does for nanowires (see Table 6.1). With increasing temperature, the role of various branches of the phonon spectrum in the thermal conductivity of films changes essentially. These changes are mainly due to the dispersion of thermal phonons, as well as the drift motion of longitudinal phonons. As is clear

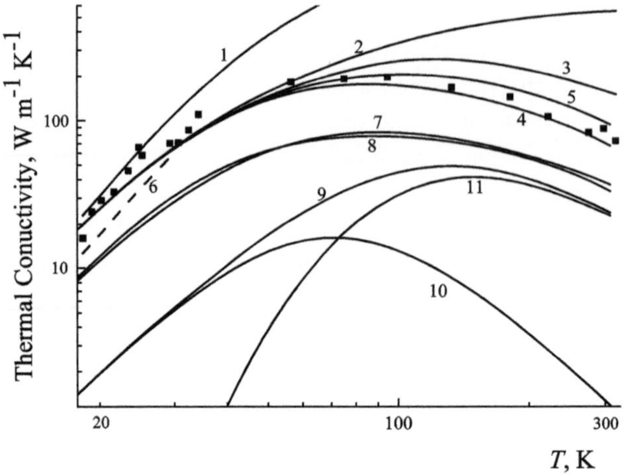

Figure 6.9: The temperature dependencies of thermal conductivity of silicon films with $D = 0.42\ \mu m$, $L = 8\ \mu m$, and $P = 0.29$ for the [100] temperature gradient direction and {100} film orientation for different phonon scattering mechanisms: (1) – boundary scattering regime, $v = \tilde{v}_{B[100]}^{\lambda\{100\}}$; (2) – $v = \tilde{v}_{B[100]}^{\lambda\{100\}} + v_{iso}$; (3) – $v = \tilde{v}_{B[100]}^{\lambda\{100\}} + v_{iso} + v_U$; (4) – diffusion contribution; (5) – total thermal conductivity; (6) – boundary scattering regime for P = 0; (7) contribution of the fast transverse mode; (8) contribution of the slow transverse mode; (9) total contribution of longitudinal phonons; (10) diffusion contribution from the longitudinal mode; (11) contribution of longitudinal phonon drift. Symbols ■ correspond to experimental data [6, 7].

from Figure 6.9, at $T < 54$ K, the slow transverse mode makes the dominant contribution to thermal conductivity (see Figure 6.9, curve 8). At $T = 20$ K, its contribution is 48%, and the contribution of the fast mode is 42%. Long flat segments in the spectrum of the slow transverse mode t_2 when $q_{\max}/2 < q < q_{\max}$ (see Section 1.2, Figure 1.5) are responsible for abnormally low values of the group velocity and, accordingly, for a great decrease in its contribution to thermal conductivity with rising temperature. In contrast, the contribution of the fast transverse mode t_1 grows faster with increasing temperature and at $T > 54$ K it becomes larger than that of the t_2 mode (see Figure 6.9, curve 8). Therefore, in the entire temperature range from 54 K to 350 K, the fast transverse mode t_1 makes the dominant contribution to thermal conductivity. For transverse phonons, the diffusion contribution to thermal conductivity over the entire temperature range significantly exceeds the contribution of the drift motion.

At $T = 20$ K, the contribution of longitudinal phonons to thermal conductivity is small; it is 10%. However, with rising temperature, this contribution increases rapidly mainly due to the drift motion and, at $T > 200$ K, it becomes comparable to the contribution of the slow transverse mode (see Figure 6.9, curve 11). At $T < 60$ K, the drift contribution to the thermal conductivity is much less than the diffusion contribution. However, in the temperature range from 100 K to 300 K, normal phonon-phonon scattering processes dominate for longitudinal phonons, and the drift contribution to the thermal conductivity significantly exceeds the diffusion one (see Figure 6.9, curves 10, 11). The contributions of the drift motion of phonons to the total thermal conductivity of films with D = 1.6 and 0.42 µm at T = 100 K are 15%. At T = 300 K, they reach 26% and 28%, respectively. With decreasing film thickness, the relative contribution of longitudinal phonons to the thermal conductivity increases. However, for films with D = 1.6 and 0.42 µm, it remains smaller than the contribution of the slow transverse mode. As can be understood from Figure 6.9, accounting for the drift motion of longitudinal phonons allows agreeing the predicted thermal conductivity values of the films with the experimental data of [6, 7].

Figure 6.10 displays the temperature dependencies of the thermal conductivity for films of different thicknesses in the regime of boundary phonon scattering ($\nu = \nu_B$, curves 2–6) and in the regime of bulk phonon relaxation mechanisms ($\nu_V = \nu_{iso} + \nu_U + \nu_N$, curve 1). The point of intersection of these curves stands for the temperatures of transition from the boundary scattering to the bulk relaxation mechanisms, T_{BV}. At these temperatures, the boundary scattering provides 50% of the total heat resistance. A numerical analysis shows that, for silicon films with D = 1.6 and 0.42 µm, the T_{BV} transition temperatures are 61 and 89 K, respectively. A decrease in the film thickness and specularity parameter gives rise to an increase in the role of the boundary scattering and to a rather rapid increase in the T_{BV} transition temperature.

So, for example, a decline in the film thickness (with other fixed parameters) to values of D = 0.1 and 0.05 µm when $P = 0.14$ gives $T_{BV} = 143$ and 196 K, respectively. For films with D = 0.02 µm when P = 0, the T_{BV} transition temperature is $T_{BV} = 345$ K. As the figure indicates, for bulk materials, the T_{BV} transition temperatures are far below. At room temperatures, the contributions of the boundary scattering for films with D = 1.6 nm, 0.42 nm, 0.1 nm, 0.05 nm, and 0.02 µm are 2%, 6%, 17%, 29%, and 58%, respectively. Thus, at room temperature, the phonon boundary scattering plays an essential role in the thermal conductivity of sufficiently thin films.

Let us compare the temperature dependencies of the thermal conductivity of silicon nanowires with circular and square-shaped cross-sections. As noted in Section 2.3, the mean free paths of phonons for circular and square-shaped cross-section infinite-length samples differ by less than 1% provided the equality of areas of their transverse cross-sections. Therefore, it is of interest to calculate the temperature dependencies of the thermal conductivity of square-shaped cross-section nanowires at temperatures varied from 20 K to 300 K. Then, it is necessary to compare them with the experimental data of [4] and the results obtained in [23, 113] for 56 nm and 115 nm diameter nanowires. Suppose the square's side D and radius R of the nanowires are defined by the condition $D = \sqrt{\pi} R$, and their lengths coincide (see Figure 6.11). The reflection specularity parameter for nanowires is taken equal to P = 0.15. As can be inferred from Figure 6.11, the calculated results of [23, 113] for nanowires with square-shaped and circular cross-sections are in

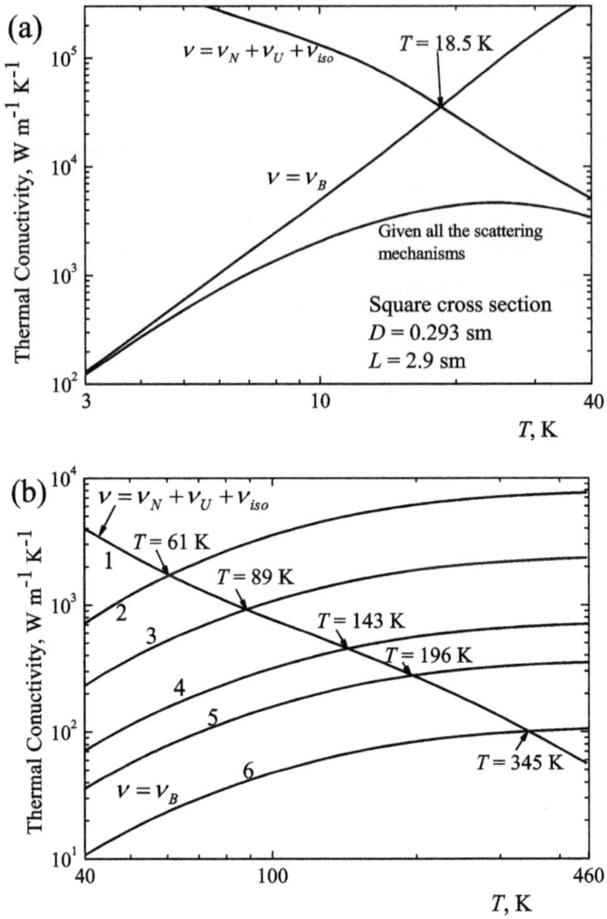

Figure 6.10: The temperature dependencies of thermal conductivity: (a) – for bulk samples [19], (b) – for films with the {100} orientation for bulk relaxation mechanisms (curve 1), and boundary scattering in films with the parameters: $D = 1.6$ μm, $L = 8$ μm, and $P = 0.48$ (curve 2); $D = 0.42$ μm, $L = 8$ μm, and $P = 0.29$ (curve 3); $D = 0.10$ μm, $L = 100\ D$, and $P = 0.14$ (curve 4), $D = 0.05$ μm, $L = 100\ D$, and $P = 0.14$ (curve 5) and $D = 0.02$ μm, $L = 100\ D$, and $P = 0$ (curve 6).

good agreement with each other and with the experimental data of [4]. At $T = 20$ K, the difference between the two calculations is 3.7%, while at $T = 300$ K it amounts to 0.6 %. This means that the above values are significantly less than the experimental error. Moreover, not only the full thermal conductivity values are well consistent, but also the contributions of all vibrational modes, as well as the drift and diffusion motion of phonons (see Figure 6.11).

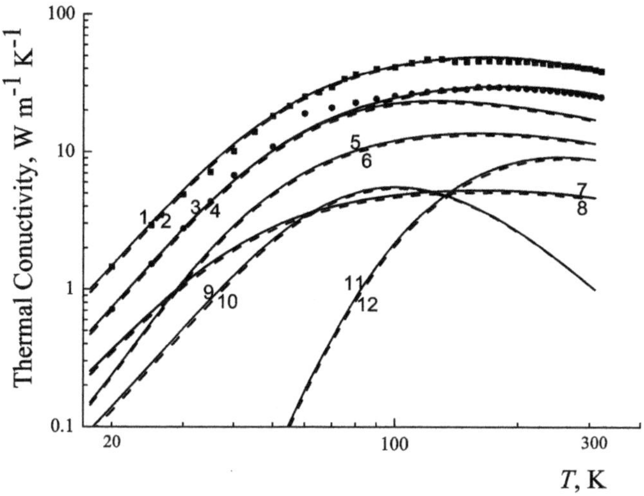

Figure 6.11: The temperature dependencies of thermal conductivity of silicon nanowires with circular (solid curves 1, 3) and square-shaped (dashed curves 2 and 4) cross-sections of the equal areas for $2R_1 = 115$ nm (1,2) and $2R_2 = 56$ nm (3,4), and the contributions of fast transverse phonons (curves 5, 6), slow transverse phonons (curves 7,8), diffusion motion of longitudinal phonon (curves 9, 10), and drift motion of longitudinal phonons (curves 11, 12) to the thermal conductivity of nanowires with $2R_2 = 56$ nm. Symbols correspond to the experimental data of [4].

These results indicate that, with diffuse boundary scattering, the analytical solutions for the phonon relaxation rates deduced in Chapter 2 are quite adequate to the real situation and can be used for interpreting the features of phonon transport in nanoscale samples of various geometries.

6.2.2 Anisotropy of the Thermal Conductivity of Silicon Films

Let us explore the change in the anisotropy of thermal conductivity of films of various widths ($\mu = 1$, 10, and 100) with increasing temperature. Suppose the reflection specularity parameter is $P = 0.14$, the film length is fixed as $L = 100\ D$, and the thickness is $D = 0.1$ μm. Further, we identify the orientations of the film planes and directions of heat flux, which provide maximum or minimum thermal conductivity in the film plane. For this, we consider three cases of changing the $\kappa^{\{J\}}_{[I(\psi)]}(T)$ angular dependencies of thermal conductivity when the heat flux rotates in the film plane:

(1) $\{J\} = \{100\}$,
(2) $\{J\} = \{110\}$,
(3) $\{J\} = \{111\}$.

For these cases, the phonon group-velocity components V^{λ}_{g1}, V^{λ}_{g2}, and V^{λ}_{g3} in a coordinate system related to the film can be extracted from formulas (4.4).

For $\mu = 1$ or $W = D$, we have a nanowire with a square-shaped cross-section. Therefore, the analytical theory developed here gives a unique opportunity to investigate the change in the anisotropy of thermal conductivity during the transition from nanowires ($\mu = 1$) to sufficiently wide films ($\mu = 100$) at a fixed temperature. Figure 6.12 shows such dependencies for temperatures of 20 K and 320 K and three film plane's orientations $\{J\} = \{100\}$, $\{110\}$, and $\{111\}$. Let us dwell on the peculiarities of phonon transport in silicon square-shaped cross-section nanowires.

Unlike circular cross-section nanowires, square-shaped cross-section nanowires exhibit thermal conductivity as depending on the orientation of the side faces. However, this dependence is weak. An analysis of cases when the nanowire's axis coincides with the axis of rotation and [111] and [100] heat flux directions shows that the anisotropy of the thermal conductivity due to a change in the orientation of the side faces of the nanowire is less than 1%. In contrast, the heat flux changing its direction relative to the axis of the crystal causes strong anisotropy of the thermal conductivity of nanowires. In this case, the results of calculating the thermal conductivity of square-shaped and circular cross-section nanowires are in good agreement with each other provided the equality of their cross-sectional areas (see Figure 6.11). A detailed analysis of the angular dependencies of thermal conductivity of silicon nanowires, given in Section 6.1.2, shows that the anisotropy of thermal conductivity is caused by focusing and defocusing of phonons.

Against nanowires, the thermal conductivity anisotropy of sufficiently wide films is mainly due to their orientation dependence. As can be seen in Figures 6.12, the angular dependencies of thermal conductivity of silicon films as the temperature gradient rotates in the {100}, {110}, and {111} planes significantly differ from both each other and those calculated for nanowires. With an increase in the film width, the values of thermal conductivity dramatically increment in comparison with those for nanowires. For example, at $T = 20$ K and a change in the μ parameter from 1 to 100, a rise of the values of the $\kappa^{\{100\}}_{[I(\psi)]}$ thermal conductivity of films oriented along the {100} plane is 3.53 and 5.35 times in the [100] and [110] directions, respectively. For the {110} plane, the values of the $\kappa^{\{110\}}_{[I(\psi)]}$ thermal conductivity in the $I = [100]$ and [110] directions rise by 2.50 and 2.51 times, respectively. An increase in the film width leads to a qualitative change in the angular dependencies of the thermal conductivity as compared with nanowires: the dependence of thermal conductivity on the heat flux direction weakens. For films with the {100} and {111} planes for $\mu = 100$ (L = 100 D), it becomes isotropic although small anisotropy still remains for $\mu = 10$ (see Figure 6.12a, b). At $T = 20$ K, for nanostructures with $\mu = 1$, 10, and 100, the ratios $\kappa^{\{100\}}_{[100]}(\mu)/\kappa^{\{100\}}_{[110]}(\mu)$ turn out to be equal to 1.52, 1.19, and 1.00, respectively (see Table 6.2).

It should be pointed out that the maximum values of the thermal conductivity are achieved for the {100} orientation, whereas the minimum ones are observed for

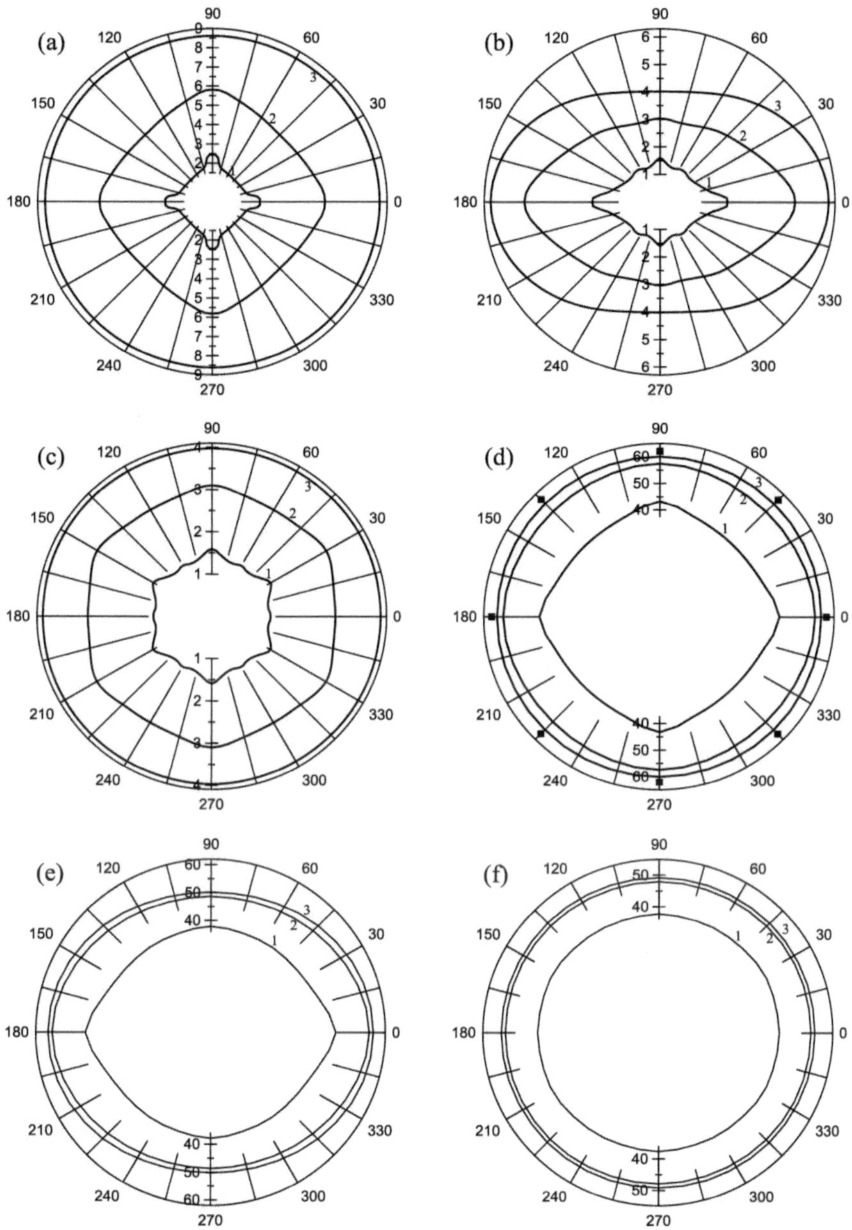

Figure 6.12: The κ(ψ) angular dependencies of thermal conductivities, (W/mK), of silicon films with $D = 0.10\ \mu$m, $L = 100\ D$, and $P = 0.14$ for temperatures of 20 K (a,c,e) and 320 K (b,d,f) for the film plane's orientations $\{J\} = \{100\}$ (a,b), {110} (c,d), and {111} (e,f), calculated for $\mu = 1$ (curves 1), $\mu = 10$ (curves 2), and $\mu = 100$ (curves 3). Symbols ■ correspond to the experimental data of [5].

Table 6.2: The thermal conductivity ratios for silicon nanofilms and square-shaped nanowires in symmetrical directions for $D = 0.10$ μm at temperatures of 20 K and 320 K.

T, K	μ	$\kappa^{\{100\}}_{[100]}/\kappa^{\{100\}}_{[110]}$	$\kappa^{\{110\}}_{[100]}/\kappa^{\{110\}}_{[110]}$	$\kappa^{\{110\}}_{[100]}/\kappa^{\{110\}}_{[111]}$	$\kappa^{\{110\}}_{[100]}/\kappa^{\{111\}}_{[110]}$
20	1	1.52	1.50	1.72	1.01
	10	1.20	1.60	1.50	1.6
	100	1.00	1.50	1.3	2.2
320	1	1.16	1.17	1.19	1.02
	10	1.01	1.16	1.12	1.16
	100	1.00	1.15	1.10	1.20

the {111} orientation (see Figure 6.12a, c). At $T = 20$ K and $P = 0.14$, the ratio of thermal conductivities of sufficiently wide Si films ($\mu = 100$) oriented along the above orientations is 2.2 (see Table 6.2). With an increase in temperature, the role of bulk scattering mechanisms grows, and the anisotropy of the thermal conductivity drops. For example, at $T = 320$ K, for nanostructures with $\mu = 1$ and 10, the ratio declines to 1.16 and 1.01, respectively (see Table 6.2). With increasing temperature, the anisotropy associated with the dependence of thermal conductivity on the film plane's orientation also falls off: the ratio of thermal conductivities for the orientations {100} and {111} is 2.2 at $T = 20$ K and $\mu = 100$, while at $T = 320$ K it diminishes to 1.20 (see Table 6.2). In this case, at room temperature, the contribution of boundary scattering to the thermal resistance of nanofilms with $D = 0.1$ μm amounts to 17%. For the {110} plane, the $\kappa^{\{110\}}_{[I(\psi)]}(\mu)$ angular dependencies in sufficiently wide films acquire an ellipsoidal form, with the long axis coinciding with the [100] direction. For this orientation, at $T = 20$ K and $\mu = 1$, 10, 100, the ratios of thermal conductivity values are equal to $\kappa^{\{110\}}_{[100]}(\mu)/\kappa^{\{110\}}_{[110]}(\mu) = 1.50,\ 1.60$, and 1.50, while at $T = 320$ K it decreases to values of 1.17, 1.16, and 1.15, respectively (see Table 5.2).

Thus, it has been shown that the angular dependencies of the thermal conductivity qualitatively change when going over from square-shaped cross-section nanowires to sufficiently wide films. Secondly, when diffuse scattering at the boundaries of Si films takes place, maximum thermal conductivity is typical of films with the {100} orientation, whereas films with the {111} orientation possess minimum thermal conductivity.

6.3 Conclusion

The main results of the sixth chapter can be formulated as follows:

1. The use of the phonon-boundary relaxation times calculated in the monograph and the three-mode Callaway model allowed to adequately describe the temperature dependences of the thermal conductivity of silicon nanowires with a diameter of more than 50 nm and films with a thickness of more than 20 nm.
2. The optimal film plane's orientations and directions of the heat flux have been found, which ensure the maximum or minimum heat removal from the elements of silicon microcircuits at both low and room temperatures. It has been shown that, with diffuse phonon-boundary scattering, Si films oriented along the {100} and {111} planes possess the highest and lowest thermal conductivity, respectively.
3. It has been shown that the dispersion of thermal phonons has a significant influence on the thermal conductivity of silicon nanowires and thin films. Long flat segments in the spectrum of slow transverse short-wave phonons cause abnormally low group-velocity values and, accordingly, a significant decrease in their contribution to the thermal conductivity with rising temperature.
4. Research on the role of boundary scattering shows that, at room temperature, the contribution of boundary scattering to the thermal resistance reaches 60% both for nanowires with a diameter of 56 nm and for films with a thickness of 20 nm.

Chapter 7
Prospects for Further Research

Currently, the theory of phonon transport in bulk and nanoscale monocrystalline samples is still far from being completed, and a number of important problems have not yet been solved. One of these challenges is the influence of anisotropy of elastic energy on the propagation of phonons, their focusing, and thermal conductivity of monocrystalline samples. The main drawback of existing theories for describing the temperature dependencies of the thermal conductivity is the need to introduce a significant number of fitting parameters for phonon boundary scattering, normal phonon–phonon scattering processes and the Umklapp processes (see, for example, [9, 35, 36, 85]). One of the achievements of the above theory is finding the analytical solution to the problem of diffuse phonon scattering at the boundaries of finite-length samples with circular, square-shaped, and rectangular cross-sections and determination of the corresponding phonon relaxation times. This makes it possible to get rid of fitting parameters of the phonon-boundary scattering, accounting for different directions of the heat flux or geometric parameters of the samples. We have shown that the method developed in [22] for taking into account phonon focusing and analytical solutions for phonon relaxation times at the boundaries of finite-length samples with circular, square-shaped, and rectangular cross-sections allows one to adequately describe the temperature dependencies of the thermal conductivity of silicon crystals [19] for all side face's orientations of the samples and directions of the temperature gradient. Moreover, in the range from 3 to 15 K, when scattering at the boundaries and by isotopic disorder dominates, the theory of [20–22] quantitatively describes the above temperature dependencies without resorting to fitting parameters. With increasing temperature, the contribution of anharmonic scattering processes increases and the discrepancy between the predicted and experimental data grows. A maximum difference between the calculated and experimental values of [19] takes place in the vicinity of the maximum thermal conductivity when the contributions of boundary scattering and bulk relaxation mechanisms to the thermal resistance become equal. Although the deviation from the experimental data in the vicinity of the maximum is not large, the calculated values for the [001] directions turn out to be 5% higher than the measured ones and below the experimental values by 6% for the [111] direction. However, the calculated values of the anisotropy are three times bigger than those from the experiment. At higher temperatures, the thermal conductivity anisotropy drops more slowly than in the experiment. As already noted, the Matthiessen rule is a possible reason for this discrepancy when going over from boundary scattering to bulk one. This can be explained by the fact that phonon boundary scattering occurs near the surface of the sample, whereas bulk processes are distributed uniformly throughout the

https://doi.org/10.1515/9783110670500-008

volume. Therefore, the relaxation rates in these scattering mechanisms cannot be summed up additively, as the Matthiessen rule claims. Obviously, the maximum deviation from the Matthiessen rule can be revealed provided the equality of the phonon relaxation rates in bulk and boundary scattering mechanisms. Analysing the role of low-energy phonons in the thermal conductivity and thermoelectric power of monocrystalline semiconductors, Herring [38, 105] also pointed out the weakness of the Matthiessen rule at temperatures near the maximum thermal conductivity. To correctly account for the combined action of the boundary and bulk phonon scattering mechanisms, the Boltzmann kinetic equation (BKE) with allowance for the inhomogeneous distribution of the heat flux over the transverse cross-section of the sample needs to be solved. With taking phonon focusing into account, this problem is relevant but rather complicated. It requires separate consideration.

Therefore, one of the promising trends for further analysis of phonon transport in elastically anisotropic films and nanowires is to study the influence of phonon focusing on competition between the boundary and bulk phonon relaxation mechanisms using the BKE. The results obtained by the BKE method for thermal conductivity of bulk materials and nanostructures must be compared with the outcomes of [20–24, 27, 69–71, 106] found within the phenomenological Casimir–McCurdy method [13, 19]. This allows one to:

1) Circumscribe the limits of applicability of the phenomenological method for analysing phonon transport in nanostructures of different geometry.
2) Verify the justification of utilizing the Matthiessen rule for boundary and bulk scattering mechanisms when studying phonon transport in nanostructures both far from and in the vicinity of the temperature of the transition from boundary scattering to bulk relaxation mechanisms.
3) Analyse the dependencies of the thermal conductivity of nanostructures on the geometric parameters, directions of the heat flux, and film plane's orientations relative to the crystal axes.
4) Make clear the role of interference of the boundary and bulk scattering mechanisms for phonon transport in nanostructures.
5) Apply the developed theory to analysing the thermal conductivity in isotopically highly enriched crystals of Ge, Si, and diamond.

It should be emphasized that research on the influence of phonon focusing covers not only phonon transport in dielectric crystals. It is relevant in investigating kinetic phenomena in metals and semiconductors, which are due to the drag of electrons by phonons and their mutual drag. At sufficiently low temperatures, when the electron–phonon interaction is a dominant electron relaxation mechanism (it must also dominate for the mutual phonon drag), kinetic coefficients such as the electrical conductivity and drag thermoelectric power become dependent on the phonon momentum relaxation mechanism [67, 114]. In this interval of temperatures, diffuse scattering at the sample boundaries is the main mechanism of phonon relaxation [67, 114]. Certainly, the phonon focusing that gives rise to the anisotropy of thermal conductivity in the regime of the Knudsen phonon gas flow must also cause anisotropy of the drag thermoelectric power. Earlier, when exploring the drag thermoelectric power in semiconductors, the model of an isotropic medium has been used for phonons, as a rule (see, e.g., [67, 114]). Therefore, the effects due to the influence of anisotropy of elastic energy on the electron-phonon drag could not be considered. It is well known [67, 114] that the model of an isotropic medium offers the interaction of electrons only with longitudinal phonons through the deformation potential. However, quasi-longitudinal and quasi-transverse phonons can propagate in elastically anisotropic crystals. Moreover, quasi-transverse phonons have a non-zero longitudinal component and can also interact with electrons through the deformation potential. An analysis of elastic waves in cubic crystals has shown that the contribution of the transverse component to quasi-longitudinal vibrations in cubic crystals is small and can be neglected [44]. The contribution of the longitudinal components to quasi-transverse modes is not small. Its maximum values for the slow transverse mode amount to 13, 14, and 16.5% for HgSe: Fe, Si, and Ge crystals, respectively. For InSb, GaAs, and KCl crystals, it reaches 16, 17, and 27%, respectively [44]. Therefore, to calculate and analyse the contributions of quasi-longitudinal and quasi-transverse phonons to electron-phonon drag and thermoelectric effects in semiconductor crystals and nanostructures based on them is of great interest. Moreover, the contribution of the transverse modes to the thermal conductivity of Si and Ge at low temperatures in the [100] directions exceeds 90%. This is due to both the higher density of phonon states and the influence of focusing of phonons on their mean free paths. Therefore, one of the tasks for theory and technical applications is to study the influence of elastic energy anisotropy on electron-phonon drag and thermoelectric phenomena in semiconductor crystals and nanostructures based on them at low temperatures.

The first step in this end has been done in [115], where the influence of the anisotropy of elastic properties on the electron–phonon drag and thermoelectric phenomena in a gapless HgSe: Fe semiconductor with degenerate carrier statistics have been studied, and a method of their calculating has been developed, as well. The influence of phonon focusing on the dependencies of the thermopower on the geometric parameters and directions of the heat flux relative to the crystal axes is analysed. The crystallographic directions that provide the maximum and minimum values of the thermopower are determined. The role of quasi-longitudinal and quasi-transverse phonons in the drag thermoelectric power in HgSe:Fe crystals at low temperatures is explained. It is shown that the contribution of longitudinal phonons to the drag thermoelectric power and their mean free paths become maximal in the [111] focusing direction, whereas their minimum values are observed in the [001] defocusing direction. The contribution of the slow quasi-transverse mode reaches a maximum in the [001] focusing direction and, for long enough samples, can exceed the contribution of quasi-longitudinal phonons. Therefore, analysing the drag thermoelectric power in elastically anisotropic crystals requires accounting for both contributions.

In light of the findings secured in Ref. [115], it seems interesting to examine the influence of phonon focusing on electron–phonon drag in monocrystalline nanowires, films, and heterostructures.

Bibliography

[1] D.G. Cahill, W.K. Ford, K.E. Goodson, G.D. Mahan, A. Majumdar, H.J. Maris, R. Merlin and S.R. Phillpot, Nanoscale thermal transport, Journal of Applied Physics. 93 (2003), pp. 793–818.

[2] A. D. McConnell and K. E. Goodson, Nanoscale thermal transport, Annual Review on Heat Transfer. 14 (2005), pp. 129–168.

[3] D. G. Cahill, P. V. Braun, G. Chen, D. R. Clarke, S. Fan, K. E. Goodson, P. Keblinski, W. P. King, G. D. Mahan, A. Majumdar, H.J. Maris, S. R. Phillpot, E. Pop and L. Shi, Nanoscale thermal transport. II. 2003–2012, Applied Physics Reviews. 1 (2014), p. 011305.

[4] D. Li, Y. Wu, P. Kim, L. Shi, P. Yang and A. Majumdar, Thermal conductivity of individual silicon nanowires , Applied Physics Letters. 83 (2003), pp. 2934–2936.

[5] W. Liu and M. Asheghi, Phonon–boundary scattering in ultrathin single-crystal silicon layers, Applied Physics Letters. 84 (2004), pp. 3819–3821.

[6] M. Asheghi, M. N. Touzelbaev, K. E. Goodson, Y. K. Leung and S. S. Wong, Temperature-dependent thermal conductivity of single-crystal silicon layers in SOI substrates, Journal of Heat Transfer. 120 (1998), pp. 30–36.

[7] M. Asheghi, Y. K. Leung, S. S. Wong and K. E. Goodson, Phonon-boundary scattering in thin silicon layers, Applied Physics Letters. 71 (1997), pp. 1798–1800.

[8] J.M. Ziman, Electrons and Phonons. The Theory of Transport Phenomena in Solids, Oxford University Press, New York (1960), 554 pp.

[9] B. M. Mogilevsky and A. F. Chudnovsky, Thermal Conductivity of Semiconductors, Nauka, Moscow (1972), 536 pp. (in Russian)

[10] V.P. Silin, Introduction to the Kinetic Theory of Gases, Nauka, Moscow (1971), 332 pp. (in Russian)

[11] M. Devien, The Flow and Heat Transfer of Rarefied Gases. Trans. from English M.: Publishing house of foreign literature (1962), 188 pp. (in Russian)

[12] M. Knudsen, Die Gesetze der Molekularströmung und der inneren Reibungsströmung der Gase durch Röhren, Annalen der Physik 333 (1909), pp. 75–130.

[13] H. B. G. Casimir, Note on the conduction of heat in crystals, Physica 5 (1938) pp. 495–500.

[14] R. Berman, F. E. Simon and J. M. Ziman, The thermal conductivity of diamond at low temperatures, Proceedings of Royal Society of London A. 220 (1953), pp. 171–183.

[15] R. Berman, E. L. Foster and J. M. Ziman, Thermal conduction in artificial sapphire crystals at low temperatures. I. nearly perfect crystals, Proceedings of Royal Society of London A. 231 (1955), pp. 130–144.

[16] B. Taylor, H.J. Maris and C. Elbaum, Phonon focusing in solids, Physical Review Letter. 23 (1969), pp. 416–419.

[17] H. J. Maris, Enhancement of heat pulses in crystals due to elastic anisotropy, Journal of Acoustics Soceity of Am. 50 (1971), pp. 812–818.

[18] J. P. Wolfe, Imaging Phonons: Acoustic Wave Propagation in Solids, Cambridge University Press, New York. (1998), 411 p.

[19] A. K. McCurdy, H. J. Maris and C. Elbaum, Anisotropic heat conduction in cubic crystals in the boundary scattering regime, Physical Review B. 2 (1970), pp. 4077–4083.

[20] I. I. Kuleyev, I. G. Kuleyev, S. M. Bakharev and A. V. Inyushkin, Relaxation times and mean free paths of phonons in the boundary scattering regime for silicon single crystals, Physics of the Solid State. 55 (2013), pp. 31–44.

[21] I. I. Kuleyev, I. G. Kuleyev, S. M. Bakharev and A. V. Inyushkin, Features of phonon transport in silicon rods and thin plates in the boundary scattering regime. The effect of phonon focusing at low temperatures, Physica B. 416 (2013), pp. 81–87.

https://doi.org/10.1515/9783110670509-009

[22] I. I. Kuleyev, I. G. Kuleyev, S. M. Bakharev and A. V. Inyushkin, Effect of phonon focusing on the temperature dependence of thermal conductivity of silicon, Physica Status Solidi B. 251 (2014), pp. 991–1000.

[23] I. G. Kuleyev, I. I. Kuleyev and S. M. Bakharev, Phonon focusing and temperature dependences of the thermal conductivity of silicon nanowires, Journal of Experimental And Theoretical Physics. 118 (2014), pp. 253–265.

[24] I. I. Kuleyev, S.M. Bakharev, I. G. Kuleyev and V. V. Ustinov, Phonon focusing and temperature dependences of thermal conductivity of silicon nanofilms, Journal of Experimental And Theoretical Physics. 120 (2015), pp. 638–650.

[25] H. J. Maris and S. Tamura, Heat flow in nanostructures in the Casimir regime, Physical Review B. 85 (2012), p. 054304.

[26] N. Mingo, Calculation of Si nanowire thermal conductivity using complete phonon dispersion relations, Physical Review B. 68 (2003), p. 113308.

[27] I.I. Kuleyev, I.G. Kuleyev and S.M. Bakharev, Low temperature anisotropy of the thermal conductivity of single crystal nanofilms and nanowires, Journal Of Experimental And Theoretical Physics. 119 (2014), pp. 460–472.

[28] Z. Aksamija and I. Knezevic, Anisotropy and boundary scattering in the lattice thermal conductivity of silicon nanomembranes, Physical Review B. 82 (2010), p. 045319.

[29] J. E. Turney, A. J. H. McGaughey and C. H. Amon, In-plane phonon transport in thin films, Journal of Applied Physics. 107 (2010), p. 024317.

[30] K. Fuchs, The Conductivity of Thin Metallic Films According to the Electron Theory of Metals, Mathematical Proceedings of the Cambridge Philosophical Society. 34 (1938), pp. 100–108.

[31] E.H. Sondheimer, The mean free path of electrons in metals, Advances in Physics. 1 (1952), pp. 1–42.

[32] M. P. Zaitlin, Boundary scattering of phonons in noncrystalline materials, Physical Review B. 12 (1975), pp. 4487–4492.

[33] W. Li, N. Mingo, L. Lindsay, D. A. Broido, D. A. Stewart and N. A. Katcho, Thermal conductivity of diamond nanowires from first principles, Physical Review B. 85 (2012), p. 195436.

[34] J. Guo, B. Wen, R. Melnik, S. Yao and T. Li, Geometry and temperature dependent thermal conductivity of diamond nanowires : A non-equilibrium molecular dynamics study, Physica E: Low-dimensional Systems and Nanostructures. 43 (2010), pp. 155–160.

[35] R. Berman, Thermal Conduction in Solids, Published Oxford: Clarendon Press. (1976), 193 p.

[36] V.L. Gurevich, Transport in phonon systems, Elsevier Science Ltd. (1988), 434 p.

[37] M. Asheghi, K. Kurabayashi, R. Kasnavi and K. Goodson, Thermal conduction in doped single-crystal silicon films, Journal of Applied Physics. 91 (2002), pp. 5079–5088.

[38] C. Herring, Role of low-energy phonons in thermal conduction, Physical Review. 95 (1954), pp. 954–965.

[39] I. G. Kuleyev, I. I. Kuleyev and S. M. Bakharev, Effect of dispersion and damping of thermal phonon states on the longitudinal ultrasonic absorption in germanium crystals, Physics of the Solid State. 53(8) (2011), pp. 1644–1657.

[40] I. G. Kuleyev, I. I. Kuleyev and I. Yu. Arapova, Interaction of collinear and noncollinear phonons in anharmonic scattering processes and their role in ultrasound absorption of fast quasi-transverse modes in cubic crystals , Journal of Physics: Condensed Matter. 22(9) (2010), p. 095403.

[41] G. Leibfried, Gittertheorie der Mechanischen und Thermischen Eigenschaften der Kristalle, Berlin: Springer-Verlag (1955), pp. 104–324.

[42] F. I. Fedorov, Theory of Elastic Waves in Crystals, Springer Science & Business Media (2013), 375 pp.

[43] C. Kittel, Introduction to Solid State Physics, Wiley (2004), 704 pp.
[44] I.G. Kuleev, I. I. Kuleev, Elastic waves in cubic crystals with positive or negative anisotropy of second-order elastic moduli, Physics of the Solid State. 49 (2007), pp. 437–444.
[45] B. Taylor, Focusing of phonons in crystalline solids due to elastic, Physical Review B. 3 (1971), pp. 1462–1472.
[46] R. Truell, C. Elbaum, B. B. Chick, Ultrasonic Methods in Solid State Physics, New York and London: Academic Press (1969), 464 p.
[47] J. W. Tucker and V. W. Rampton, Microwave Ultrasonics in Solid State Physics, Elsevier Science Publishing Co Inc.,U.S. (1972), 428 p.
[48] W. Voigt, Lehrbuch der Kristalphysik, Springer-Verlag (2014), 979 p.
[49] S. Simons, The Absorption of Very High Frequency Sound in Dielectric Solids, Mathematical Proceedings of the Cambridge Philosophical Society. 53(7) (1957), pp. 702–716.
[50] Y. K. Yogurtcu, A. J. Miller and G.A. Saunders, Elastic behaviour of YAG under pressure, Journal of Physics C: Solid State Physics. 13 (36) (1980), pp. 6585.
[51] W.C. Overton and R. T. Swim, The adiabatic elastic constants of rock salt, Physical Review. 84 (1951), pp. 758–762.
[52] H. J. McSkimin and W. L. Bond, Elastic moduli of diamond, Physical Review. 105 (1957), pp. 116–121.
[53] H. J. McSkimin and P. Andreatch, Elastic moduli of silicon vs hydrostatic pressure at 25.0 C and 195.8 C, Journal of Applied Physics. 35(7) (1964), pp. 2161–2165.
[54] D. R. Huffman, M. H. Norwood, Specific heat and elastic constants of calcium fluoride at low temperatures, Physical Review. 117 (1960), pp. 709–711.
[55] J. K. Galt, Mechanical properties of NaCl, KBr, KCl, Physical Review. 73 (1948), pp. 1460–1462.
[56] J.R. Drabble and A.J. Brammer, Third order elastic constants of gallium arsenide, Solid State Communications. 4(9) (1966), pp. 467–468.
[57] C. V. Briscoe and C. F. Squire, Elastic constants of LiF from 4.2 K to 300 K by ultrasonic methods, Physical Review. 106 (1957), pp. 1175–1177.
[58] A. F. Wright, Elastic properties of zinc-blende and wurtzite AlN, GaN, and InN, Journal of Applied Physics. 82(6) (1997), pp. 2833–2839.
[59] A. Lehoczky, D. A. Nelson and C. R. Whitsett, Elastic constants of mercury selenide, Physical Review. 188 (1969), pp. 1069–1073.
[60] G. Nilsson and G. Nelin, Phonon dispersion relations in Ge at 80 K, Physical Review B. 3(2) (1971), pp. 364.
[61] G. Nilsson and G. Nelin. Study of the homology between silicon and germanium by thermal-neutron spectrometry, Physical Review B. 6 (1972), pp. 3777–3786.
[62] H. Bilz, W. Kress, Phonon Dispersion Relations in Insulators, New York: Springer Ser. in Solid-State Sci.10, Springer-Verlag Berlin-Heidelberg (1979), 241 p.
[63] I.I. Kuleyev, I.G. Kuleyev, S.M. Bakharev and A.V. Inyushkin, Effect of dispersion on the phonon focusing and anisotropy of thermal conductivity of silicon single crystals in the boundary scattering regime, Physics of the Solid State. 55 (2013), pp. 1545–1556.
[64] G.G. Devyatykh, A. V. Gusev, A.M. Gibin and O.V. Timofeev, Heat capacity of high-purity silicon, Inorganic Materials. 33(12) (1997), pp. 1206–1209.
[65] P. Flubacher, A. J. Leadbetter and J. A. Morrison, The heat capacity of pure silicon and germanium and properties of their vibrational frequency spectra, Philosophical Magazine. 4, 39 (1959), pp. 273–294.
[66] J. M. Ziman, Electrons and Phonons: The Theory of Transport Phenomena in Solids, Oxford: Oxford University Press (1962), 554 p.

[67] A. I. Anselm, Introduction To Semiconductor Theory, Mir Publishers, Moscow (1981), 646 p.
[68] G. F. Miller and M. J. P. Musgrave, On the propagation of elastic waves in aeolotropic media. III. Media of cubic symmetry, Proceedings of the Royal Society of London A: Mathematical, Physical and Engineering Sciences. 236 (1206), (1956), pp. 352–383; M. J. P. Musgrave, Proc. Roy. Soc. A 226(1206)(1954), pp. 339–356.
[69] I. I. Kuleev, S.M. Bakharev, I.G. Kuleev and V. V. Ustinov, Effect of phonon focusing on Knudsen flow of phonon gas in single-crystal nanowires made of spintronics materials, Physics of Metals and Metallography. 118 (2017), pp. 10–20.
[70] I. I. Kuleyev, S. M. Bakharev, I. G. Kuleyev and V.V. Ustinov, Influence of phonon focusing on the Knudsen flow of phonon gas in single-crystal nanofilms of spintronic materials, Physics of Metals and Metallography. 118 (2017), pp. 316–327.
[71] I.I. Kuleyev, S.M. Bakharev, I.G. Kuleyev and V.V. Ustinov, The influence of phonon focusing on density of states and the Knudsen phonon gas flow in nanowires with different types of anisotropy of elastic energy, Physical status solidi C. 14(3-4) (2017), p. 1600263.
[72] A. G. Every, Formation of phonon-focusing caustics in crystals, Physical Review B. 34(4) (1986), p. 2852.
[73] M. Lax and V. Narayanamurti, Phonon magnification and the Gaussian curvature of the slowness surface in anisotropic media: detector shape effects with application to GaAs, Physical Review B. 22(10) (1980), pp. 4876–4897.
[74] J.A. Shields, J.P. Wolfe and S.I. Tamura, Channeling of acoustic phonons in silicon: the polarization dependence in elastic scattering, Zeitschrift fur Physik B. 76 (1989), pp. 295–301.
[75] Cz. Jasiukiewicz, T. Paszkiewicz and D. Lehmann, Phonon focussing patterns: Calculation of response of finite area detectors to pulsed ballistic beams of dispersive and dispersionless phonons, Zeitschrift fur Physik B. 96 (1994), pp. 213–222.
[76] J. de Klerk and M. J. P. Musgrave, Internal conical refraction of transverse elastic waves in cubic crystal, Proceedings of the Physical Society. Section B. 68 (1955), p. 81.
[77] P. C. Waterman, Orientation dependence of elastic waves in single crystals, Physical Review. 113 (1959), pp. 1240–1253.
[78] K.B. Vlasov and A.B. Rinkevich, Internal conical refraction of elastic waves, Acoustical Journal 41 (1)(1995) pp. 67–71. (in Russian)
[79] E. Held, W. Klein and R.P. Huebener, Characterization of single-crystalline GaAs by imaging with ballistic phonons, Zeitschrift fur Physik B: Condensed Matter. 75 (1989), pp. 17–29.
[80] G.A. Northrop and J. P. Wolfe, Ballistic phonon imaging in germanium, Physical Review B. 22 (12) (1980), p. 6196.
[81] J. Philip and K. S. Viswanathan, Phonon magnification in cubic crystals , Physical Review B. 17(12) (1978), p. 4969.
[82] A. G. Every, Ballistic phonons and the shape of the ray surface in cubic crystals , Physical Review B. 24(6) (1981), p. 3456.
[83] A.V. Pogorelov, Differential Geometry, Noordhoff, Groningen (1960), 172 p.
[84] I. G. Kuleev and S. M. Bakharev, Phonon Flux Enhancement Factors in Crystals with Different Elastic Energy Anisotropy Types, Physics of the Solid State 60(7) (2018), pp. 1263–1272.
[85] P. Carruthers, Theory of thermal conductivity of solids at low temperatures, Reviews of Modern Physics. 33 (1961), pp. 92–138.
[86] Y. P. Joshi, Effect of phonon focussing on thermal conductivity of silicon, Pramana. 18 (1982), pp. 461–472.
[87] J. P. Harrison and J. P. Pendrys, Thermal conductivity of cerium magnesium nitrate, Physical Review B. 7(8) (1973), p. 3902.

[88] H. Lundt, M. Kerstan, A. Huber and P. O. Hahn, Subsurface damage of abraded silicon wafers, Semiconductor Silicon/1994, Proceedings of the 7th International Symposium on Silicon Materials Science and Technology. – The Electrochemical Society (1994), pp. 218–224.

[89] J.R. Drabble and H.J. Goldsmid, Thermal Conduction in Semiconductors, Pergamon press, Oxford, London, New York, Paris (1961), p. 235.

[90] S. Uma, A. McConnell, M. Asheghi, K. Kurabayashi and K. Goodson, Temperature-dependent thermal conductivity of undoped polycrystalline silicon layers, International Journal of Thermophysics. 22 (2001), pp. 605–616.

[91] J. Callaway, Model for lattice thermal conductivity at low temperatures, Physical Review. 113(4) (1959), p. 1046.

[92] J. A. Krumhansl, Thermal conductivity of insulating crystals in the presence of normal processes, Proceedings of the Physical Society. 85(5) (1965), p. 921.

[93] B. H. Armstrong, N-processes, the relaxation-time approximation, and lattice thermal conductivity , Physical Review B. 32(6) (1985), pp. 3381–3390.

[94] I.G. Kuleev and I.I Kuleev, Normal phonon-phonon scattering processes and the thermal conductivity of germanium crystals with isotope disorder, Journal of Experimental and Theoretical Physics. 93(3) (2001), pp. 568–578; The effect of normal phonon-phonon scattering processes on the maximum thermal conductivity of isotopically pure silicon crystals, Journal of Experimental and Theoretical Physics. 95(3) (2002), pp. 480–490

[95] M. Asen-Palmer, K. Bartkowski, E. Gmelin, M. Cardona, A. P. Zhernov, A. V. Inyushkin, A. N. Taldenkov, V. I. Ozhogin, K. M. Itoh and E. E. Haller, Thermal conductivity of germanium crystals with different isotopic compositions, Physical review B. 56 (15) (1997), p. 9431.

[96] A. P. Zhernov and A. V. Inyushkin, Effect of isotopic composition on phonon modes. Static atomic displacements in crystals, Physics-Uspekhi. 44 (2001), pp. 785–811; Kinetic coefficients in isotopically disordered crystals, Phys. Usp. 45 (2002), pp. 527–552.

[97] M. G. Holland, Analysis of lattice thermal conductivity , Physical Review. 132(6) (1963), p. 2461.

[98] L. Landau and G. Rumer, Absorption of sound in solids, Physikalishce Zeitschrift Sowjetunion. 11 (1937), pp. 18–25.

[99] S. Simons, On the mutual interaction of parallel phonons, Proceedings of the Physical Society. 82(3) (1963), p. 401.

[100] I. G. Kuleyev, I. I. Kuleyev and I. Yu. Arapova, Anharmonic processes of scattering and absorption of slow quasi-transverse modes in cubic crystals with positive and negative anisotropies of second-order elastic moduli , Journal of Physics: Condensed Matter. 20 (2008), p. 465201; Interaction of collinear and noncollinear phonons in anharmonic scattering processes and their role in ultrasound absorption of fast quasi-transverse modes in cubic crystals, Journal of Physics: Condensed Matter. 22 (2010), p. 0945403.

[101] P. G. Klemens, The scattering of low-frequency lattice waves by static imperfections, Proceedings of the Physical Society. Section A. 68(12) (1955), p. 1113.

[102] S. I. Tamura, Isotope scattering of dispersive phonons in Ge, Physical Review B. 27(2) (1983), pp. 858–866.

[103] A.P. Zhernov and A.V. Inyushkin, Isotopic Effects in Solids, Russian Research Center, Kurchatovski Institut, Moscow (2001). (in Russian)

[104] I. G. Kuleyev, I. I. Kuleyev and I. Y. Arapova, Quasi-transverse ultrasound absorption due to point defects and anharmonic scattering processes in cubic crystals with positive and negative anisotropies of the second-order elastic moduli , Journal of Physics: Condensed Matter. 19(40), (2007), p. 406216.

[105] C. Herring, Theory of the thermoelectric power of semiconductors, Physical Review. 96 (1954), pp. 1163–1187.

[106] I.I. Kuleyev, I.G. Kuleyev and S.M. Bakharev, Phonon focusing and features of phonon transport in the silicon nanofilms and nanowires at low temperatures, Physical Status Solidi B. 252 (2015), pp. 323–332.

[107] M. Smoluchowski, Zur kinetischen Theorie der Transpiration und Diffusion verdünnter Gase, Annalen der Physik. 338(16) (1910), pp. 1559–1570.

[108] Y. F. Zhu, J. S. Lian and Q. Jiang, Re-examination of Casimir limit for phonon traveling in semiconductor nanostructures , Applied of Physical Letter. 92 (2008), p. 113101.

[109] I. I. Kuleev, Anisotropy of the mean free paths of phonons in single-crystal films of germanium, silicon, and diamond at low temperatures, Physics of the Solid State. 59 (2017), pp. 682–693.

[110] Y. Kruglyak, Landauer-Datta-Lundstrom generalized transport model for nanoelectronics, Journal of Nanoscience. 3 (2014), 02NAESF01(4pp).

[111] W. O'Mara, R. B. Herring and L. P. Hunt, Handbook of Semiconductor Silicon Technology. 1 edition. William Andrew (1990), 795 p.

[112] S. B. Soffer, Statistical model for the size effect in electrical conduction, Journal of Applied Physics. 38(4) (1967), pp. 1710–1715.

[113] I. G. Kuleyev, I. I. Kuleyev and S. M. Bakharev, Anisotropy and temperature dependences of thermal conductivity for silicon nanowires , Bulletin of the Russian Academy of Sciences: Physics. 78 (2014), pp. 905–907.

[114] V.M. Askerov, Electron Transport Phenomena in Semiconductors, World Scientific (1994), 394 p.

[115] I. G. Kuleyev, I. I. Kuleyev, S. M. Bakharev and V. V. Ustinov, Phonon focusing and electron–phonon drag in semiconductor crystals with degenerate charge-carrier statistics, Journal of Experimental And Theoretical Physics. 123 (2016), pp. 489–495.

[116] I. G. Kuleyev, I. I. Kuleyev, and S. M. Bakharev, McCurdy's Effects in the Thermal Conductivity of Elastically Anisotropic Crystals in the Mode of Knudsen Phonon-Gas Flow, Semiconductors 52(13) (2018), pp. 1643–1652.

[117] I. I. Kuleev, Influence of focusing on phonon propagation and thermal conductivity in single crystal films with different types of anisotropy of elastic energy, Physics of the Solid State. 60(5) (2018), pp. 870–876.

Index

https://doi.org/10.1515/9783110670509-010

Also of interest

Coherent Quantum Physics.
A Reinterpretation of the Tradition
Arnold Neumaier, 2019
ISBN 978-3-11-066729-5, e-ISBN (PDF) 978-3-11-066738-7,
e-ISBN (EPUB) 978-3-11-066736-3

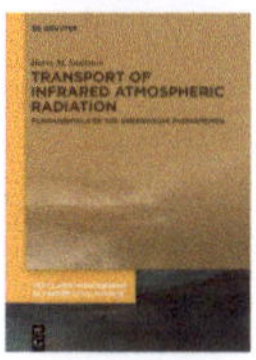

Transport of Infrared Atmospheric Radiation.
Fundamentals of the Greenhouse Phenomenon
Smirnov, Boris M., 2020
ISBN 978-3-11-062765-7, e-ISBN (PDF) 978-3-11-062875-3,
e-ISBN (EPUB) 978-3-11-062903-3

Metrology and Theory of Measurement.
Slaev, Valery A. / Chunovkina, Anna G. / Mironovsky, Leonid A., 2019
ISBN 978-3-11-065094-5, e-ISBN (PDF) 978-3-11-065250-5,
e-ISBN (EPUB) 978-3-11-065131-7

Quantum Systems, Channels, Information.
A Mathematical Introduction
Alexander S. Holevo, 2019
ISBN 978-3-11-064224-7, e-ISBN (PDF) 978-3-11-064249-0,
e-ISBN (EPUB) 978-3-11-064240-7

Phase-Field Crystals.
Fast Interface Dynamics
Galenko, Peter / Ankudinov, Vladimir / Starodumov, Ilya, 2019
ISBN 978-3-11-058597-1, e-ISBN (PDF) 978-3-11-058809-5,
e-ISBN (EPUB) 978-3-11-058653-4

Wilson Lines in Quantum Field Theory
Cherednikov, Igor Olegovich / Mertens, Tom / Van der Veken,
Frederik, 2019
ISBN 978-3-11-065092-1, e-ISBN (PDF) 978-3-11-065169-0,
e-ISBN (EPUB) 978-3-11-065103-4